Agricultural Research Policy

Vernon W. Ruttan

*Department of Agricultural and Applied Economics
and Department of Economics
University of Minnesota*

University of Minnesota Press Minneapolis

Published by the University of Minnesota Press,
2037 University Avenue Southeast, Minneapolis, MN 55414
Printed in the United States of America.

Library of Congress Cataloging in Publication Data

Ruttan, Vernon W.
   Agricultural research policy
   Includes index.
      1. Agricultural research. 2. Agricultural
research—Government policy. I. Title.
S540.A2R87   630'.72   81-16396
ISBN 0-8166-1101-7   AACR2
ISBN 0-8166-1102-5 (pbk.)

The University of Minnesota
is an equal-opportunity
educator and employer.

There are many interesting research problems.
Some of them are important.

Richard Bradfield

Acknowledgments

In writing this book I have drawn on the counsel and guidance and on the insights and experience of a large number of my colleagues in the national research agencies and the international agricultural research system. I have attempted to acknowledge their contributions in the specific chapters to which they contributed.

Of all those who have contributed to my thinking about agricultural research policy I would like most to express my special appreciation to Robert E. Evenson, Albert H. Moseman, Theodore W. Schultz, and Sterling Wortman. I must also thank Stephen Hoenack, who read and criticized the entire manuscript.

I have appreciated the strong interest and support for my work on agricultural research policy of Keith Huston (former director) and Richard J. Sauer (present director) of the Minnesota Agricultural Experiment Station. At several points in the book, I have drawn upon research conducted under Minnesota Agricultural Experiment Station Project 14-067.

I am grateful to Don Hadwiger of Iowa State University for allowing me to read the manuscript of *The Politics of Agricultural Research* (to be published by the University of Nebraska Press) and to Per Pinstrup-Anderson for allowing me to read the manuscript of *The Role of Agricultural Research and Technology in Economic Development* (to be published by Longman).

Finally, I am indebted to Mary Strait, Mary Jane Baumgart, and Elizabeth J. Butler for their efforts in translating my handwritten draft into a manuscript that I could transmit to the University of Minnesota Press. I was also saved from numerous errors of fact and interpretation, as well as spelling and grammar, by Marcia Bottom's copyediting. Several of the figures were drawn or redrawn by Karen Leonard.

Contents

Acknowledgments   vii

Chapter 1   Introduction . . . 3
A Personal Perspective   4
The Book's Plan   13
A Philosophical Perspective   15

Chapter 2   Technical Change and Agricultural Development . . . 17
Agricultural Growth   17
Agricultural Development Models   20
An Induced Innovation Model   26
Disequilibrium in Agriculture   36
Perspective   41

Chapter 3   The Agricultural Research Institution . . . 45
The Experiment Station   45
The Research and Development Process   50
The Role of Skill and Insight in a Biological
    Innovation: The Rice Blast Nursery Case   54
Linkages between Science and Technology   56
Technology and Society   60

Chapter 4   National Agricultural Research Systems . . . 66
Organizing Agricultural
    Research in the United Kingdom   67
Institutionalizing Agricultural Science in Germany   71
The Federal-State System of
    Agricultural Research in the United States   76

Establishing an Agricultural
Research System in Japan 83
Building Agricultural
Research Institutions in India 90
Reforming the Brazilian
Agricultural Research System 96
Agricultural Research Institutes in Malaysia 100
Some Perspectives on the
Organization of Agricultural Research 107

Chapter 5 The International Agricultural Research System . . . 116
The International
Agricultural Research Institute Model 117
Some Organization and Management Issues 124
Future Policy and Program Directions 132
Pre- and Postharvest Mechanical Technology 136
Small-Farm Technology 138
A Perspective on the Future 142

Chapter 6 Reviewing Agricultural Research Programs . . . 147
The Research Review Process 149
The Research Program 151
The Training Program 153
The Outreach Program 155
Perspective 157

Chapter 7 Location and Scale in Agricultural Research . . . 161
The Location of Agricultural Research Facilities 161
Size and Productivity in Agricultural Research 166
Structural Rigidity 171
The Small Country Problem 173

Chapter 8 The Private Sector in Agricultural Research . . . 181
Public Research and Private Research 182
How Much Does the Private Sector Spend? 183
Public Funding of Mechanization Research 186
Development and Protection of Plant Varieties 192
Innovation and Regulation in Insect Control 200
Perspective 210

Chapter 9 Institutional and Project Funding of Research . . . 215
The Research Scientist 216
The Research Administrator 219

Some Research
   Management and Policy Implications 221
The USDA's Competitive Research Grant Program 225
An Appraisal of the Two
   Systems of Research Support 229
Research Strategies for an Imperfect World 231

Chapter 10 The Economic Benefits
   from Agricultural Research . . . 237
Productivity Growth 237
The Contribution of Research to Productivity 241
The Character of Public
   Agricultural Research in the United States 249
Why Does Agricultural
   Research Remain Undervalued? 251
Some Implications 258

Chapter 11 Research Resource Allocation . . . 262
The Parity Model of Research Resource Allocation 264
Allocation of Resources to Research 267
Allocation of Resources in Research 275
Linking Project Selection to
   Resource Allocation Objectives 286
The Limits and Potential of
   Research Planning Methodologies 292

Chapter 12 The Social Sciences in Agricultural Research . . . 298
The Emergence of Rural Social Science 299
The Demand for Social Science Knowledge 304
Social Science Research in the
   Agricultural Research Institute 308
Social Science Research
   in the College of Agriculture 313
Social Science Research
   in the Ministry of Agriculture 320
A Perspective on the Demand
   for Social Science Knowledge 327

Chapter 13 Responsibility and Agricultural Research . . . 331
The Agricultural Scientist as Hero and Villain 332
Responsibility for Research Results 335
Toward Some Guidelines to Moral Responsibility 340

Technologies, Institutions, and Reforms 343
The Technology Assessment Movement 348
Agricultural Research and the Future 350
Indexes . . . 355

Tables

2.1 Accounting for Intercountry
Differences in Labor Productivity . . . **38**

2.2 Accounting for Intercountry Differences in Labor
Productivity between the U.S. and Selected Countries . . . **39**

5.1 The International Agricultural Research Institutes . . . **120**

8.1 Estimate of Industry Expenditures
for Farming and Postharvest Efficiency . . . **185**

8.2 Historical Events in the
Development of the Tomato Harvester . . . **189**

8.3 Evolution of Pesticide Regulation, 1900-1975 . . . **202**

8.4 Percentages of Research and Development
Expenditures Allocated According to Function by
American Pesticide Manufacturers . . . **208**

8.5 Time Required for Registration and Number of New
Registrations of Agricultural Chemicals, 1969-1978 . . . **208**

9.1 Amount and Relative Importance of Research Performed
by Funding Sources for State Agricultural Experiment Stations,
the 1980 Colleges and Tuskegee Institute, and Forestry
Schools, and USDA Agencies, 1967 and 1977-1979 . . . **226**

9.2 Funds Appropriated for Agricultural Research
under Cooperative Research Programs of the
USDA Science and Education Administration . . . **227**

10.1 Average Annual Rates of Change in Total Output, Inputs,
and Productivity in U.S. Agriculture, 1870-1979 . . . **241**

10.2 Average Annual Change in Total Output, Inputs,
and Productivity in Japanese Agriculture, 1880-1975 . . . **241**

10.3 Summary Studies of
Agricultural Research Productivity . . . **242**

10.4 Estimated Impacts of Research and
Extension Investments in U.S. Agriculture . . . **248**

11.1 Classification Codesheet for
      Report of Agricultural Research . . . **268**

11.2 Criteria and Weights Used for Establishing
      Relative Program Projections in the *National Program
      of Research for Agricultural Study* . . . **277**

11.3 Criteria for Evaluating Research Problem Areas at the
      North Carolina Agricultural Experiment Station . . . **280**

11.4 Alternative Projections and Realized Farm
      Output and Factor Input Indexes for 1960 and 1975 . . . **288**

Figures

2.1 Historical Growth Paths of Agricultural Productivity
     in the United States, Japan, Germany, Denmark,
     France, and the United Kingdom, 1880-1970 . . . **19**

2.2 Estimated Area Planted to High-Yielding Varieties
     of Wheat and Rice in Asia (East and South) . . . **26**

2.3 International Comparision of Labor/
     Output and Land/Output Ratios in Situations
     Characterized by Different Land/Labor Ratios . . . **27**

2.4 Historical Growth Path of Labor Productivity in
     Relation to Land Productivity and Cultivated Area
     per Worker in Philippine Agriculture, 1948-1968 . . . **30**

2.5 Relationship between Fertilizer Input per Hectare of Arable
     Land and the Fertilizer/Arable Land Price Ratio . . . **32**

2.6 Relationship between Tractor Horsepower per Male Worker
     and the Price of Machinery Relative to Labor, 1970 . . . **33**

2.7 Intercountry Cross-sectional Comparison of Changes
     in Tractor Horsepower per Male Worker and in
     Fertilizer Consumption per Hectare, 1960-1970 . . . **34**

3.1 Systems Model of Experiment Station's
     Performance and Development . . . **46**

3.2 The Emergence of Novelty in the Act of Insight . . . **52**

3.3 The Process of Cumulative Synthesis . . . **53**

3.4 The Interaction between Advances
     in Scientific and Technical Knowledge . . . **57**

3.5 Stages in the Development of Sugarcane Varieties . . . **59**

3.6 Supply and Demand for
     Technological and Institutional Innovations . . . 61
4.1 Scientist-years in SAES-USDA Program
     by Research Goal for FY1965 and FY1977 Compared
     to Those Recommended for 1977 . . . 80
4.2 Comparison of Research Expenditures
     with the Producer Price of Rice, 1950-1976 . . . 89
4.3 Evolution of Rubber Planting Materials
     through Breeding and Selection . . . 104
5.1 International Agricultural Research Centers'
     Annual Core Expenditures and Core
     Operating Expenditures, 1960-1980 . . . 124
5.2 International Winter and Spring
     Wheat Research Network, 1979 . . . 127
5.3 Cumulative Percentage of Farms in Three Size Classes
     Adapting Modern Varieties in 30 Villages in Asia . . . 139
8.1 Research and Development Funds
     for the U.S. Food Research Systems, 1976 . . . 184
8.2 Alternative Methods for Developing Seed Enterprises . . . 193
9.1 Current Grants Process . . . 223
10.1 Partial and Total Productivity Measures
     for U.S. Agriculture, 1950-1978 . . . 240
10.2 Congruence between Research Expenditures
     and Value of Commodity Production in 1975 . . . 252
11.1 Modular Form of Cost-Benefit Estimation Model . . . 285
12.1 The Concept of Yield Gaps between
     an Experiment Station's Rice Yield, the Potential
     Farm Yield, and the Actual Farm Yield . . . 311

Agricultural Research Policy

Chapter 1

Introduction

Throughout most of human history, increases in agricultural output have been achieved almost entirely from increases in area cultivated— by the expansion of the cultivated margin of existing villages, the establishment of new villages, and the opening of frontier lands to cultivation. The few exceptions to this generalization—in East Asia, the Fertile Crescent (Iraq and Egypt), and parts of Western Europe— were more important as indicators of the gains that would be achieved in the 20th century than as significant contributions to the world's food supplies.

By the end of the 20th century, almost all of the increase in agricultural production will have occurred as a result of increases in yield—in output per unit land area. The few areas still to be brought into agricultural production, such as the campos cerrado and the llanos in tropical America and the tsetse fly-infested plains of Africa, will be more important as reminders of an earlier pattern of development than as significant sources of growth in the world's food supply.

During the 20th century, agriculture has been undergoing a transition from a resource-based sector to a science-based industry. Growth in agricultural output is increasingly based on development of scientific and technical capacity to invent new mechanical, chemical, and biological technologies. It is increasingly dependent on growth of the capacity of the industrial sector and a technology-based subsector of agriculture to embody advances in agricultural science and technology in new and more productive inputs (seeds, fertilizers, herbicides, insecticides, machines, and equipment). And its impact is enhanced by increased investment in the formal and informal education of rural people that enlarges their capacity to discriminate among the new technologies that have become available and to employ efficiently the new technology and the new practices under the wide variations in the physical, economic, and cultural environments in which they practice agriculture.

Prior to the 20th century, much of the new knowledge on which advances in agricultural technology and agricultural practices were based occurred as a by-product of scientific curiosity or the accumulation of practical experience. New crops became available as by-products of exploration and migration. By the beginning of the 20th century, only a few countries had developed the institutional capacity—in the form of the agricultural experiment station, the industrial laboratory, or the engineering development staff—necessary to produce a continuous stream of new agricultural technology. By the middle of the 20th century, even the poorest developing countries had begun to establish the institutional foundations for the conduct of agricultural research. During the last quarter of the 20th century, the world was moving rapidly toward the establishment of an integrated network of national and international agricultural research institutions.

The issues related to the strategy and organization of agricultural research have, however, rarely been the subject of scientific inquiry. There is no school or discipline of agricultural research management. This is probably as it should be. Effective research management is the product of a unique combination of experience, insight, skill, and personality. I doubt that it could be taught. I would argue, however, that these qualities could be enhanced and refined by drawing on the accumulated experience of research organization, management, and strategy that has accumulated since agricultural research became institutionalized.

This book consists of a series of essays on the strategy, organization, and management of agricultural research. The essays attempt to explore some of the concepts and hypotheses that have evolved over three decades of association with, and writing about, agricultural research. It may be useful, in this introduction, to present something of the personal history behind the evolution of the concepts and perspectives discussed at greater length in subsequent chapters.

A PERSONAL PERSPECTIVE

My first professional employment (in 1951) was with the economics staff in the office of the general manager of the Tennessee Valley

Authority (TVA). The TVA had special responsibility for agricultural, water, and power development in the Tennessee Valley region. My responsibilities involved working in close liaison with the agricultural and chemical engineering programs. The agricultural program had as one of its major objectives the development and dissemination of technologies appropriate for the small farms that prevailed in the bottomland along the rivers and streams and in the upland areas of the southern Appalachians and Cumberland highland areas of the Tennessee Valley. A major thrust of the program was to utilize the new fertilizers being developed by the chemical engineering division as a basis for the development of grassland farming systems that would be less destructive to the soil resources of the region than the traditional row-crop farming systems that prevailed in the upland areas. A secondary thrust was the development of machinery and equipment suitable for the small farms of the region.

These objectives had seemed entirely appropriate in the environment of the mid-1930s, when the TVA program got under way. The Great Depression of the 1930s had reinforced a view that the American economy had reached a mature stage in which growth of nonagricultural employment opportunities would be limited. The appropriate objective of agricultural development policy was, therefore, to improve the productivity of the land and of the labor on the region's small farms.

By the late 1940s, the presumptions on which the agricultural development program of the Tennessee Valley Authority had been based were no longer valid. The American economy, including the economy of the South, had entered a period of rapid growth in industrial output and employment. Rising wage rates and expanding employment opportunities pulled labor, particularly new entrants into the labor force, off the small farms and into the urban areas of the South and the North. Progress in the development of higher-yielding crops, such as hybrid corn, combined with falling fertilizer prices to raise productivity in the more favored areas of The corn Belt in the Midwest and in the black-soil areas of the South more rapidly than on the upland and hill farms. There was little demand for the improved agricultural practices and the small-scale equipment that had been developed by the TVA and under TVA sponsorship. The problems of soil erosion and low labor productivity were being solved by the transfer of resources out of agriculture rather than by the adoption of the agricultural technologies developed for the small farms of the region.

It was out of this experience that I developed a perspective that the value of a research program must be continuously reviewed and evaluated in terms of the consistency of its objectives with changes in the economic environment where it expects to make a contribution. It is not enough to judge a research program simply on the basis of its scientific and/or technical merits. If, in fact, the U.S. economy had continued to stagnate—if it had been necessary to push the margin of cultivation farther up the slopes of the Appalachians and the Cumberlands and if farm size had continued to decline and the labor force in rural areas had continued to rise—the TVA agricultural program would have continued to be relevant. Continuation of the original program thrust into the 1950s, however, reflected a failure to bring economic knowledge to bear on problems of research resource allocation and on rural development program formulation.

The TVA chemical engineering program, which later was designated the National Fertilizer Development Center, is one of the few examples in the United States of effective public-sector contribution to the development of industrial technology. The center was located on the site of a synthetic nitrogen plant that was built at Muscle Shoals, Alabama, during World War I in order to take advantage of the power-generating capacity of the Tennessee River. The TVA management arrived at an early decision to focus the direction of the chemical engineering program on the development of high-analysis phosphate fertilizer materials and on the processes and equipment needed to produce the new materials. This focus was consistent both with the judgments of the agricultural development potential of the region, particularly the proposed shift from row-crop to grassland farming, and with evaluations of the opportunities for technical change in fertilizer production.

During its initial decade, the TVA fertilizer research and development program was highly productive of new knowledge, new technology, and new materials. Highly skilled and strongly motivated chemists and engineers with diverse backgrounds and experience were brought together. Their interaction around a common set of research and development programs was highly productive. New, high-analysis phosphate fertilizer, multiple-nutrient carriers of phosphorus and nitrogen, and innovations in the equipment needed to produce the new products were developed. By the early 1950s, much of the excitement and dynamism of the program had been lost. The scientific effort had settled down to filling in the gaps in the literature on phosphate chemistry, and the engineering program focused on minor modifications in fertilizer-manufacturing equipment.

As I reflected on my experience in the TVA chemical engineering program and attempted to learn something about other industrial and agricultural research and development programs, I began to formulate a hypothesis about the natural history of the research institute. My hypothesis followed these lines: the establishment of a new research institute creates an opportunity to bring together a unique combination of scientific and professional capacity characterized by diverse training and experience. The excitement of establishing a new research enterprise is attractive to the more venturesome and imaginative scientists and engineers. If the management of the institute is skillful at encouraging effective intellectual interchange and avoiding excessive interdisciplinary aggression, the new institute may experience a period of exceptional productivity that may last a decade or longer. Experience seems to suggest a decline in intellectual excitement and a rise in managerial bureaucracy sometime before the end of the second decade. As a result, the institute will arrive at a period of maturity in which the research program becomes more routine and less productive. At this point, a major problem becomes how to renew the intellectual vigor of the mature research institute. This view of the natural history of the research institute was reinforced during my period as an economist in the International Rice Research Institute in the Philippines (1963-1965) and during my continued association with the international agricultural research system.

In 1954 I left the TVA to join the Agricultural Economics Department at Purdue University. I recall rather vividly an early conversation with the director of the agricultural experiment station about research strategy and priorities, particularly his comment to the effect that "my job is to attempt to see that the faculty is able to pursue the research that interests them." It struck me even at that time, when as a young scientist I was anxious about preserving control over my own research agenda, that this was not a responsible position for a research director to take. It has concerned me increasingly over the years as it has become clear that most agricultural experiment station directors had given up any pretension about exercising significant intellectual leadership over the research activities that were funded by the stations and had lost whatever inclination they once had to engage the research station scientific staff in discussions of their research activities. These functions were left to the heads or chairpersons of the disciplinary departments. The result was fragmentation of the experiment station research along disciplinary lines and only limited capacity to focus research resources on significant problems. The data that are available on research resource allocation in U.S. agricultural experiment stations suggest that resources tend to be allocated among departments on the basis of parity and that whatever reallocation does take place occurs primarily within departments—such as a shift in research in the soils department from the effects of fertilizer on crop productivity to a concern with the environmental effects of nitrogen infiltration into ground water. Within the last several years, however, I have noted a tendency among some of the younger, more vigorous agricultural experiment station directors to reassert their responsibility for research direction and resource allocation.

In 1963 I joined the staff of the International Rice Research Institute (IRRI) in the Philippines. The IRRI was the first unit in what has now become an international system of agricultural research institutes. (See chapter 5.) The objective that the director and the associate director of the institute set before the staff was stated, not simply in terms of the quality of the research product, but "to raise rice yields in Asia." This struck me as audacious in the extreme. How could a small research institute with fewer than 20 senior scientists aspire to such an objective? In retrospect, of course, it is clear that the new technology developed at the IRRI did have a major impact on rice production in Asia. The modern rice varieties developed at the IRRI had a direct impact on rice production in parts of the Philippines and in several other areas in Asia. Even more important, the model of the biologically efficient rice plant that served as a conceptual model for the varieties developed at IRRI influenced the direction of national breeding programs and the varieties produced by these breeding programs throughout the rice-growing areas of the world.

I am now convinced that the articulation of the IRRI objective in terms of impact on rice yield rather than in terms of contribution to scientific knowledge about rice was a key factor in directing the scientific effort of the IRRI staff. The logic of the institute's objective was more effective in inducing interdisciplinary collaboration than what could have been achieved by the exercise of administrative guidelines or authority.

In the fall of 1965, I returned to the United States to become head of the Department of Agricultural Economics at the University of Minnesota. After my very positive experience in interdisciplinary research at the IRRI, I felt very strongly about the importance of building effective interdisciplinary linkages with other departments in the University of Minnesota Institute of Agriculture. I failed completely in this objective. One of the reasons for my failure was, I believe, the centripetal forces that encourage intra- rather than inter departmental collaboration in a large department consisting, at Minnesota, of over 40 staff members at the professorial level (assistant professors, associate professors, and professors), and between 60 and 80 graduate students holding research assistantships. At the IRRI, where I had been the only agricultural economist, the only alternative to interdisciplinary collaboration was isolation. Achievement of the same objective at a large, research-oriented university would, I believe, require the establishment of several interdisciplinary centers in which a group of scientists and their graduate students would have their offices and laboratories. Interdisciplinary collaboration can, I believe, only be achieved at the expense of intradisciplinary collaboration.

During the late 1960s, I began to develop an interest in the potential application of more formal analytical approaches to the allocation of research resources. The United States had evolved a federal-state system of agricultural research. The federal system was highly centralized. The state system was decentralized. Federal funds made available to the states under the Hatch Act and related legislation involved rather rigorous financial accountability requirements. Although a system of regional research committees facilitated communication among researchers working on related areas, there was little central direction of research effort. During the late 1960s, there were several studies dealing with agricultural research priorities. New management systems, particularly the Program Planning and Budgeting System (PPBS), were being introduced. The methodology that had been developed for the historical study of agricultural research productivity was being adapted to produce *ex ante* estimates of the costs and benefits of agricultural research. It seemed only a matter of time until the new management tools such as the PPBS, zero-based budgeting, or other methodologies that could be cast in a cost-benefit or systems-analysis framework would be applied to agricultural research. Experimental research attempting to explore the possibility of such applications was initiated at Iowa State University and at the University of Minnesota.

In the summer of 1968, I helped to organize an informal session at the American Agricultural Economics Association to assess some of the work both on the *ex post* measurement of the economic returns to agricultural research and on the measurement of the potential returns to the allocation of resources to research. This was followed by a major conference held at the University of Minnesota in the winter of 1970, which brought together most of the people from agricultural science and economics who had been working on the methodology of assessing and planning agricultural research. It seemed to me at the time that the issues involved in the new methodologies could be classified under two broad headings. First, what possible advances in agricultural science and technology could be made available to farms within the foreseeable future? The question, could only be answered by the most imaginative scientists in each discipline or problem area. Second, of those advances that could be visualized, which would be of sufficient potential value in production to warrant a claim on the allocation of financial and scientific resources? The answer to this question required the skills of economists. At the conclusion of the Minnesota symposium, I was quite optimistic about the potential for the development of formal methodologies as guides for research resource allocation.[1]

The 1970 Minnesota symposium was followed in 1974 by an Airlie House conference sponsored by the Agricultural Development Council and the World Bank on research resource allocation and productivity in national and international research systems. The conference included key participants from the Minnesota symposium plus several scholars who had worked on similar problems in developing countries. It also included several research entrepreneurs and administrators who had been involved in organizing and directing the new international agricultural research system. (See chapter 5.)[2] By the end of the Airlie House conference, I had become much more skeptical, though certainly not completely agnostic, about the contributions of both the *ex post* analysis and the *ex ante* planning methodologies to resource allocation in agricultural research. My current perspective on these issues is outlined in chapter 10.

By the mid-1970s, my perspective on research resource allocation had been strongly influenced by the research that Yujiro Hayami and I had conducted on agricultural development in the United States, Japan, and a number of other developed and developing countries.[3] Japan and the United States began to build their agricultural research and extension systems in the last quarter of the 19th century. Both countries drew rather heavily on Germany for their inspiration and example. Yet, the two countries followed quite different paths of technological change. The United States followed a mechanical technology path, and Japan followed a biological technology path. Prior to the 1950s, the United States achieved most of its productivity gains by increasing output per worker—by inventing technologies that enabled the successful cultivation of more land per worker. During the same period, Japan achieved most of its gains in productivity by increasing output per hectare—by developing production practices and crop varieties that resulted in continuous increases in output per hectare.

This comparative research on technical change in agriculture, combined with the insight into agricultural research strategies and potentials I gained through my experience in North America and Asia, led me to a view that nature is relatively "plastic." I became convinced that alternative paths to technological change in agriculture, based on a nation's physical and human resource endowments, are not only feasible but that research strategies that do not take such differences into account are likely to be unproductive. An appropriate research strategy involves directing research resources to releasing the physical and institutional constraints on production that are most inelastic.

As I became convinced of the importance of each nation's developing an agricultural research capacity suited to its own resource and institutional endowments, I also became more concerned with the issue of the organization of national research systems. What steering mechanisms or focusing devices are needed to direct scientific effort toward releasing the constraints on production in a particular nation or region? It appeared that in both Japan and the United States the development of prefectural and state agricultural research systems, with substantial local funding and direction, represented an important steering mechanism. There was a short feedback loop between the productivity and the financial support from the state and the prefectural stations. Potato production in Aroostook County (Maine) might not be very important on the national scale of things, but it is important to Maine!

I also became convinced that the productivity of the decentralized prefectural and state systems was enhanced by a national research system that could provide institutional support and capacity for research on issues of broader national and regional problems, such as the 1970 corn blight in the United States. The new system of international agricultural research institutes plays a similar role relative to the smaller national agricultural research systems of many developing countries. The issue of location, scale, and organization in agricultural research have, however, been the subject of very little formal economic analysis.

It should be evident by now that an issue of continuing concern throughout my association with agricultural research has been how to maintain and enhance the effectiveness of agricultural research institutions and systems in response to internal and external developments. At the level of the individual research institute or disciplinary department, a common device for assuring institutional productivity is a combination of internal and external program reviews. Over the years I have participated in a number of internal and external reviews. These experiences have convinced me that the internal reviews have, in general, been more effective than the external reviews in bringing about change in research programs. External reviews have often been more productive of complacency or conflict than of reform or redirection. Since, however, there seems to be no alternative to the external review in establishing and maintaining credibility with funding agencies, I have attempted, on the basis of some experience, to reflect on how such reviews can be made more productive.

A further issue of continuing concern to me has been the role of social science research in agricultural research institutes and agricultural universities and ministries. My perspective on this issue has evolved as a result of my experience as a junior scientist at the Tennessee Valley Authority and at Purdue University, as a staff economist working on issues of agricultural policy at the President's Council of Economic Advisors, as the only agricultural economist in a biological research institute at the International Rice Research Institute, as a departmental administrative officer at the University of Minnesota, and as president of the Agricultural Development Council.

In the United States, the social sciences—rural sociology, agricultural economics, and communications—became institutionalized in the experiment station at a later stage than the biology- and engineering-based disciplines. The process involved considerable tension and interdisciplinary aggression. This tension has largely disappeared in the United States. It did not emerge at some of the international institutes, such as IRRI and International Crop Research Institute for the Semi-arid Tropics (ICRISAT), where agricultural economics was built into the research program of the institutes at the very beginning. The tension still exists in some of the new national systems that have evolved from strong colonial research traditions.

It is clear that this new knowledge resulting from social science research does not result in the production of new technology. If, however, social science research is appropriately focused, it can contribute to more effective institutional performance. This means, at the individual research institute, collaboration with other scientists to identify the value of technical changes designed to remove technical constraints on production. At the university level it means, in addition, research directed to and understanding the interactions between technical and institutional change. At the ministry and planning commission levels, it means focusing on policy and planning in order to achieve consistency among agricultural objectives and general development objectives. The administration of an agricultural research institute, of an agricultural faculty, or of an agricultural ministry that fails to build the social science capacity needed to assist in substituting informed judgment for trial and error in research resource allocation or in program development and management must be prepared to defend the economic cost and to bear the political cost of weak institutional performance. (See chapter 12.)

THE BOOK'S PLAN

The concerns that have led to the organization of this book have developed out of my particular institutional and research experience. The book attempts to deal with these issues in a more systematic manner than is indicated in the personal perspective outlined above. It is organized into 13 chapters, including this introduction.

In chapter 2 I explore the implications of factor endowments for agricultural research strategy and planning. The implications of both alternative resource endowment at a particular point in history as well as changing resource endowments over time are examined. The research implications of shifting from a land- or resource-based strategy of agricultural development are explored. The last section in this chapter deals with the challenge of making a transition from a strategy based on low-cost energy to one based on high-cost energy.

Chapter 3 is devoted to the behavioral and institutional aspects of the research and development process. The problem of creativity and novelty in thought and action for the organization of research is discussed. A behavioral model of the agricultural research institute or experiment station is outlined in order to clarify the relationship between the internal knowledge-generating processes and the external scientific, economic, and political environment in which the institute functions.

In chapter 4, I discuss the historical development of several important national systems in developed countries—such as the United States, Germany, the United Kingdom, and Japan—as well as the more recent history of several national systems in developing countries—such as India, Malaysia, and Brazil. Particular attention is given to the role of cultural and political factors in the development of national research systems. Alternative forms of organization and stages in the evolution of research management are identified. What are the implications for the growing demand that agricultural research serve multiple objectives? In addition to productivity, equity and nutritional objectives are increasingly being sought.

Chapter 5 will be devoted to a discussion of the development, the contributions, and the problems of the new system of international agricultural research institutes. The system has been highly productive. It is difficult to think of any international aid effort that has achieved such favorable returns. Yet, the system is not without problems. How does an international system solve the problem of maintaining continuing relevance to the problems of national systems? How can it deal with the problem of institutional maturity?

One institutional response to the problem of maintaining the quality of staff and the productivity of research effort in mature systems has been the internal and external research review. Yet, such reviews often suffer from either a lack of competence or bias. The results are often ignored. Chapter 6 will discuss making the research review process a positive rather than a divisive experience.

There are a number of neglected issues relating to the economic organization of the agricultural research industry. These include such issues as the location and size of agricultural research facilities. They also include the factors that bring about changes in the boundaries between, or the mixture of, public- and private-sector agricultural research. In chapters 7 and 8 an attempt is made to clarify some of these issues.

Chapter 9 deals with the implications of institutional and grant funding of research. Public-sector agricultural research typically has been funded from institutional sources. In recent years, both in the United Kingdom and in the United States, there has been an attempt to design funding institutions that improve the responsiveness of the agricultural research system to changing social (and political) objectives. In this chapter it is argued that some of the methods used to obtain greater responsiveness may also have a negative impact on the productivity of the research system.

In chapter 10, I examine the results of a large body of research on the contribution of agricultural research to output and productivity growth. The results are summarized in terms of rates of return to investment in agricultural research. These rates of return typically have been high when judged by rates of return necessary to induce investment in the private sector. Some of the reasons for the continued high rates of return are examined.

Chapter 11 is devoted to a review and an evaluation of efforts to develop and apply increasingly formal research resource allocation procedures in agricultural research. The roles of cropping-systems research, crop-growth models, and systems-simulation approaches are evaluated. Guidelines are suggested for the flexible use of both *ex post* data on research productivity and *ex ante* simulations in allocating research resources.

In chapter 12, the primary source of demand for social science research on problems of agricultural and rural development is identified as derived from the demand for better institutional performance. The criteria are employed to identify the objectives of social science research activities at the agricultural research institute, the agricultural ministry, and the university level.

The productivity of modern agriculture is the result of a remarkable fusion of science and technology. In the last several decades, there has been a concern that the power released by this fusion is dangerous to the modern world and to the future of humankind. The consequences of the use of agricultural technology and its impact on the productive power of the land, on the health and safety of producers and consumers, on environmental amenities, and on rural communities have added new items to the agricultural research agenda. Chapter 13 is devoted to a discussion of social and moral responsibility in agricultural research.

A PHILOSOPHICAL PERSPECTIVE

In concluding this introductory chapter, I would like to emphasize that this is a book about agricultural research policy. It is not a scientific treatise on agricultural research. Nor is it limited to the economics, the sociology, or the politics of agricultural research. An attempt is made to go beyond formal analysis. Judgments are made about agricultural research policy. These judgments are not confined by any formal philosophical system. They arise out of a combination of scientific analysis, professional experience, and my personal perception of responsibility for the results of scientific inquiry and technological development.

NOTES

1. The results of the Minnesota symposium were published in Walter L. Fishel, ed., *Resource Allocation in Agricultural Research* (Minneapolis: University of Minnesota Press, 1971).

2. The results of the Airlie House Conference were published in Thomas G. Arndt, Dana G. Dalrymple, and Vernon W. Ruttan, eds., *Resource Allocation and Productivity in National and International Agricultural Research* (Minneapolis: University of Minnesota Press, 1977).

3. The results of this research are reported in Yujiro Hayami and Vernon W. Ruttan, *Agricultural Development: An International Perspective* (Baltimore: Johns Hopkins University Press, 1971). See also Hans P. Binswanger and Vernon W. Ruttan, *Induced Innovation: Technology, Institutions and Development* (Baltimore: Johns Hopkins University Press, 1978).

Chapter 2

Technical Change and Agricultural Development

The capacity to develop and to manage technology in a manner consistent with a nation's physical and cultural endowments is the single most important variable accounting for differences in agricultural productivity among nations.[1] The development of such capacity depends on many factors. These include the capacity to organize and to sustain the institutions that generate and transmit scientific and technical knowledge, the ability to embody new technology in equipment and materials, the level of husbandry skill and the educational accomplishments of rural people, the efficiency of input and product markets, and the effectiveness of social and political institutions.

The purpose of this chapter is to provide a historical interpretation and a theoretical explanation for the evolution of alternative paths of technical change in agriculture in the process of development. This induced innovation perspective will provide the conceptual framework for much of the discussion of agricultural research policy in subsequent chapters.[2]

AGRICULTURAL GROWTH

The 1970s were not kind to the reputations of either the prophets who have attempted to plot the course of world food production or the scholars who have attempted to understand the processes of agricultural development. Perspectives have shifted from a sense of impending catastrophe engendered by the world food crisis of the mid-1960s; to the euphoria of the new potentials opened up by the Green Revolution; to the crunch on world grain supplies resulting from poor harvests in South Asia, parts of sub-Saharan Africa and the Soviet Union between 1972 and 1974. The resurgence of food production (except in Eastern Europe and the Soviet Union) in 1975 and again in 1976 was followed by a renewed complacency about the prospects for world food supplies and agricultural development. Yet, the simple mathematics of inelastic food demand, uncertain weather, and improvident food stock policies virtually assures that there will be at least one food crisis of global significance during the next decade.

Uncertainty about the longer-term prospects for the growth of agricultural production and the economic welfare of rural people stems, however, from more fundamental concerns than the dramatic behavior of agricultural commodity markets. There has been a convergence of scientific opinion and ideological perspective to the effect that the world is fast approaching both the physical and cultural limits of growth. The theme that "progress breeds not welfare, but catastrophe" has again emerged from the underworld of social thought as a serious theme in scientific and philosophical inquiry.[3] Yet, the last two decades have been highly productive in advancing both our analytical capacity and our empirical knowledge of the role of technical change in agricultural development and of the sources of productivity growth in agriculture. The dating of "modern" agricultural growth in the now-conventional model, or paradigm, of agricultural development begins with the emergence of a period of sustained growth in total productivity—a rise in output per unit of total input.

When growth is based on a more intensive use of traditional inputs, little surplus becomes available to improve the well-being of rural people or to be transferred to the rest of the economy.[4] Little surplus has been generated by simple resource reallocation within farms, communities, or regions in the absence of technical change embodied in less expensive and more productive inputs. Only when the constraints on growth imposed by the primary reliance on indigenous inputs—those produced primarily within the agricultural sector— are released by new factors whose productivity is augmented by the use of new technology is it possible for agriculture to become an efficient source of growth in a modernizing economy.

Initially, growth in productivity—output per unit of total input— typically has been dominated by growth in a single partial productivity ratio. In the United States and the other developed countries of recent settlement, growth in labor productivity—output per worker—typically has "carried" the initial burden of growth in total

productivity. In countries characterized by relatively high population/
land ratios at the beginning of the development process, Germany
and Japan, for example, growth in land productivity—output per
hectare—has largely been responsible for growth in total productivity during the early years of modernization. As modernization
has continued, there has emerged a tendency for total productivity
growth to be fed by a more balanced growth in the partial productivity ratios, that is, by a growth in output per worker and per
hectare. (See figure 2.1.)

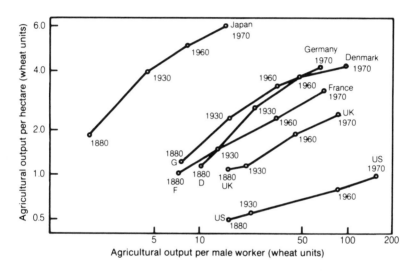

Figure. 2.1. Historical Growth Paths of Agricultural Productivity in the United
States, Japan, Germany, Denmark, France and the United Kingdom, 1880-
1970. (Used with permission from Vernon W. Ruttan, Hans P. Binswanger,
Yujiro Hayami, William Wade, and Adolf Weber, "Factor Productivity and
Growth: A Historical Interpretation," in Hans P. Binswanger and Vernon W.
Ruttan, *Induced Innovation: Technology, Institutions, and Development*
[Baltimore: Johns Hopkins University Press, 1978], p. 55.)

For a number of countries, however, the model outlined above has
little meaning. The 20th century has been characterized by a massive,
and continuously widening, disequilibrium between rich and poor
countries in the efficiency of resource use and the welfare of rural
people. Since World War II, output per hectare has been growing
at approximately the same rate of about 2 percent per year in
the developing countries as in the developed countries. But output per worker in the developing countries has been growing at
only one-third the rate of that in the developed countries—about

1.5 percent per year compared to about 4.5 percent per year. And, for large numbers of developing countries and for the lagging regions in many others, even those rates remain outside the personal experience of most farm families. In these lagging regions, output per hectare is growing at rates that are barely perceptible and output per worker has experienced no measurable change, not only between years but between generations. In the "developing world" food output per person is not significantly higher today than it was in the mid-1930s, and in many areas of Asia and Africa the number of kilograms of food grain that can be purchased with a day's labor has declined since the early 1960s.[5]

AGRICULTURAL DEVELOPMENT MODELS

During the rest of this century, it will be imperative that we develop and implement more effective agricultural development strategies than have been followed in the past. A useful step in thinking about this issue is to review the approaches to agricultural development that were available to us in the past and that will remain part of our intellectual equipment as we attempt to build on existing knowledge in the future.

Any attempt to evolve a meaningful perspective on the process of agricultural development must abandon the view of agriculture in premodern and traditional societies as being essentially static. Historically, the problem of agricultural development is not one of transforming a static agricultural sector into a modern dynamic sector but one of accelerating the rate of growth in agricultural output and productivity consistent with the growth of other sectors in a modernizing economy. Similarly, a theory of agricultural development should provide insight into the dynamics of agricultural growth (that is, into the changing sources of growth) in economies ranging from those in which output is growing at a rate of 1 percent or less to those in which agricultural output is growing at an annual rate of 4 percent or more.

The traditional literature on agricultural development can be represented by five general models:

- The frontier model
- The conservation model
- The urban-industrial impact model
- The diffusion model
- The high-payoff input model.[6]

The frontier model

Throughout most of history, expansion of the area cultivated or grazed has represented the main way of increasing agricultural production. The most dramatic example in Western history was the opening up of the new continents, North and South America and Australia, to European settlement during the 18th and 19th centuries. With the advent of cheap transport during the latter half of the 19th century, the countries of the newly opened continents became increasingly important sources of food and agricultural raw materials for the metropolitan countries of Western Europe.

In earlier times, similar processes had proceeded, though at a less dramatic pace, in the peasant and village economies of Europe, Asia, and Africa. The first millennium AD saw the agricultural colonization of Europe north of the Alps, the Chinese settlement of the lands of south of the Yangtze, and the Bantu occupation of Africa south of the tropical forest belts. Intensification of land use in existing villages was followed by pioneer settlement, the establishment of new villages, and the opening up of forest or jungle land to cultivation. In Western Europe there was a series of successive changes from neolithic forest fallow to systems of shifting cultivation on bushland and grassland followed first by short fallow systems and in recent years by annual cropping.

Where soil conditions were favorable, as in the great river basins and plains, the new villages gradually intensified their systems of cultivation. Where soil resources were poor, as in many of the hill and upland areas, new areas were opened up to shifting cultivation or to nomadic grazing. Under conditions of rapid population growth, the limits to the frontier model were often quickly reached. Crop yields typically were low, measured in terms of output per unit of seed rather than in output per unit of crop area. Output per hectare and per person hour tended to decline, except in the delta areas such as those in Egypt and South Asia and the wet rice areas of East Asia. In many areas the result was the worsening of the wretched conditions of the peasantry.

There are relatively few remaining areas of the world where development along the lines of the frontier model will represent an efficient source of growth during the last quarter of the 20th century. The 1960s and 1970s saw the "closing of the frontier" in most areas of Southeast Asia. In Latin America and Africa, the opening up of new lands awaits the development of technologies for the control of pests (such as the tsetse fly in Africa) and diseases or for the release and maintenance of productivity in problem soils.

This century can be seen as the transition from a period when most of the increases in world agricultural production occurred as a result of expansion in areas cultivated to a period when most of the increase in crop and animal production will come from increases in the frequency and intensity of cultivation. In the future, growth in agricultural production must come from changes in land use that make it possible to crop a given area of land more frequently and more intensively and hence to increase the output per unit area and per unit of time.

The conservation model

The conservation model of agricultural development evolved from the advances in crop and livestock husbandry associated with the English agricultural revolution and the concepts of soil exhaustion suggested by the early German chemists and soil scientists. It was reinforced by the concept, in the English classical school of economics, of diminishing returns to labor and capital applied to land. The conservation model emphasized the evolution of a sequence of increasingly complex land- and labor-intensive cropping systems, the production and use of organic manures, and labor-intensive capital formation in the form of physical facilities to more effectively use land and water resources.

Until well into the 20th century, the conservation model of agricultural development was the only approach to the intensification agricultural production that was available to most of the world's farmers. Its application can be effectively illustrated by the development of the wet-rice culture systems that emerged in East and Southeast Asia and by the labor- and land-intensive systems of integrated crop-livestock husbandry that increasingly characterized European agriculture during the 18th and 19th centuries. During the English agricultural revolution, more intensive crop-rotation systems replaced the open three-field system in which arable land was allocated between permanent cropland and permanent pasture. This involved the introduction and more intensive use of new forage and green manure crops and an increase in the availability and use of animal manures. This "new husbandry" permitted the intensification of crop-livestock production through the recycling of plant nutrients, in the form of animal manures, to maintain soil fertility. The inputs used—the plant nutrients, the animal power, the land improvements, the physical capital, and the agricultural labor force—were largely produced or supplied by the agricultural sector itself.

Agricultural development, within the framework of the conservation model, clearly was capable in many areas of the world of sustaining rates of growth in agricultural production at around 1 percent per year over relatively long periods of time. This rate is not compatible, however, with modern rates of growth in the demand for agricultural output, which typically fall between 3 and 5 percent in the developing countries.

The urban-industrial impact model

According to the conservation model, locational variations in agricultural development were related primarily to differences in environmental factors. It stands in sharp contrast to models that interpret geographical differences in the level and rate of economic development primarily in terms of the level and rate of urban-industrial development.

Initially, the urban-industrial impact model was formulated by von Thunen in Germany to explain geographical variations in the intensity of farming systems and in the productivity of labor in an industrializing society. In the United States, it was extended to explain the better performance of the input and product markets, which link the agricultural and nonagricultural sectors, in regions characterized by rapid urban-industrial development than in regions where the urban economy had not made a transition to the industrial stage. In the 1950s, interest in the urban-industrial impact model reflected a concern with the failure of agricultural resource development and price policies adopted in the 1930s to remove the persistent regional disparities in agricultural productivity and rural incomes.

The rationale for this model was developed in terms of more effective factor and product markets in areas of rapid urban-industrial development. Industrial development stimulated agricultural development by expanding the demand for farm products, by supplying the industrial inputs needed to improve agricultural productivity, and by drawing away surplus labor from rural areas. The empirical tests of the model have repeatedly confirmed the importance of a strong nonfarm labor market as a stimulus to higher labor productivity in agriculture.

The importance of urban-industrial development for agricultural development is frequently overlooked in the debates about the priority of agricultural and industrial development. Rapid technical change is sometimes a sufficient condition for expanding agricultural production. But it is rarely a sufficient condition for improving the incomes of agricultural workers. Failure to generate a strong demand for labor in the nonagricultural sector is often accompanied by a growing number of landless workers in rural areas and by a continued lag in the incomes of agricultural producers. The slow growth of the French economy between 1870 and World War II was a major constraint on productivity growth and economic prosperity in French agriclulture. During this period, the French peasant provided the urban-industrial sector with more food per capita at lower real prices. But neither the demand for commodities nor the demand for labor in the nonagricultural sector was expanding rapidly enough to generate significantly higher incomes in rural areas.[7]

The diffusion model

The diffusion approach to agricultural development rests on the empirical observation of substantial differences in land and labor productivity among farmers and regions. The route to agricultural development, in this view, is through more effective dissemination of technical knowledge and a narrowing of the productivity difference among farmers and among regions. The diffusion of better husbandry practices was a major source of productivity growth even in premodern societies. Prior to the development of modern agricultural research systems, substantial effort was devoted to crop exploration and introduction. Even in nations with well-developed agricultural research systems a significant effort is still devoted to the testing and refinement of farmers' innovations and to the testing and adaptation of exotic crop varieties and animal species.

This model provided the major intellectual foundation of much of the research and extension effort in farm management and production economics since the emergence, during the latter years of the 19th century, of agricultural economics and rural sociology as separate subdisciplines linking the agricultural sciences and the social sciences. The developments that led to the establishment of active programs of farm management research and extension occurred at a time when experiment station research was making only a modest contribution to agricultural productivity growth. A further contribution to the effective diffusion of known technology was provided by the research of rural sociologists on the diffusion process. Models that emphasized the relationship between diffusion rates and the personality characteristics and educational accomplishments of farm operators were developed.

The insights into the dynamics of the diffusion process, when coupled with the observation of wide agricultural productivity gaps among developed and developing countries and a presumption of inefficient resource allocation among "irrational, tradition-bound" peasants, produced an extension or a diffusion bias in the choice of agricultural development strategy in many developing countries during the 1950s. The limitations of the diffusion model as a foundation for the design of agricultural development policies became increasingly apparent as technical assistance and community development programs, based explicitly or implicitly on the diffusion model, failed to generate either rapid modernization of traditional farms and communities or rapid growth in agricultural output.

The high-payoff input model

The inadequacy of policies based on the conservation, urban-industrial impact, and diffusion models led to a new perspective in the 1960s. In the high-payoff input model the key to transforming a traditional agricultural sector into a productive source of economic growth is investment designed to make modern, high-payoff inputs available to farmers in poor countries. Peasants, in traditional agricultural systems, were viewed as rational, efficient resource allocators. They remained poor because, in most poor countries, there were only limited technical and economic opportunities to which they could respond. The new, high-payoff inputs were classified into three categores.

- The capacity of public- and private-sector research institutions to produce new technical knowledge
- The capacity of the industrial sector to develop, produce, and market new technical inputs
- The capacity of farmers to acquire new knowledge and to use new inputs effectively

The enthusiasm with which the high-payoff input model has been accepted and translated into economic doctrine has been due in part to the proliferation of studies reporting high rates of return to public investment in agricultural research. (See chapter 10.) It was also due to the success of efforts to develop new, high-productivity grain varieties suitable for the tropics. New, high-yielding wheat varieties were developed in Mexico in the 1950s and new, high-yielding rice varieties were developed in the Philippines in the 1960s. These varieties were highly responsive to industrial inputs, such as fertilizer and other chemicals, and to more effective soil and water management.

The high returns associated with the adoption of the new varieties and the associated technical inputs and management practices have led to rapid diffusion of the new varieties among farmers in several countries in Asia, Africa, and Latin America. (See figure 2.2.)

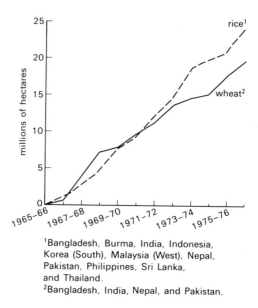

[1]Bangladesh, Burma, India, Indonesia, Korea (South), Malaysia (West), Nepal, Pakistan, Philippines, Sri Lanka, and Thailand.
[2]Bangladesh, India, Nepal, and Pakistan.

Figure 2.2. Estimated Area Planted to High-Yielding Varieties of Wheat and Rice in Asia (East and South). (From Dana G. Dalrymple, *Development and Spread of High-Yielding Varieties of Wheat and Rice in Less Developed Nations* [Washington, D.C.: U.S. Department of Agriculture, in cooperation with U.S. Agency for International Development, September 1978], p. 117.)

AN INDUCED INNOVATION MODEL

The high-payoff input model remains incomplete as a theory of agricultural development. The major contributors to higher agricultural productivity—education and research—are public goods that generally are not traded through the market place. The mechanism by which resources are allocated to education and research was not fully incorporated into the model. The high-payoff input model does not explain how resource endowments induce the development of efficient technologies for a particular society. Nor does it attempt to specify the processes by which input and product price relationships induce investment in research in a direction consistent with a nation's particular resource endowments.

These limitations in the high-payoff input model led to efforts to

develop a model of agricultural development in which technical
change is treated as endogenous to the development process, rather
than as an exogenous factor that operates independently of other
development processes.[8] The induced innovation perspective was
stimulated by historical evidence that different countries had followed alternative paths of technical change in the process of agricultural development (figure 2.1) and by a consideration of the wide
productivity differentials among countries (figure 2.3).

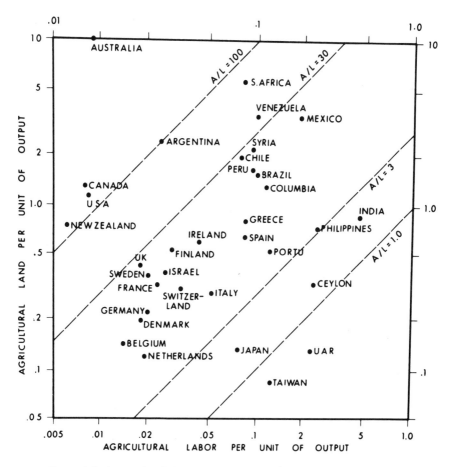

Figure 2.3. International Comparison of Labor/Output and Land/Output
Ratios in Situations Characterized by Different Land/Labor Ratios. The
diagonal lines represent constant land/labor ratios. (From Saburo Yamada and
Vernon W. Ruttan, "International Comparisons of Productivity in Agricul-
ture," in *New Developments in Productivity Analysis*, John W. Kendrick and
Beatrice N. Vaccara, eds., Conference on Research in Income and Wealth:
Studies in Income and Wealth, vol. 44 [Chicago: University of Chicago Press,
for the National Bureau of Economic Research, 1980], p. 534.)

The productivity levels achieved by farmers in the most advanced countries in each productivity grouping (figures 2.1 and 2.3) can be seen as arranged along a productivity frontier. This frontier reflects the level of technical progress and factor inputs achieved by the most advanced countries in each resource endowment classification. In order to make these productivity levels available to farmers in low productivity countries, it is necessary to make investments in the agricultural research capacity required to develop technologies that are appropriate to the natural and institutional environments of the low productivity countries. Furthermore, there must be investments in the physical and institutional infrastructures needed to realize the new production potential opened up by advances in technology.

Alternative paths of technological development

There is clear evidence that technology can be developed to facilitate the substitution of relatively abundant and hence cheap factors for relatively scarce and hence expensive factors of production. The constraints imposed on agricultural development by an inelastic supply of land have, in economies such as Japan's and Taiwan's, been offset by the development of high-yielding crop varieties designed to facilitate the substitution of fertilizer for land. The constraints imposed by an inelastic supply of labor, in countries such as the United States, Canada, and Australia, have been offset by technical advances leading to the substitution of animal and mechanical power for labor. In both cases, the new technologies—embodied in new crop varieties, new equipment or new production practices—may not always be substituted for land or labor by themselves. Rather, the new technologies may serve as catalysts that facilitate the substitution of the relatively abundant factors for the relatively scarce factors.

In agriculture, two kinds of technologies generally correspond to this taxonomy: mechanical technology to "labor-saving" and biological (or biological and chemical) technology to "land-saving." The primary effect of the adoption of mechanical technology is the facilitation of the substitution of power and machinery for labor. Typically, this results in a decline in labor use per unit of land area. The substitution of animal or mechanical power for human labor enables workers to extend their efforts over larger land areas. The primary effect of the adoption of biological technology is the facilitation of the substitution of labor and/or industrial inputs for land. This may occur through increased recycling of soil fertility by more labor-intensive conservation systems; through use of chemical fertilizers; and through husbandry practices, management systems, and inputs such as insecticides that permit an optimum yield response.

Historically, there has been a close association between advances in output per unit of land area and advances in biological technology and between advances in output per worker and advances in mechanical technology. These historical differences have given rise to the cross-sectional differences in productivity and factor use illustrated in figure 2.3. Advances in biological technology may also result in increases in output per worker if the rate of growth per hectare exceeds the rate of growth of the agricultural labor force.

In the Philippines, for example, growth in output per worker prior to the mid-1950s was due primarily to expansion in the area cultivated per worker. Since the early 1960s, growth in output per worker has been due to increase in the output per unit of land area. (See figure 2.4.)

Induced technical innovation

An examination of the historical experience of the United States and Japan illustrates the theory of induced technical innovation. In the United States it was primarily the progress of mechanization, first with animal and later with tractor motive power, that facilitated the expansion of agricultural production and productivity by increasing the area operated per worker. In Japan, it was primarily the progress of biological technologies, such as varietal improvement leading to increased yield response to higher levels of fertilizer application, that permitted rapid growth in agricultural output in spite of severe constraints on the supply of land. These contrasting patterns of productivity growth and factor use can best be understood in terms of a process of dynamic adjustment to changing relative factor prices.

In the United States the long-term rise in wage rates relative to the prices of land and machinery encouraged the substitution of land and power for labor. This substitution generally involved progress in the application of mechanical technology to agricultural production. The more intensive application of mechanical technology depended on the invention of technology that was more intensive in its use of equipment and more extensive in its use of land relative to labor.

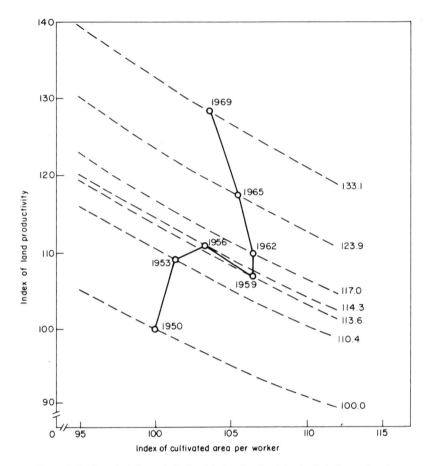

Figure 2.4. Historical Growth Path of Labor Productivity in Relation to Land Productivity and Cultivated Area per Worker in Philippine Agriculture, 1948-1968. (Used with permission from Cristina Crisostomo David and Randolph Barker, "Agricultural Growth in the Philippines, 1948-1971," in *Agricultural Growth in Japan, Taiwan, Korea and the Philippines*, Yujiro Hayami, Vernon W. Ruttan, and Herman Southworth, eds. [Honolulu: University Press of Hawaii, an East-West Center Book, 1979], p. 134.)

For example, the Hussy and McCormick reapers in use during the 1860s and 1870s required, over a harvest period of about two weeks, five workers and four horses to harvest 140 acres of wheat. When the binder was introduced, it was possible for a farmer to harvest the same acreage of wheat with two workers and four horses. The process illustrated by the substitution of the binder for the reaper has been continuous. As the limits to horse-powered mechanization were reached in the early part of this century, the process was continued by the introduction of the tractor as the primary source of motive power. The process has continued with the substitution of larger and more highly powered tractors and the development of self-propelled harvesting equipment.

In Japan, the supply of land was inelastic and its price rose relative to wages. It was not, therefore, profitable to substitute power for labor. Instead, the new opportunities arising from a continuous decline in the price of fertilizer relative to the price of land were exploited through advances in biological technology. Varietal improvement was directed, for example, toward the selection and breeding of more fertilizer-responsive varieties of rice. The enormous changes in fertilizer input per hectare that have occurred in Japan since 1880 reflect not only the effect of the response by farmers to lower fertilizer prices but the development by the Japanese agricultural research system of "fertilizer-consuming" rice varieties in order to take advantage of the decline in the real price of fertilizer.

The effect of relative prices in the development and choice of technology is illustrated with remarkable clarity for fertilizer in figure 2.5, in which the U.S. and Japanese data on the relationship between fertilizer input per hectare of arable land and the fertilizer/land price ratio are plotted for the period 1880 to 1960. In both 1880 and 1960, U.S. farmers were using less fertilizer than Japanese farmers. However, despite enormous differences in both physical and institutional resources, the relationship between these variables has been almost identical in the two countries. As the price of fertilizer declined relative to other factors, scientists in both countries responded by inventing crop varieties that were more responsive to the lower prices of fertilizer. American scientists, however, always lagged behind the Japanese by a few decades in the process because the lower price of land relative to fertilizer in the United States resulted in a lower priority being placed on yield-increasing technology.

It is possible to illustrate the same process with cross-sectional data for mechanical technology. Variations in the level of tractor horsepower per worker among countries is very largely a reflection of the price of labor relative to the price of power. (See figure 2.6.) Even in countries with small farms, such as Japan and Taiwan, increases in wage rates have induced the adaptation of mechanical technology to the size of the farm.

The effect of a rise in the price of fertilizer relative to the price of land or of the price of labor relative to the price of machinery has been to induce advances in biological and mechanical technology.

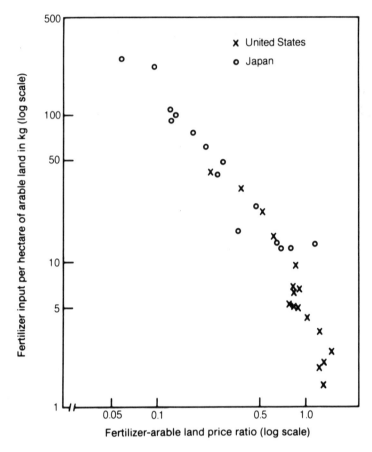

Figure 2.5. Relationship between Fertilizer Input per Hectare of Arable Land and the Fertilizer/Arable Land Price Ratio. (Used with permission from Yujiro Hayami and Vernon W. Ruttan, *Agricultural Development: An International Perspective* [Baltimore: Johns Hopkins University Press, 1971], p. 127.)

The effect of the introduction of lower cost and more productive biological and mechanical technology has been to induce farmers to substitute fertilizer for land and mechanical power for labor. These responses to differences in resource endowments among countries and to changes in resource endowments over time by agricultural research institutions, by the farm supply industries, and by farmers have been remarkably similar in spite of differences in culture and tradition.

During the last two decades, as wage rates have risen rapidly in Japan and as land prices have risen in the United States, there has

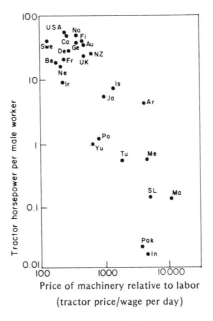

Figure 2.6. Relationship between Tractor Horsepower per Male Worker and the Price of Machinery Relative to Labor, 1970. (From Saburo Yamada and Vernon W. Ruttan, "International Comparisons of Productivity in Agriculture," in *New Developments in Productivity Measurement and Analysis*, John W. Kendrick and Beatrice N. Vaccara, eds., Conference on Research in Income and Wealth: Studies in Income and Wealth, vol. 44 [Chicago: University of Chicago Press, for the National Bureau of Economic Research, 1980], p. 540.

been a tendency for the patterns of technological change in the two countries to converge. (See figure 2.7.) Between 1960 and 1970 fertilizer consumption per hectare rose more rapidly in the United States than in Japan and tractor horsepower per worker rose more rapidly in Japan than in the United States. Both countries appear to be converging toward the European pattern of technical change in which increases in output per worker and increases in output per hectare occur at approximately equal rates.

There will be further changes in the future. In the early and mid-1970s, the price of energy has risen. This has affected both the price of fuel and the price of fertilizer. It is unlikely that declining fertilizer prices in the future will be as important a factor in determining the direction of biological technology as during the past century. Higher fertilizer prices have already induced a substantial increase in the research resources devoted to the investigation of potential biological and organic sources of plant nutrition. It is possible that

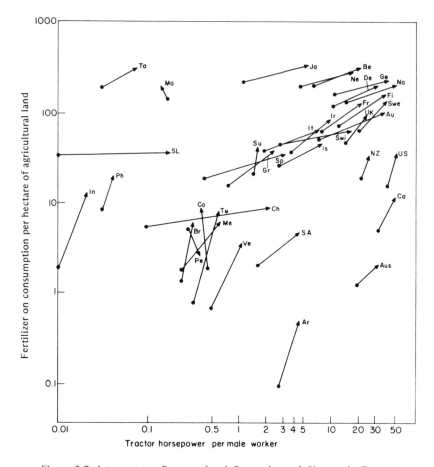

Figure 2.7. Intercountry Cross-sectional Comparison of Changes in Tractor Horsepower per Male Worker and in Fertilizer Consumption per Hectare, 1960-1970 (log scale). (From Saburo Yamada and Vernon W. Ruttan, "International Comparisons of Productivity in Agriculture," *New Developments in Productivity Measurement and Analysis,* John W. Kendrick and Beatrice N. Vaccara, eds., Conference on Research in Income and Wealth: Studies in Income and Wealth, vol. 44 [Chicago: University of Chicago Press, for the National Bureau of Economic Research, 1980], p. 546.

momentum of advance in biological technology during the next several decades will be affected by the necessity of a transition comparable to the shift from the horse to the tractor as a source of motive power in the area of mechanical technology.

In most developing countries, however, fertilizer prices at the farm level should continue to decline relative to the price of land for some time. In these countries, the expansion of the domestic fertilizer-production capacity and the improvement of the marketing and transport systems should result in a decline in the costs of moving fertilizer from the factory to the farm. Increased fertilizer use per hectare will continue to be an important source of productivity growth in most developing countries during the next several decades.

Induced institutional innovation

In discussing the theory of induced technical change in agriculture, we need to consider the personal behavior of individual research scientists and the institutional behavior of the agricultural experiment stations and research institutes that support their research.[9] In most countries that have successfully achieved rapid rates of technical progress in agriculture, the "socialization" of agricultural research has been deliberately employed as an instrument of modernization in agriculture. The induced innovation model of technical change in agriculture implies that both research scientists and research administrators are responsive to differences in resource endowments and to changes in the economic environment in which they work.

The response of research scientists and administrators represents the critical link in the inducement mechanism. The model does not imply that it is necessary for individual scientists and research administrators in public institutions to respond consciously to market prices or directly to farmers' demands for research results in the selection of research objectives. They may, in fact, be motivated primarily by a drive for professional achievement and recognition. Or they may view themselves as responding to an "obvious and compelling need" to remove the constraints on growth of production or on factor supplies. It is only necessary that there exists an effective incentive mechanism to reward the scientist or administrators, by material benefits or prestige, for their contributions to the solution of problems that are of social or economic significance.

The importance of effective responses to economic and social priorities is not limited to the field of applied science. If appropriate incentives lead applied researchers to respond to the needs of society and if research workers in the basic sciences are sensitive to the needs of applied researchers for new theories and new methodologies, basic researchers in effect are also responding to the needs of society. It is not uncommon that major breakthroughs in basic science or supporting science are created through the process of solving the problems raised by research workers in the more-applied fields.

The response by the scientific community to the recent rise in the price of fossil fuel-based inputs represents a dramatic example of the induced innovation process. Increases in the price of nitrogen fertilizer have induced a shift in scientific resources toward more intensive research and development activity on the biological and organic sources of plant nutrition. The low productivity of agricultural scientists in many developing countries is due to the fact that many societies have not yet succeeded in developing incentives that lead to the focusing of scientific effort on the significant problems of domestic agriculture. Under such conditions, scientific skills atrophy or are directed to the reward systems of the international scientific community.

It is not argued, however, that technical change in agriculture is wholly of an induced character. The supply of new technology has an exogenous dimension that stems from autonomous development in the basic sciences as well as an endogenous dimension that can be influenced by demand. Technical change in agriculture reflects, in addition to the effects of resource endowments and growth in demand, the progress of general science and technology. Progress in general science that lowers the "cost" of technical change may influence the direction of technical change in agriculture in a manner that is unrelated to changes in factor proportions and product demand. Similarly, advances in science and technology in the developed countries, in response to their own resource endowments, may result in a bias in the technical opportunities that become available in the developing countries. Even in these cases, the rate of adoption and the impact on productivity will be strongly influenced by the conditions of resource supply and product demand since these forces are reflected through input and product markets.

DISEQUILIBRIUM IN AGRICULTURE

During the last two decades, the institutional capacity to generate technical changes adapted to national and regional resource endowments has been established in many developing countries. More recently, these emerging national systems have been buttressed by a new system of international crop and animal research institutes. (See chapter 5.) These new institutes have become both important sources of new knowledge and technology and increasingly effective communication links among the developing national research systems.

Both the new international system and many of the new national systems have been highly productive. The evidence cited in chapter 10 shows that in India, for example, investment in agricultural research has generated annual rates of return in the range of 40 to 60 percent. Rates of return in this range are, however, not an entirely valid source of self-congratulation. Though they testify to the efficient allocation of the research resources that society has made available to the agricultural science community, they also indicate a continuing underinvestment in agricultural research.

At the global level, there has been a continuing disequilibrium in agricultural productivity and in the well-being of rural people. This disequilibrium has taken the form of the widening disparities in the rate of growth of total agricultural outputs, in labor and land productivity, and in income and wage rates among regions. It is clear that a fundamental source of such disequilibrium has been the lag in shifting from a natural resource-based to a science-based agriculture. The effects of lags in the application of knowledge are also important sources of regional disequilibria in many countries, such as Mexico and India.

It seems increasingly clear that the elimination of both the international and domestic disequilibria in agricultural productivity will require increases in the allocation of research resources and of development investment to the agricultural sector and toward rural areas. It was a major step forward when the allocation of research resources in developing countries broke away from the mold that had been established in the developed countries and began to emphasize the development of biological technologies designed to raise output per unit of land area in order to release the constraints imposed by an inelastic supply of land. It is now time to begin to make the next step and look directly at the most abundant resource available in most poor countries—people—and the low productivity of human resources in rural areas.

The importance of this refocusing can be illustrated by the data presented in tables 2.1 and 2.2.[10] The sources of labor productivity differences among countries are classified into three broad categories —resource endowments, technical inputs, and human capital. Land and livestock serve as proxy variables for resource endowments; machinery and fertilizer, for technical inputs; and general education in agriculture, for human capital.

Land and livestock represent a form of long-term capital formation embodying inputs supplied primarily from within the agricultural sector. In traditional systems of agriculture, indigenous, labor-intensive capital formation represents almost the only source of growth in labor productivity. Fertilizer, as measured by nutrient

Table 2.1. Accounting for Intercountry Differences in Labor Productivity

Difference	Between 11 Developing Countries[a] and 4 Recently Developed Countries[b]		Between 11 Developing Countries and 9 Older Developed Countries[c]		Between 9 Older Developed Countries and 4 Recently Developed Countries	
	%	Index	%	Index	%	Index
Difference in output per male worker	93.6	100	83.5	100	61.5	100
Difference explained						
Total	**90.0**	**96**	**71.1**	**85**	**50.5**	**82**
Resource endowments	*32.6*	*35*	*17.5*	*21*	*29.1*	*47*
Land	9.7	10	1.8	2	9.7	16
Livestock	22.9	25	15.7	19	19.4	31
Technical inputs	*24.5*	*26*	*24.3*	*29*	*10.4*	*17*
Fertilizer	14.6	16	14.5	17	3.9	6
Machinery	9.9	10	9.8	12	6.5	11
Human capital	*32.9*	*35*	*29.4*	*35*	*10.9*	*18*
General education	19.5	21	17.6	21	3.3	6
Technical education	13.4	14	11.7	14	7.6	12

Source: Adapted from Yujiro Hayami and Vernon W. Ruttan, *Agricultural Development: an International Perspective.* (Baltimore: Johns Hopkins University Press, 1971), pp. 96-101.
a. Brazil, Sri Lanka, Colombia, India, Mexico, Peru, Philippines, Syria, Taiwan, Turkey, United Arab Republic.
b. Australia, Canada, New Zealand, United States.
c. Belgium, Denmark, France, Germany, Netherlands, Norway, Sweden, Switzerland, United Kingdom.

Table 2.2. Accounting for Intercounty Differences in Labor Productivity
between the United States and Selected Countries

Difference	India	Japan	UK	Argentina	Canada
Difference in output per male worker					
%	97.8	89.2	55.8	60.0	24.0
Index	100	100	100	100	100
Difference Explained					
Total Index	104	74	89	76	98
Resource endowments					
Index	33	33	33	−8	20
Technical inputs index	26	25	24	40	51
Human capital index	45	10	33	44	28

Source: Adapted from Yujiro Hayami and Vernon W. Ruttan, *Agricultural Development: An International Perspective*. (Baltimore: Johns Hopkins University Press, 1971). pp. 96-101.

consumption in commercial fertilizer, and machinery, as measured by tractor horsepower, are employed as proxies for the whole range of inputs in which modern mechanical and biological technologies are embodied. The proxies for human capital include measures of both the general educational level of the rural population and specialized education in the agricultural sciences and technology. General education is viewed as a measure of the capacity of a population to utilize new technical knowledge. Graduates in the agricultural sciences and technological fields represent the major source of scientific and technical personnel for agricultural research and extension.

The difference in the average agricultural output per worker between the 11 developing countries and the 9 older developed countries was 83.5 percent. Differences in human capital investment alone account for over one-third of the difference. Differences in land resources per worker account for only 2 percent of the difference. It seems apparent that, in spite of the limitations of land resources in the developing countries, they could achieve levels of output per worker comparable to the European levels of the early 1960s through a combination of investment in human capital, investment in experimental stations and industrial capacity to make modern technical inputs available to their farmers, and investment in the labor-intensive capital formation characterized by livestock and perennial crops and by land and water development.

The difference in the average agricultural output per worker between the nine older developed countries and the four recently developed countries was 61.5 percent. The results are quite different from the comparison between the developing countries and the older developed countries. Technical inputs and human capital account for only slightly more than one-third of the difference. Resource endowments account for close to half. It appears that output per worker in the older developed countries would have great difficulty in approaching the levels of the recently developed countries in the absence of substantial adjustments in labor/resource ratios. However, the older developed countries have clearly failed to take full advantage of the growth opportunities available to them through greater investment in technical manpower and in agricultural science capacity. The comparisons of individual countries tend to reinforce the inferences based on the group comparisons. Failure to take full advantage of the potential growth from human capital and technical inputs is significantly more important than limitations in resource endowments in accounting for differences in output per worker.

It is clear that a fundamental source of the widening disequilibrium in world agriculture has been the lag in shifting from a natural resource-based agriculture to a science-based agriculture. In the developed countries, human capital and technical inputs have become the dominant sources of output growth. Differences in the natural resource base have accounted for an increasingly less significant share of the widening productivity gap among nations. Productivity differences in agriculture are increasingly a function of investments in the education of rural people and in scientific and industrial capacity rather than in natural resource endowments.

The role of education as a factor affecting the productivity of agricultural labor is particularly important during periods in which a nation's agricultural research system is introducing a continuous stream of new technology into the agricultural system. In an agricultural system characterized by static technology, there are few gains to be realized from education in rural areas. Rural people who have lived for generations with essentially the same resources and the same technology have learned from long experience what their efforts can get out of the resources that are available to them. Children acquire the skills that are worthwhile from their parents. Formal schooling has little economic value in agricultural production.

As soon as new technical opportunities start becoming available, this situation changes. Technical change requires the acquisition of new husbandry skills; additional resources such as new seeds, new chemicals, and new equipment have to be acquired from nontraditional sources. New skills in dealing both with natural resources and with factor and product markets have to be acquired. New and more efficient factor and product market institutions linking agriculture with the nonagricultural sector have to be developed. The economic value of education to farmers and farm workers, and to the larger society, experiences a sharp rise as a result of the disequilibria introduced by new technical opportunities.[11]

The underutilization of labor resources in rural areas poses a serious challenge both to agricultural scientists and administrators whose training and experience are in the natural sciences and to those whose training and experience are in the social sciences. They must begin to view the existence of poor or underutilized labor resources as an opportunity for development just as they have in the past viewed poor or underutilized land and water resources as opportunities for development. The challenge to make productive use of the underutilized labor in Brazil's Northeast must be given at least as high a priority as the challenge to make more productive use of the problem soils of the llanos or the Amazon basin. The new high-payoff agricultural technologies will increasingly be those that have the effect of increasing the demand for underutilized labor resources—that put people to work more days per year and more productively. The high-payoff institutional innovations will be those that enable the people living in rural areas both to upgrade and to achieve more effective command over the physical and human resources that are available to them in order to expand their capacity to respond to the new technical opportunities that are becoming available.

PERSPECTIVE

The induced innovation perspective does not imply that the progress of agricultural technology can be left to an "invisible hand"—to the undirected market forces that will direct technology along an "efficient" pattern determined by "original" resource endowments or relative factor and product prices. The production of the new knowledge leading to technical change is the result of a process of institutional development. The invention of the public-sector agricultural research institute—that is, the socialization of agricultural research— was one of the great institutional innovations of the 19th century.

Technological change, in turn, represents a powerful source of demand for institutional change. The processes by which new knowledge can be brought to bear to alter the rate and direction of technical change in agriculture is, however, substantially greater than our understanding of the processes by which resources are brought to bear on the process of institutional innovation and transfer. The developing world is still trying to cope with the debris of nonviable institutional innovations, with extension services that have no capacity to extend knowledge or little useful knowledge to extend, with cooperatives that serve to channel resources to village elites, with price stabilization policies that have the effect of amplifying commodity price fluctuations, and with rural development programs that are incapable of expanding the resources available to rural people.

The need for viable institutions capable of supporting more rapid agricultural growth and rural development is even more compelling today than a decade ago. As the technical constraints on the growth of agricultural productivity become less binding, there is an increasing need for institutional innovation that will result in a more effective realization of the new technical potential. The trial-and-error approaches involved in ad hoc production campaigns and rural development programs have been costly in terms of human resources and have rarely been effective in building rural institutions that have prevailed beyond the enthusiasm of the moment.

One implication of the induced innovation perspective is the growing interdependence between advances in knowledge in the natural sciences and the social sciences as they relate to agricultural and rural development. In the absence of new technical opportunities—new sources of disequilibrium in the productivity of physical and human resources—there would be little demand for new knowledge about the institutional dimensions of agricultural and rural development processes. Similarly, unless social science research can generate new knowledge leading to viable institutional innovation and more effective institutional performance, the potential productivity growth made possible by scientific and technical innovation will be underutilized.

A major challenge to the legislative and executive agencies that are responsible for the allocation of resources to research, and to the administrators responsible for research program development and management, is how to direct the limited research resources available to them to those areas that have high rates of return when evaluated in terms of the economic and social priorities of the societies of which they are a part. The analytical methods that are available to the individuals and organizations who are responsible for research policy and management are discussed in chapter 11.

The final point I want to make in this chapter is that the historical evidence suggests that those countries that have been successful in generating rapid technical change in agriculture have found it necessary to develop the institutional capacity for agricultural research and development that has enabled them to follow a path of technical change consistent with their own resource and cultural endowments. Those countries that have attempted to rely primarily on borrowed technology have rarely developed the capacity to adapt and manage the borrowed technology in a manner capable of sustaining agricultural development. The induced innovation model provides a powerful analytical perspective through which to interpret the unique demands of both developed and developing societies for technical and institutional inovation.

NOTES

1. This chapter represents a revised version of Vernon W. Ruttan, "Induced Innovation and Agricultural Development," *Food Policy*, 2 (3) (August 1977), pp. 196-216 (published by IPC Science and Technology Press Ltd., Guildford, UK).

2. A major limitation of much of the appropriate technology literature is its lack of adequate conceptualization of the relationship between technical change and changes in resource and cultural endowments. As a result, the appropriate technology concept has remained subjective, undefined, qualitative, and idiosyncratic. The induced innovation model outlined in this chapter suggests one method of arriving at a definition of appropriate technology. The basic reference in the literature of the appropriate technology movement is E.F. Schumacher, *Small Is Beautiful: Economics as if People Mattered* (New York: Harper and Row, 1973). For a review and an evaluation of the appropriate technology literature, see Richard S. Eckaus, *Appropriate Technologies for Developing Countries* (Washington, D.C.: National Academy of Sciences, 1977).

3. For a perspective on the limits to growth literature, see Vernon W. Ruttan, "Technology and the Environment," *American Journal of Agricultural Economics*, 53 (December 1971), pp. 707-17; and Robert M. Solow, "The Economics of Resources or the Resources of Economics," *American Economic Review*, 64 (May 1974), pp. 1-14.

4. This argument is developed most fully in Theodore W. Schultz, *Transforming Traditional Agriculture* (New Haven, Conn.: Yale University Press, 1964). See particularly pp. 36-52. Following the publication of Schultz's book, a view of the peasant as "poor but efficient" rapidly emerged as a new orthodoxy, replacing the earlier view that the major limit on agricultural development was the burden of custom and tradition that bore down on the peasant producer in traditional agricultural systems.

5. See Michael Lipton, "Urban Bias and Food Policy in Poort Countries," *Food Policy*, 1 (1) (November 1975), pp. 41-52.

6. A more detailed development of the conservation, urban industrial impact, diffusion, and high-payoff input models is presented in Yujiro Hayami and Vernon W. Ruttan, *Agricultural Development: An International Perspective* (Baltimore: Johns Hopkins University Press, 1971), pp. 27-43. The discussion of the frontier model presented here draws very heavily from D. P. Grigg, *The Agricultural Systems of the World: An Evolutionary Approach* (Cambridge: Cambridge University Press, 1974), and Shigeru Ishikawa, *Economic Development in Asian Perspective* (Tokyo: Kinokuniya Bookstore Co., 1967).

7. Vernon W. Ruttan, "Structural Retardation and the Modernization of French Agriculture: A Skeptical View," *Journal of Economic History*, 37 (September 1978), pp. 714-28.

8. Hayami and Ruttan, *Agricultural Development*, pp. 53-63 and 111-35.

9. The theory of induced institutional innovation is developed more fully in Hans P. Binswanger and Vernon W. Ruttan, *Induced Innovation: Technology, Institutions and Development* (Baltimore: Johns Hopkins University Press, 1979). The discussion of research resource allocation in this section draws on Walter L. Fishel, ed., *Resource Allocation in Agricultural Research* (Minneapolis: University of Minnesota Press, 1971), and Thomas M. Arndt, Dana Dalrymple, and Vernon W. Ruttan, eds., *Resource Allocation and Productivity in National and International Agricultural Research* (Minneapolis: University of Minnesota Press, 1977).

10. The accounting for intercountry differences in labor productivity utilizes coefficients obtained from estimating an intercountry "metaproduction function" of the Cobb-Douglas form. The percentage differences in output per worker can be expressed as the sum of percentage differences in conventional and unconventional factor inputs per worker, weighted by their respective production elasticities. The coefficients used in the growth accounting were derived from data centered on 1960. Preliminary results obtained by Saburo Yamada from data centered on 1970 are generally consistent with the results presented in tables 2.1 and 2.2. See Saburo Yamada and Vernon W. Ruttan, "International Comparisons of Productivity in Agriculture," in *New Developments in Productivity Measurement and Analysis*, John W. Kendrick and Beatrice N. Vaccara, eds., Conference on Research in Income and Wealth: Studies in Income and Wealth, vol. 44 (Chicago: University of Chicago Press for the National Bureau of Economic Research, 1980), pp. 509-94.

11. There is substantial evidence over a wide range of developed and developing countries that, under conditions of rapid improvement in agricultural technology, a 1 percent increase in the level of general education (measured in terms of schooling or literacy ratios) has approximately the same impact on agricultural output as a 1 percent increase in the agricultural labor force. See, for example, Zvi Griliches, "Research Expenditures, Education, and the Aggregate Agricultural Production Function," *American Economic Review*, 54 (December 1964), pp. 961-74; Hayami and Ruttan, *Agricultural Development*, pp. 90-96. There is also some empirical evidence that intensive extension activity can serve as a partial substitute for formal education at lower levels of education but that at higher levels of education schooling and extension are complementary. See, for example, Abdul Halim, "The Economic Contribution of Schooling and Extension to Rice Production in Laguna, Philippines," *Journal of Agricultural Economics and Development*, 7 (January 1977), pp. 33-46. The return to education is, of course, sensitive to the level of other inputs. Under conditions of static technology, it is possible to overinvest in education. See, for example, Arnold C. Harberger, "Investment in Men versus Investment in Machines: The Case of India," in *Education and Economic Development*, C. Arnold Anderson and Mary Jean Bowman, eds. (Chicago: Aldine, 1965), pp. 11-50. The literature on the contribution of farmer education to agricultural production has been reviewed by Marlaine E. Lockhead, Dean T. Jamison, and Lawrence J. Lau, "Farmer Education and Farm Efficiency: A Survey," *Economic Development and Cultural Change*, 29 (October 1980), pp. 37-76. For a skeptical perspective on the "human capital" literature, see Mark Blaug, "Human Capital Theory: A Slightly Jaundiced Survey," *Journal of Economic Literature*, 14 (September 1975), pp. 850-55.

Chapter 3

The Agricultural Research Institution

The basic functional unit in any research system is the experiment station, the research institute, or the laboratory. This chapter presents a model of the process by which a research institution transforms the resources available to it into new knowledge. Next, the processes by which new knowledge is generated by individual scientists and research teams are discussed. The final section of the chapter presents a framework for analyzing the returns from the new knowledge produced by a research institution and considers the process by which the research system itself is transformed by the demands that society places on it.

THE EXPERIMENT STATION

The productivity of the human and material resources that are devoted to the development and operation of an agricultural research institution—whether operated as part of a centralized national research system, as a state unit in a decentralized federal-state system, as a commodity research center in the international system, or as a research laboratory or station by a firm in the private sector—must be evaluated in terms of its contribution to economic and social objectives.[1] The experiment station or research institute can be viewed as a system for transforming intellectual and physical capital into new knowledge and new technology. This knowledge is made available in research papers, books, bulletins, and information releases and in consultations with other scientists, science administrators, technicians, extension workers, and producers of agricultural and industrial products. It is frequently embodied in blueprints, formulas, models, seeds, and chemicals. And its social and economic impact is ultimately realized in the form of technical or institutional change.

In attempting to understand the responses of the modern agricultural research institute or experiment station to the institutional and physical environment in which it finds itself, it is helpful to conceptualize the experiment station along the lines suggested in figure 3.1. The chart is an attempt to relate the internal production processes of the station to the external environment in which it operates. An experiment station's production processes involve the transformation of stock and flow resources into intermediate products. These products are, in turn, transformed into outputs. The outputs can be categorized under three headings—information, capacity, and influence.

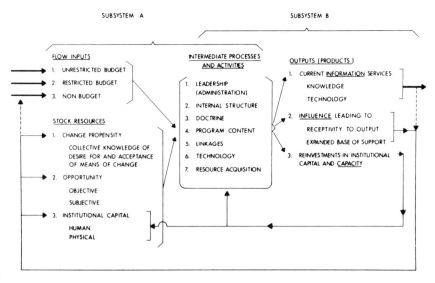

Figure 3.1. Systems Model of Experiment Station's Performance and Development. (Adapted from Melvin G. Blase and Arnold Paulson, "The Agricultural Experiment Station: An Institutional Development Perspective," *Agricultural Science Review,* 10 [Second Quarter, 1972], pp. 11-16.)

The most important and visible output of an experiment station or a research laboratory is the information, in the form of new knowledge or new technology, that is generated and released. In some fields, plant breeding, for example, the new technology may be embodied in higher-yielding or pest-resistant crop varieties (cultivars). In other fields the knowledge may be embodied in published reports on farm management, cropping practices, or animal nutrition. Over the long run, the use of resources for agricultural research must be justified in terms of the value of the new knowledge or new technology that it produces.

If a research system is to remain a valuable social asset, it must also devote resources to reinvestment in institutional capacity, to the enlargement of its physical and intellectual capital. This means diverting some resources from the production of information that does not have immediate application. This also means expanding the capacity of its scientific staff through time devoted to graduate education, study leaves, and supporting or basic research. The facilities, administrative structure, and ideology that serves as a rationale for the research program must also be modernized. The frequent arguments about the relative emphasis that should be devoted to "basic" and "applied" research often reflect a deeper disagreement concerning the relative emphasis that should be given to the production of current services relative to the expansion of capacity—particularly to the capacity embodied in the professional staff of the station. They may also reflect the fact that there are very large spillover effects to investment in expansion of intellectual capacity. In general, only a small part of the benefits of the basic research leading to capacity expansion may be realized by the individual research institution, particularly if such benefits lead to greater mobility of the staff members in which the expanded intellectual capacity is embodied.

Resource acquisition

Research institutions typically devote significant resources to increasing their influence. In order to establish a successful claim on current and future resources, a research institution usually finds it necessary to maintain effective relationships with funding agencies— legislative bodies, operating bureaus and divisions, and private foundations. The institution may also find it useful to devote resources to building a positive image as a valuable public resource. Although the resources devoted to the production of influence may have little direct value to society, such activities are essential to the maintenance and continuity of both public and private research institutions. However, institutions that have outlived their social function as producers of information and technology frequently devote excessive amounts of resources to organizational maintenance activities.

The intermediate or transactive processes and activities identified in figure 3.1 are for internal use. They have little direct value to society; their value is derived from their contribution to the output of the research institution. But they are indispensable to the productivity of the research institution itself. They represent the "engine" or the "production function" that determines the efficiency with which the research system makes use of the resources available to it. By and large, these intermediate services must be produced by the research institution itself rather than be purchased from external sources.

Linkage

The linkage services include the contacts and relationships with individuals and institutions outside the research station or laboratory —those with other scientists, the clients who use the services of the station, and sources of support. The linkages carry messages in both directions. It is through the linkages with the outside that a research institution influences technical and institutional change and is, in turn, influenced by the external changes in society and in science and technology. In the United States, the state agricultural experiment stations are characterized by an exceedingly complex set of linkages with the external environment—those with state crop improvement associations; community and regional development councils, producer- and consumer-interest groups, extension services; and the entire hierarchy of local, state, and federal administrative and political institutions. Maintaining effective communication with outside groups without becoming either a captive or an adversary is an exceedingly difficult process. Maintenance of effective linkages between technology-oriented and science-oriented research is also very important. It has been argued that the failure of the agricultural science community to maintain adequate links with the general science community has contributed to the sterility of many agricultural research efforts.[2]

Leadership

Leadership is an extremely important intermediate product. The idea that all a research director needs to do is to hire good people and let them "do their own thing" has only minimal relevance at a time when the solution to many significant technical and social problems requires concerted research effort. Leadership must be sensitive to changing social goals, and it must effectively transmit their implications to the scientific staff.

Leadership must be able to conceptualize and effectively communicate the potential contribution of the experiment station or the research laboratory to the solution of emerging problems. In addition, it must be capable of mobilizing and allocating both financial and scientific resources in such a manner as to produce the high returns to scientific and technological effort that society has come to expect. This means not only acquiring the necessary human and financial resources but also performing the more difficult task of creating an institutional environment in which these resources can become productive.

Research technology

The technology or the methodology of research is in continuous flux. The research program must be organized in such a way that the research staff is aware of, and contributes to, the advances in its own and closely related fields. Resources devoted to the production of research technology represent a capital investment in the capacity of the individual research worker, the research institute that makes the investment, and the broader research system of which it is a part. It has often been noted that research institutes that devote a relatively high proportion of their efforts to either basic or "frontier" supporting research are often able to pay lower salaries to comparable scientific personnel because institute staff members are willing to accept part of their "pay" in the form of investment in their own capacity.

Institutional doctrine

Doctrine is reflected in the articulation of institutional goals and philosophy and in operating style of an institution. For example, during the 1960s, the traditional production-oriented doctrine of the U.S. state and federal agricultural research system experienced severe stress under an increasing pressure to give heavier weight in the formulation of research priorities to studying the environmental spillover effects of technical change in agriculture, the problems of human capital formation, and the socioeconomic dimensions of community development. The international agricultural research institutes have been urged to place greater weight on the geographic and social distributional implications of the technology they produce. The operational manifestation of a shift in doctrine is the reformulation of a research station's program. Modifications in doctrine are sterile unless accompanied by a realization of the revised priorities.

Internal structure

Program specifications often imply drastic modifications in the internal structure of a research station. Problem-oriented inter-disciplinary "centers" or "teams" erode the decision-making authority of discipline- or commodity-oriented departments. During periods of stress, the reformulation of doctrine, the redirection of program, and the reorganization of internal structure may absorb substantial resources and seriously compete with the production of information. These efforts must be justified primarily in terms of their impact on the future productivity and viability of the experiment station.

The international agricultural research institutes and centers organized under the sponsorship of the Consultative Group on International Agricultural Research (CGIAR) operate in an even more complex environment than national or state (provincial) agri-cultural experiment stations. The functions of leadership, resource acquisition, and linkages are particularly complex. The responsi-bilities of the institute directors and of the CGIAR in the generation of support for core budget, special projects, and outreach are still evolving. Linkages with scientists and other professional workers, research institutions, and funding sources in both developed and developing countries are at times characterized by political, as well as scientific and economic, considerations. An institute's leadership and staff must establish working relationships in an environment characterized by widely different perceptions of the role of political, administrative, and scientific considerations.

THE RESEARCH AND DEVELOPMENT PROCESS

What are the processes by which an experiment station or a research institute generates new knowledge and information?[3] The generation of information and its embodiment in the form of a new crop variety, a synthetic fiber, or an improved model of commodity demand and supply relationships is a complicated process. It requires the coopera-tion of large numbers of individuals and organizations. The work of educated people—scientists, engineers, technicians, and literate farmers and laborers—is critical to the process that produces tech-nological change and economic growth.

A striking feature of modern research and development is the extent to which it has become institutionalized. In applied research and development, the industrial laboratory and the research insti-tute have increasingly replaced the individual inventor and the academic scientist who did research in the free time they could divert from teaching.

The process of innovation

The design of effective institutions for research and development requires a clear understanding of the processes by which new things (innovations) emerge in science and technology. This subject has received much attention in applied technology, sociology, and history. In economics it has occupied only a peripheral, although expanding role.

The distinction between the role of insight and the role of skill in the emergence of a new concept or technique has posed major difficulties in developing a theory of innovation. A. P. Usher proposed the following distinction.

> Acts of skill include all learned activities whether the process of learning is an achievement of an isolated adult individual or a response to instructions by other individuals; . . . acts of insight are unlearned activities that result in new organizations of prior knowledge and experience. . . . Such acts of insight frequently emerge in the course of performing acts of skill, though characteristically the act of insight is induced by the conscious perception of an unsatisfactory gap in knowledge or mode of action.[4]

Usher distinguished among the several concepts of how innovation occurs on the basis of the relative prominence given to insight and skill.

At one extreme is the transcendentalist approach, which attributes the emergence of invention to the inspiration of the occasional genius who from time to time achieves new insight through intuition. This theory lies behind the great inventor concept—the assertion that Fulton invented the steamship or that Marconi invented the radio. At the other extreme is the mechanistic process approach, which assumes that "necessity is the mother of invention." The individual inventor is one of a number of individuals who has the appropriate skills to make the final modification that is acclaimed as an invention.

A more comprehensive view regards major innovations as emerging from the cumulative synthesis of relatively simple innovations, each of which requires an individual "act of insight" by a person who also possesses a high degree of skill acquired through training and experience in the particular area of investigation. In this view, the individual invention involves four steps.

1. *Perception of the problem,* in which an incomplete or unsatisfactory pattern or method of satisfying a want is perceived.
2. *Setting the stage,* in which the elements or data necessary for a solution are brought together through some particular configuration of events or thoughts. Among the elements of the

solution is an individual who possesses sufficient skill in manipulating the other elements.

3. *The act of insight,* in which the essential solution to the problem
   is found. Large elements of uncertainty surround the act of
   insight. It is this uncertainty that makes it impossible to predict
   the timing or the precise configuration of an innovation in
   advance.
4. *Critical revision,* in which the newly perceived relations become
   more fully understood and effectively worked into the entire
   context to which they belong. This may again call for new
   acts of insight.

A major or strategic invention represents the cumulative synthesis
of many individual inventions. It will usually involve all the separate
steps that may be found in the case of the individual invention. Many
of the individual inventions do no more than set the stage for the
major invention, and new acts of insight are again essential when the
major invention requires substantial critical revision to adapt it to

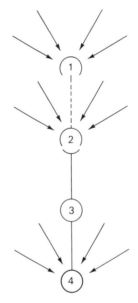

Figure 3.2. The Emergence of Novelty in the Act of Insight. Synthesis of
familiar items: (1) perception of an incomplete pattern; (2) setting the stage;
(3) the act of insight; and (4) critical revision and full mastery of the new
pattern. (After Abbott P. Usher, *A History of Mechanical Inventions* (Cambridge, Mass.: Harvard University Press, 1954].)

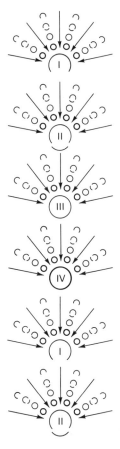

Figure 3.3. The Process of Cumulative Synthesis. A full cycle of strategic invention and part of a second cycle. Roman numerals I through IV represent steps in the development of a strategic invention. Small figures represent individual elements of novelty. Arrows represent familiar elements included in the new synthesis. (After Abbott P. Usher, *A History of Mechanical Inventions* [Cambridge, Mass.: Harvard University Press, 1954].)

particular uses. A schematic presentation of the elements of the individual act of insight and the cumulative synthesis as visualized by Usher are presented in figures 3.2 and 3.3.

A major advantage of the cumulative-synthesis approach is that it clarifies the points at which conscious efforts to speed the rate or alter the direction of innovation can be effective. The possibility of affecting the rate or the direction of innovation is obscured by the transcendentalist approach, with its dependence on the emergence of "the great inventor," and is denied by the mechanistic process approach, with its dependence on broad historical trends and forces.

The focus of much conscious effort to affect the speed or direction of innovations centers on the second and fourth steps in the process, in setting the stage and in critical revision. By consciously bringing together the elements of a solution—by creating the appropriate research environment—the stage can be set in such a manner that fewer elements are left to chance. It would be inaccurate to suppose that it is yet possible to set the stage in such a manner as to guarantee a breakthrough in any particular area, although the probability of a breakthrough may be increased as more resources are devoted to research on a given problem. As more is learned about the effectiveness of various research environments, the probability that breakthroughs will be achieved should increase. The establishment of the International Rice Research Institute set the stage for the insight that led to the development of high-yielding rice varieties for the tropics, but there was no way of assuring, in 1960, that a breakthrough would actually occur. For the present, it can only be emphasized how little is known about the administration of basic research.

At the level of critical revision, considerable progress has already been made in bringing economic and administrative resources to bear. Many of the elements of critical revision require acts of skill rather than acts of insight. Once the initial development of a high-yielding prototype rice variety had been developed, the incorporation of disease and insect resistance was much more predictable than the original breakthrough. The effectiveness of modern research procedures in shortening the time span from the test tube to the production line testifies to our ability to exert conscious direction at the applied research level.

THE ROLE OF SKILL AND INSIGHT IN A BIOLOGICAL INNOVATION: THE RICE BLAST NURSERY CASE

The four stages in the process of innovation—(1) perception of the problem; (2) setting the stage, which brings together the elements of a solution; (3) the act of insight, which permits solution of the problem; and (4) critical revision, which includes the refinement and improvement of the original solution—are illustrated ideally in the development of a relatively simple biological innovation, the rice blast nursery.[5] The example also illustrates the complex interactions that frequently accompany technological change.

Rice blast is an airborne fungus disease caused by *Piricularia oryzae*, which attacks the leaves and the stems of rice plants and substantially reduces yield. It emerged as a significant disease problem in Southeast Asia during the 1950s. Before that time, a combination of natural selection and farmer selection resulted in the use of rice varieties that were relatively resistant to local strains of the rice blast fungus. Most of these varieties were also adapted to conditions of low fertility, but, unfortunately, their yields were usually low and they were not responsive to improved fertility. Attempts to increase yields by the use of nitrogen fertilizer and by the introduction of new varieties were accompanied by the increased virulence of the rice blast fungus in those areas where farmers had adopted new varieties and cultural practices most extensively.

The first step, perception of the problem, was the recognition of the rising incidence of rice blast disease after the introduction of new rice varieties. The increased recognition that rice blast disease was a significant factor limiting attempts to achieve higher rice yields in tropical areas led to the search for an economically and biologically efficient technique for screening rice varieties to be used in rice-breeding programs and the new varieties produced by such programs.

The object of the screening was to select varieties of the host (the rice variety) that were highly resistant to infection by the fungus. Initial screening efforts involved inefficient laboratory techniques or inexact field observations. To complicate matters further, the fungus spores were found to be present in the atmosphere in concentrations that exceeded those that could be effectively produced under laboratory conditions.

The second step, setting the stage, involved a high degree of technical skill but relatively little new knowledge. The problem was to create an optimum environment for the growth of the rice blast fungus spores that provided, at the same time, favorable conditions for the mass production of rice seedlings. The components of such an environment, well known at the time, were: (1) adequate moisture in the environment, (2) high levels of nutrition, (3) upland nursery conditions, and (4) tropical temperatures (20° to 30° C).

The third step, the act of insight, was facilitated by the increasing concern with the damage caused by the rice blast disease in Thailand during the mid-1950s and by the presence of a pathologist (S. H. Ou) in Thailand who possessed the professional skill necessary to combine the elements of the solution into the design for a blast disease nursery that would permit the effective identification of blast-resistant varieties and the identification of the races of the blast pathogen.

The fourth step, critical revision of the solution, involved the addition of refinements aimed at improving the effectiveness of the solution—for example, the addition of border rows of susceptible cultivars and the installation of automatic moisture-control devices to increase the concentration of spores and to maintain the required moisture conditions.

LINKAGES BETWEEN SCIENCE AND TECHNOLOGY

The Usher cumulative synthesis model is incomplete in its representation of the linkages between science and technology. Usher used the model to clarify the relationship between skill and insight in the process of technical innovation. The cumulative synthesis model is equally valid as a description of the role of skill and insight in scientific innovation. The model is a reasonable representation of the innovation process in an area like mechanical technology, Abbott Usher's primary interest, in which advances in technology often have not been intimately linked with advances in science. But it is less adequate as a description of the innovation process in which there is an intimate relationship, for example, between advances in biological science and technology or between advances in chemistry and chemical technology.

The orthodox view often implies a rather simplistic linear relationship between advances in science and technology. "During the past fifty years the relationship between understanding and doing—between science and technology—has changed. Before, technology had almost always preceded science. In the twentieth century, however, understanding caught up, and science became the major force behind technological progress. Basic science developed theory and understanding; technology took that knowledge and provided blueprints for a change." Within the context of figure 3.3, one could, from this perspective, think of stages I and II as involving primarily scientific innovation and stages III and V as involving primarily tehcnological innovation.

In practice the interactions tend to be much more complex. Instead of a single path running from scientific discovery or innovation through applied research to development, it is more representative to think of science-oriented and technology-oriented research as two parallel but interacting paths that both lead from, and feed back into, a common pool of scientific and technical knowledge. (See figure 3.4.) The two paths are connected through the pool of existing scientific and technical knowledge from which both lead and into which both feed back advances in scientific and technical knowledge.[6]

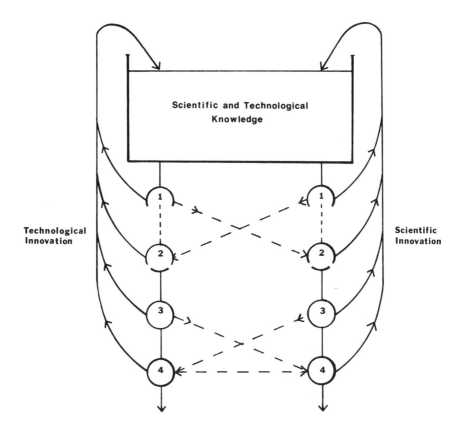

Figure 3.4. The Interaction between Advances in Scientific and Technical Knowledge.

The link to the common pool of existing knowledge, however, is not the only channel of interaction. In many instances, there are direct linkages or interactions that occur at the leading edge of both paths. In some cases, a single individual or research team may occupy a leading position in advancing knowledge along both scientific and technical paths. An example is the case of William Shockley and the solid state research group at the Bell Laboratories who both advanced the theory of semiconductors and made the initial advances in transistor technology. The process outlined in figure 3.4 is also descriptive of the interactions between George H. Schull of the Carnegie Institution and Donald Jones of the Connecticut Agricultural Experiment Station in the development and extension of the theory of hybrid vigor and the invention of the double-cross method of hybrid seed production. It also appears to represent a valid interpretation of the interrelationship between scientific and technological advances in genetics and genetic engineering that are under way at the present time.[7]

The process of interaction between scientific and technical innovation illustrated in figure 3.4 has been described in more analytical terms by Robert E. Evenson and Yoav Kislev. They described the process of research and development leading to new technology in terms of two types of activities. The first is a search for superior performance, with respect to factors such as yield potential and pathogen resistance, within a given distribution. The second activity involves an effort to expand the size of the population or the form of the distribution within which the search for superior performance is carried out.[8]

The model developed by Evenson and Kislev appears to be quite descriptive of the processes involved in plant and animal improvement and in the development of pesticides. The breeding of open-pollinated crops, for example, can be thought of as sampling from the distribution of all possible crossings. Herbicide research involves sampling from a population of chemical entities that are potentially active with respect to their impact on plant growth. In plant breeding, expansion of the distribution of the population within which the search for superior characteristics may be sought takes place in several ways. One is by the expansion of the size of the germ plasm collection available to plant breeders through more intensive plant exploration efforts. Another is by the development of genetic-engineering techniques that increase genetic diversity. In herbicide development, an attempt may be made to expand the population by synthesizing new chemical entities within a broad class of chemicals that are expected to exhibit herbicidal activity. Advances in scientific knowledge that expand the distribution can increase the productivity of technology-oriented research. Technical advances in scientific instrumentation can open up new avenues for scientific exploration. Furthermore, the market response to advances in technology serves to indicate those areas where expanding the distributions is likely to have the highest social and economic returns.

Search and distribution expanding activities can be illustrated more explicitly with the history of sugarcane varietal development.[9] (See figure 3.5.) The first stage involved selection of vegetatively produced canes from occasional natural crossings. The population sampled was heavily concentrated around the existing yield level. The second stage began with the independent discovery in Java

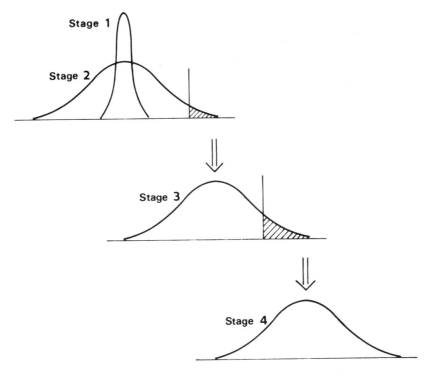

Figure 3.5. Stages in the Development of Sugarcane Varieties. (From Yoav Kislev, "A Model of Agricultural Research," *Resource Allocation and Productivity in National and International Agricultural Research*, Thomas M. Arndt, Dana G. Dalrymple, and Vernon W. Ruttan, eds. [Minneapolis: University of Minnesota Press, 1978], p. 267.)

(1887) and Barbados (1888) of the conditions necessary to induce flowering and sexual reproduction. This enormously increased the size of the distribution within which the search for characteristics associated with higher yields could be sought. The third stage involved deliberate crossing to augment desired traits, such as disease resistance. A major breakthrough involved the advances in breeding techniques in Java and India needed to successfully cross 188-chromosome disease-resistant wild canes with the 80-chromosome commercial varieties. The fourth stage involved the development of capacities in experiment stations in each major cane-producing region adequate to develop varieties suited to specific soil, climate, and disease conditions. At this stage, advances in scientific knowledge in the area of plant pathology, plant physiology, and soil science became important complements to the earlier advances in genetics in the productivity of sugar breeding.

The effects of the advances, which widened the distribution within which the search for improved varieties could be conducted, has been documented in the case of the Barbados cane-research station. Work on stage 2 varieties was carried out from 1902 to 1939. The stations released 10 important commercial varieties from 1902 to 1912, one from 1914 to 1928 and one important and three minor varieties from 1928 to 1939. When work on stage 3 varieties began in 1929, research productivity expanded sharply. Fourteen important stage 3 varieties were released between 1929 and 1939. During the period between 1929 and 1939 when both stage 2 and stage 3 programs were being pursued, the ratio of commercial varietal discoveries to seedlings brought to the final testing stage was 1:13,000 for the stage 2 program and 1:1,800 for the first five and 1:2,700 for the next nine stage 3 varieties.

As the area within a given distribution available for search (the shaded area under the stage 2 and 3 curves in figure 3.5) declines, the productivity of the search process declines. It is premature to argue that similar productivity-dampening effects are inherent in efforts to shift the distribution. There is, however, a substantial basis for hypothesizing that greater research effort will be necessary to maintain existing yield levels as yields increase. Maintenance research is the research needed to prevent yield declines as a result of the evolution of pests and pathogens, the decline in soil fertility and structure, and other factors. If the research effort required to maintain productivity is a positive function of the productivity level, it seems apparent that maintenance research will rise as a share of the research budget—that maintenance research will constitute a higher share of the research budget when the yield of rice is 8 metric tons per hectare than when it was 4 metric tons. If this hypothesis is valid, it suggests that the time may come when a relatively large share of the agricultural research effort will have to be devoted to maintenance research rather than to advancing productivity.

TECHNOLOGY AND SOCIETY

Clearly, the agricultural research system cannot, and should not, function in isolation from the broader society of which it is a part.[10] When science depends on society for patronage, society can be expected to insist on responsibility or at least on accountability. (See chapter 13.)

The linkages that impinge on the internal operation of the research station or research system, as illustrated in figure 3.1, are shown in greater detail in figure 3.6, which borrows in part from de Janvry. Figure 3.1 can be thought of as a more-detailed exposition of the area called the innovation-producing institution on the left side of figure 3.6. In figure 3.6, the generation of technical change is conceptualized as a process of "circular and cumulative causation." In this process, both the socioeconomic structure and the politico-bureaucratic structure are given explicit roles.

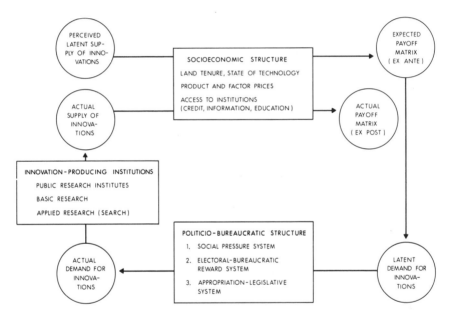

Figure 3.6. Supply and Demand for Technological and Institutional Innovations. (Adapted from Alain de Janvry, "Social Structure and Biased Technical Change in Argentine Agriculture," in *Induced Innovation: Technology Institutions and Development,* Hans P. Binswanger and Vernon W. Ruttan (Baltimore: Johns Hopkins University Press, 1978), p. 302.

Again borrowing from de Janvry, the central node of the model is a "payoff matrix" that identifies the gains and losses (economic, social, political) to each interest group from each possible choice within the technologies that might be developed. Alternative technological choices concern options among regions and commodities (e.g., irrigated versus dry land, food versus commercial crops, beef versus beans) and among technological biases (land-saving biochemical versus labor-saving mechanical devices and capital-intensive versus more labor-intensive mechanical methods). Interest groups include commercial farmers, traditional landed elites, subsistence farmers, landless agricultural workers, industrial employers, urban workers, private-sector suppliers of inputs and marketing services, and government agencies, including the research bureaucracy itself. Each relevant social group expects to realize income gains or losses from agricultural research.

In national research systems, the supply-and-demand mechanism for technical innovations in agriculture is centered on the payoff matrix and is conditioned by the socioeconomic structure on the one hand and the politicobureaucratic structure on the other. The relative power of different economic and social groups over the politico-bureaucratic structure is an important factor in getting their own demands eventually translated into a supply of new knowledge or new technology. In the case of technology, pressure on the politico-bureaucratic structure results in a specific allocation of funds and scientific effort to research institutions and, within these, to particular lines of research. The state of scientific and technical knowledge, determined in part by past allocations of physical and human capital to basic research and related fields of applied research will also influence what it is possible to achieve in each line of research. The resulting allocation of resources to applied research will influence the intensity of the search for new knowledge and new technology. The manner in which research systems are organized, funded, and staffed is an important determinant of the way in which they respond to the set of demands brought to bear on them. (See chapter 4.)

The resulting supply of new knowledge and new technology creates, through the socioeconomic structure, specific payoffs for each social group. For agricultural technology, the payoffs and their incidence among social groups are influenced by (1) the physical impact of the innovation in terms of yield effect and/or resource-saving and resource-substitution effects, (2) the diffusion effect that is conditioned by both the nature of the technological innovation and the particular position of social groups in the socioeconomic structure, and (3) the relative prices that influence the economic value of the physical and diffusion effects. Thus, the payoff matrix summarizes the economic gains and losses from the innovation for each socioeconomic group. The factors that determine the flow of gains and losses among interest groups is discussed in greater detail in chapter 11.

Anticipated payoffs induce additional demands for new knowledge and new technology. In activating the process, the organization of the research system is, again, of major importance as it influences—through the dialectical interactions or linkages it maintains with social groups—the formation of such new expectations. The payoff matrix thus reveals the nature of the dynamics—or lack of dynamics—of the inducement and diffusion of technological innovations. It also indicates that care must be taken to identify the specific groups who gain or lose from technical change in order to make meaningful statements about its economic and social significance.

The expected productivity of a research system should be measured in terms of the entries of the payoff matrix relative to the costs incurred by the research system. When other costs are incurred to create the observed payoff effect—for example, to promote diffusion of new technology or to implement institutional changes—care must be taken to charge these costs against the resulting payoffs. If the aggregate payoff is used as a criterion of net social gain from technology, care still must be taken to identify possible negative payoffs—such as the loss of employment and income for some social groups—in order to deduct them from the total payoff and to determine needed economic compensations. Indeed, the commonly used net-social-gain measure is but a global approximation of the payoff matrix that largely hides the specific socioeconomic effects of technology and the political pressures leading to support for, or opposition to, specific technological innovations.

One important implication of the perspective outlined in this section for the program of national or international agricultural experiment stations or research institutes is that the output of their research and development activities represents only part of the technical and institutional resources needed to bring about productivity growth in the production of individual commodities or in complex cropping (or farming) systems. The research output of the institute or the experiment station does not represent a complete package of practices but rather new knowledge or improvements in the components of more productive crop or livestock production systems. The payoff to the supply of innovations that represents the output of an agricultural research institution is also conditioned by advances made by other private and public suppliers of new materials and new knowledge and by the socioeconomic structure into which the new technology is introduced.

In the United States, private and public interests in agricultural commodity and research policy, at both the federal and state levels, were traditionally channeled through the U.S. Department of Agriculture (USDA) and land-grant-university research bureaucracies; the legislative leadership of the agricultural and appropriation committees;

and the organized agricultural interests consisting of general farm organizations, commodity organizations, and cooperatives. Political scientists have called this coalition the "iron triangle" because of its capacity to dominate both agricultural commodity and agricultural research policy. During the 1950s agribusiness interests were incorporated into the triangle. But during the 1970s a new coalition of food and environmental activists pried the triangle apart and forced the older participants to accede to, if not embrace, a broader agricultural research agenda.[11]

NOTES

1. This section is adapted from Melvin G. Blase and Arnold Paulson, "The Agricultural Experiment Station: An Institutional Development Perspective," *Agricultural Science Review*, 10 (second quarter, 1972), pp. 11-16. See also Jules Janick, Robert W. Schery, Frank W. Woods, and Vernon W. Ruttan, *Crop Science*, 2nd ed. (San Francisco: W. H. Freeman and Company, 1973), pp. 667-70; Vernon W. Ruttan, "Reviewing Agricultural Research Programs," *Agricultural Administration*, 5 (1978), pp. 3-7; Hans P. Binswanger and Vernon W. Ruttan, *Induced Innovation: Technology, Institutions and Development* (Baltimore: Johns Hopkins University Press, 1978), pp. 204-8.

2. National Research Council, *Report of the Committee on Research Advisory to the U.S. Department of Agriculture* (Springfield, Va.: National Technical Information Service, 1972); Andre Mayer and Jean Mayer, "Agriculture: The Island Empire," *Daedalus*, 103 (spring 1974), pp. 83-96.

3. This section is based on material presented in Vernon W. Ruttan, "Usher and Schumpeter on Invention, Innovation and Technical Change," *Quarterly Journal of Economics*, 73 (November 1959), pp. 596-606. See also Janick, et al., *Crop Science*, pp. 670-73.

4. Abbott P. Usher, *A History of Mechanical Inventions* (Cambridge, Mass.: Harvard University Press, 1954), pp. 523, 526.

5. This section was developed with the assistance of Dr. S. H. Ou, a plant pathologist with the International Rice Research Institute. For a description of the blast nursery, see S. H. Ou, "Some Aspects of Rice Blast and Varietal Resistance in Thailand," *International Rice Commission Newsletter*, 10 (3) (September 1961), pp. 11-17. See also "Report of the Committee on the Establishment of Uniform Blast Nurseries," *International Rice Report Commission Newsletter*, 11 (1) (March 1962), pp. 23-32.

6. Thomas S. Kuhn, "Scientific Discovery and the Rate of Invention: Comment" in *The Rate and Direction of Inventive Activity: Economic and Social Factors*, Richard R. Nelson, ed. (Princeton, N.J.: Princeton University Press, for the National Bureau of Economic Research, 1962), pp. 450-57.

7. For a discussion of the transistor case, see Richard R. Nelson, "The Link between science and Invention: The Case of the Transistor," in *Rate and Direction of Inventive Activity*, pp. 549-83. For an examination of the case of hybrid corn, see Paul C. Mangelsdorf, "Hybrid Corn," *Scientific American*, 185 (August 1951), pp. 39-47. For a discussion of genetic engineering, see Office of Technology Assessment, *The Impacts of Genetic Applications to Microorganisms, Animals and Plants* (Washington: USGPO, 1981).

8. See Robert E. Evenson and Yoav Kislev, *Agricultural Research and Productivity* (New Haven, Conn.: Yale University Press, 1975), pp. 140-55; Robert Evenson and Yoav Kislev, "A Stochastic Model of Applied Research," *Journal of Political Economy*, 84

(March/April 1976), pp. 265-81; and Yoav Kislev, "A Model of Agricultural Research," in *Resource Allocation and Productivity in National and International Agricultural Research*, Thomas M. Arndt, Dana G. Dalrymple, and Vernon W. Ruttan, eds. (Minneapolis: University of Minnesota Press, 1978), pp. 265-77.

9. For a technical discussion of the history of sugarcane development, see J. J. Ochse, M. J. Soule, Jr., M. J. Dijkman, and C. Wehlburg, *Tropical and Sub-Tropical Agriculture* (2 vols.) (New York: Macmillan, 1966), pp. 1197-1251.

10. This section draws on Alain de Janvry, "Inducement of Technological and Institutional Innovations: An Interpretive Framework," in *Resource Allocation and Productivity in National and International Agricultural Research*, Thomas M. Arndt, Dana G. Dalrymple, and Vernon W. Ruttan, eds. (Minneapolis: University of Minnesota Press, 1971), pp. 551-63; and Ruttan, "Reviewing Agricultural Research Programs," pp. 7-10.

11. The classic treatments of the politics of agricultural commodity and research policy in the United States are Charles M. Hardin, *The Politics of Agriculture* (Glencoe, Ill.: Free Press, 1952), and Charles M. Hardin, *Freedom in Agricultural Education* (Chicago: University of Chicago Press, 1955). For a contemporary account that focuses on the weakening of the old "iron triangle" coalition, see Don F. Hadwiger, *The Politics of Agricultural Research* (Lincoln, Neb.: University of Nebraska Press, 1981).

Chapter 4

National Agricultural Research Systems

Does the way in which a national agricultural research system is organized and funded affect its productivity or its effectiveness in generating a continuous stream of new technology capable of enhancing agricultural growth?[1] In this chapter I present a series of case studies in order to illustrate how several countries have attempted to organize their agricultural research systems. I also attempt to draw some lessons from the experiences of these research systems for the relationship between structure and performance.

Histories of the development of seven national research systems— those of the United Kingdom, Germany, the United States, Japan, India, Brazil, and Malaysia—are presented. These cases were not randomly selected. They were selected as illustrations of significant historical patterns of development and organization. The United Kingdom and Germany were selected because they were influential as early models for the development of agricultural research. Both Japan and the United States drew on the United Kingdom's and Germany's experiences and modified what they borrowed to suit their own agricultural and institutional environments. The Indian agricultural research system evolved under the influence of British colonialism. It was reformed and modernized under the influence of an American aid program. Brazil has undertaken the most massive effort of any developing country to create a modern agricultural research system. The history of agricultural research in Malaysia has been dominated by the importance of a single crop, rubber, and by the development of the Rubber Research Institute of Malaya (now Malaysia), one of the world's great research institutes.[2]

This chapter takes as a starting point a perception that the organization of a productive national agricultural research institute is not independent of the broader historical forces that have fashioned the nation's cultural, political, and economic systems. The implications of this perspective for institutional design has been ignored by a number of consultants associated with international lending and aid agencies and with private foundations, who have often recommended institutional patterns of organization and coordination without adequate regard for the traditions and experience of the country that they are advising.

ORGANIZING AGRICULTURAL RESEARCH IN THE UNITED KINGDOM

The modern agricultural revolution, it is often asserted, began in England with the planting of clover and turnips in rotation with grain crops.[3] The institutionalization of an effective agricultural research system with public funding and direction did not, however, emerge in the United Kingdom until the first decade of the 20th century.

Origins

Prior to the 20th century, agricultural research typically was initiated by private individuals, usually innovative farmers or owners of large estates. References are often made to the work of Sir William Weston, who, in the 1660s, promoted the use of turnips for winter animal feed; of Lord Townshend, who introduced the Norfolk rotation of clover, wheat, turnips, and barley; of Robert Bakewell, who attempted, in the latter half of the 18th century, to select beef cattle and sheep for improved performance; and of Arthur Young, who tried, at the beginning of the 19th century, to publicize the innovations made by "improving" landowners.

Modern agricultural historians take a less dramatic view of these events than did their predecessors, and they tend to view the events as evolutionary rather than revolutionary. Nevertheless, the success of the pattern that was established in the 18th century was sufficient to provide a basis for the English era of "high farming" in the middle of the 19th century. Its success tended to confirm a system of pragmatic applied research by innovative farmers, livestock breeders, and mechanics as the dominant source of new agricultural technology in the United Kingdom until well into the 20th century.

By the middle of the 19th century, efforts began to establish agricultural research on a more formal basis and to utilize the emerging discipline of agricultural chemistry. A laboratory was established in Edinburgh in 1842 by the Agricultural Chemistry Association of Scotland, a voluntary agricultural society. The laboratory was dissolved in 1848 because of its inability to respond to the association members' demands for immediate practical results. In 1843 an experiment station was established at Rothamsted, near London, by Sir John Bennet Lawes on his ancestral estate. Lawes, who had been engaged in the manufacture of phosphate fertilizer from bones, also began the manufacture of superphosphate in a nearby village in 1843. Sir Henry Gilbert, who had studied in Germany under Justus von Liebig, was placed in charge of the experimental work. The Rothamsted Agricultural Experiment Station is the oldest agricultural experiment station in the world today. It was supported by the profits of the Lawes phosphate enterprise until 1889, when, shortly before his death, Lawes endowed it through the Lawes Agricultural Trust. As the research program expanded and became more costly, government funds were increasingly sought and obtained to finance the work at Rothamsted, which today is almost wholly funded by the British government through the Agricultural Research Council.

Development

Government resources for the support of agricultural research in the United Kingdom first became available through the Development and Road Improvement Act of 1901. The Development Commission was established to coordinate the use of the funds that became available initially under the act and, after 1911, through the agricultural departments of England and Scotland. Grants were made to various institutes and university departments. In 1931 the Agricultural Research Council (ARC) was established; its members were to serve as scientific advisers to the Development Commission and the agricultural departments. It was also given a small budget that could be used to fill in the research gaps not covered by the several institutes. In 1954 responsibility for agricultural research was taken from the Development Commission and placed with the two agricultural departments, who were charged with the responsibility for administering and supporting the state-aided institutes on the advice of the Agricultural Research Council. In 1956 the administration of all state-aided institutes in England and Wales was transferred to the ARC. In 1959 food research plus several agriculturally related institutes that had been administered by the Department of Scientific and Industrial Research were also transferred to the ARC. In 1959 responsibility for the ARC's budget was transferred to the

Department of Education and Science, which was advised on its budget by the Council for Scientific Policy.

By the early 1960s, then, the ARC's mandate was to review and facilitate research in progress, to promote new research where it was necessary, and to ensure as far as possible that personnel and resources were used to best advantage. It supervised the state-aided research institutes, administered its own institute and units, supported fundamental and applied research in universities, and trained recruits in the agricultural service. Its style was low-key and diffuse. According to an ARC document, "No central direction of research is attempted because the Council believes that able research workers must be allowed freedom to develop their individual interests within the general requirements of the service."[4]

Attempts were made, however, to assure a balance between individual freedom in research direction and relevance of research effort for agricultural improvement. The mechanisms used for this purpose included the meetings and seminars held by the standing and technical committees of the ARC. Technical committees were established on such topics as poultry research, diseases of sheep and cattle, environment in glasshouses, potato production problems, and others. Visiting groups of distinguished scientists and agriculturalists reviewed the research programs of the institutes and units. Agricultural improvement councils, on which farmers and growers were represented, also brought the problems of agriculture to the attention of the ARC. Cooperation in the establishment and evaluation of field trials between institute scientists and specialists, advisory (extension) officers of the Advisory Services, and individual farmers has also been stressed in ARC reports. It was not felt necessary or desirable, however, for all research planning and results to receive formal review. Formal project instruments were not employed on a system-wide basis.

It seems reasonably accurate to characterize the 50 years between 1910 and 1960 as a transition from a time when agricultural research was largely directed by the interests and resources of the private sector (farmers and industrialists) to a time when agricultural research was supported primarily by the public sector and research decision making was primarily in the hands of individual scientists and scientist-administrators. The research program contained a large element of strategic and fundamental research. It was regarded by many as being unresponsive to the interests of farmers or even of the Ministry of Agriculture.

By the end of the period, a distinct division of research responsibilities had emerged. Much of the research supported by the Agricultural

Research Council was removed from the immediate problems of agricultural production. The more short-term or adaptive research was conducted by the Agricultural Development and Advisory Service (ADAS) of the Ministry of Agricultural Fisheries and Food. A distinction was often made between research (an ARC activity) and experimentation and development (the ministry's responsibilities).

Reorganization and reform

By the late 1960s, there was increasing concern in the United Kingdom that the administration of research, including agricultural research, had not kept pace with the growth of personnel and resources. In 1971 a report by Lord Rothschild, head of the Central Policy Review Staff in the Cabinet Office, which advises the Cabinet on long-term policy, proposed a drastic restructuring of the research activities carried on by the several research councils. The basic reform recommended in the Rothschild report was the establishment of a customer-contractor principle in the funding of some of the work carried out by the research councils. Tilo Ulbricht has indicated that the debate over the report focused on three issues:

(1) The principle that scientists favored by the government should be accountable for what they do and have a responsibility to meet the needs of the country; (2) the application of the customer-contractor principle as a particular means of achieving that accountability; and (3) the wider science policy issue: How can scientists through their research, help to meet the country's needs when the government has no clearly defined, long-term policies.[5]

The effect of the acceptance of the proposals on the Agricultural Research Council was that in the future over half of its funds would come in the form of contracts or commissions for specific applied research from the Ministry of Agriculture, Fisheries, and Food (MAFF).

This reorganization of funding arrangements, along with other concerns about the weakness of research management, led to the organization of a planning section in the Agricultural Research Council in 1971. The new planning section rapidly discovered that the accounting section of the council—though very precise in terms of expenditures for staff, equipment chemicals, and capital expenditures—was not capable of providing information on the research program. It could not, for example, identify the resources devoted to research on potatoes at the eight institutes that carried out work on potatoes. A major reform was the introduction of a project system and a computerized information and costing system to facilitate the analyses of research effort in relation to objectives.

What have been the effects, positive and negative, of the Rothschild reforms? Have the fears of some of the scientists that the introduction of contract research would result in an unhealthy overemphasis on short-term applied research and a neglect of strategic and fundamental research been borne out? Has the premonition of the reformers that a cozy bilateral monopoly relationship between customer (MAFF) and supplier (ARC) would emerge to contain the efforts for reform been realized? The best judgment at this time is that the Rothschild reform has resulted in a much-needed shake-up and has injected new vitality into the system. It has also been accompanied by an increase in bureaucratic constraints on research entrepreneurship.

New directions

Critics have suggested that it is now time for a new review of the United Kingdom's agricultural research system. The Rothschild reforms improved administration but did little to improve management. A new review should be directed toward the substance of the research program. How should research resources be reallocated to solve the problems of the future rather than of the past?

But agricultural research policy is itself dependent on the direction of land use, food production, and nutrition policy. For example, should agriculture in the United Kingdom continue to develop along a capital- and energy-intensive path? What will be the role of the dairy industry in food and nutrition policy? How should the nation respond to the competing agricultural and nonagricultural demands for the use of rural resources? Only as these issues are resolved will it become possible to establish a clear focus for the direction of scientific and technical effort in agricultural research.

INSTITUTIONALIZING AGRICULTURAL SCIENCE IN GERMANY

At the beginning of the 19th century, Great Britain was regarded by those interested in agricultural improvement as the "school for agriculture."[6] By the end of the century, leadership in the application of science to problems of agriculture had passed to Germany. During the latter half of the 19th century, it was almost obligatory for anyone with a serious interest in agricultural science to study in Germany, and the German experience in the organization of agricultural research and education provided models for institution building in other countries. Special attention is given in this section to the development of German agricultural science in the 19th century.

Science for development

The rapid development of agricultural science in Germany during the latter part of the 19th century was part of a more comprehensive effort by the German nation to take advantage of advances in science, technology, and education to overcome the gap in resource endowments, industrial and agricultural technology, and economic and political power between Germany and Great Britain.

Public support for education and research as instruments of economic progress was a major institutional innovation in the 19th century. In Great Britain, the strong laissez faire tradition was a major obstacle to the institutional innovations needed to take advantage of these new sources of growth. In contrast, "the German states generously financed a whole gamut of institutions, erecting buildings, installing laboratories, and above all, maintaining competent and, at the highest level, distinguished faculties."[7]

In the latter part of the 19th century, Great Britain fell behind Germany in human capital formation, including (1) the ability to read, write, and calculate; (2) the engineer's knowledge of scientific principles and applied training; and (3) the institutional support for the advancement of theoretical and applied scientific knowledge. The working skills of the artisan and mechanic may have been the only area of human capital in which Great Britain did not fall behind. As a result, there was a rapid closing of the gap between Great Britain and Germany in the 1860s and 1870s. Germany emerged as an industrial leader in fields such as chemicals and electrical machinery. In agriculture, the German approach to the development and application of science and technology are illustrated by Liebig's work in the field of agricultural chemistry and by the establishment of a publicly supported agricultural experiment station.

The publication in 1840 of *Organic Chemistry in Its Relation to Agriculture and Physiology* by Justus von Liebig is regarded by many as the dividing line in the evolution of agricultural research. Liebig's greatest accomplishment as a scientist was "to bring together and interpret the very considerable mass of chemical and related data pertaining to plants and soils that had accumulated up to that time."[8] His refutation of the humus theory of plant nutrition and his proposed mineral theory represented a major success of his approach.

Even his errors were argued so forcefully that they set a generation of researchers to work testing their validity. Liebig's greatest achievement as a teacher was the establishment at Giessen of a laboratory for training research students in organic chemistry. It was the first laboratory of its kind. Students from all over the world were attracted to it.

The demonstration of the power and value of Liebig's approach to the organization of scientific research led directly to the establishment of specialized agricultural research laboratories and experiment stations. The first publicly supported agricultural experiment station was organized in Germany (at Mockern, Saxony) in 1852 as an answer to "the search, stirring in the German provinces since the publication of Liebig's treatise in 1840, for methods of applying science to agriculture."[9] The Saxon farmers drafted a charter for the station, which the Saxon government legalized by statute, and secured an annual appropriation from the government to finance the experiment station's operations. During the next 25 years, 74 publicly supported agricultural experiment stations were established in Germany. The German model was also adapted in Austria, Italy, the United States, and Japan.

The personalities and politics of institution building

The present strucuture of agricultural research in Germany is a product of the forceful personalities of German scientific leadership in the 19th century and of the political forces that shaped the development of the German nation.

The modern research university was a German invention. The founding of the Friedrich-Wilhelm-Universität of Berlin (now Humboldt Universität) by Wilhelm von Humboldt and his associates in 1809 represented the beginning of a new type of university. The traditional European university was devoted to education, primarily in the classical professions such as theology, medicine, and law. Humboldt's objective was to create a university that could nurture the development of the new laboratory-based sciences such as chemistry, physics, and biology.

Although Albrecht Thaer, a doctor of medicine, was appointed professor of agriculture in Berlin and became a member of the Berlin Academy of Sciences, the agricultural sciences did not become institutionalized in the new university system. Thaer promoted the development of separate agricultural academies. These academies, which were primarily teaching institutions, were established in all of the German states during the first half of the 19th century.

This development was vigorously opposed by Liebig. Liebig strongly supported Humboldt's views on the unity of teaching and research. He urged that the academies be closed and that the agricultural sciences be brought within the framework of the university system. Liebig's criticism of the agricultural academies was somewhat overzealous. Several (later designated Landwirtschaftliche Hochschulen) evolved strong, unified teaching-research programs along the same lines as the universities' programs. Liebig's advocacy led to the establishment, between 1863 and 1880, of agricultural faculties in the universities at Berlin, Breslau, Göttingen, Halle, Königsberg, Leipzig, and Munich. Several other universities, such as Giessen and Kiel, developed strong faculties in agriculture after World War II. Although the agricultural academies lost some of their preeminence in the German agricultural education system, several remained as leading contributors to agricultural research and education until the late 1930s. The agricultural academy (later college) at Hoenheim remained an independent research and training institution until the early 1960s, when faculties in natural science and economics were added and the academy was raised to university status.

The new agricultural faculties did not, however, fill the need for research directly focused on the problems of agricultural production. The solution to this problem promoted by most German states was the establishment of agricultural experiment stations, which were similar to the Saxon station referred to above. These experiment stations were established outside the university system and did not have teaching obligations. They were strongly supported by local farmers' organizations and chambers of agriculture.

The decentralized agricultural research system that evolved in Germany was a direct consequence of the federal structure of the German political system. There was no central ministry of agriculture during the period of the empire. The Ministry of Agriculture, Food, and Forestry was first established under the Weimar Republic in 1919.

A move toward centralization began in the early 1900s. In 1905 the Biologische Anstalt für Land- und Forstwirtschaft for plant protection was founded. It developed a network of stations throughout Germany. The establishment of the Kaiser-Wilhelm-Gesellschaft (now the Max-Plank-Gesellschaft) in 1910 also contributed to the more centralized direction of agricultural research. Although it was initially founded to support research in the basic sciences, it later incorporated fields such as genetics, plant breeding, and agricultural engineering.

Reconstruction of agricultural science capacity

The preeminence of Germany in agricultural research was a casualty of the National Socialist period. Many of the independent agricultural schools (Landwirtschaftliche Hochschulen) were reorganized and incorporated into the university agricultural faculties. Emphasis on meeting production goals led to the formation of a series of new technology institutes and to a deemphasis on basic research and support for science related to agriculture. Agricultural economics, a field in which Germany had exercised considerable leadership, was also deemphasized. At the end of World War II, Germany was faced with the task of rebuilding its agricultural science capacity.

A major change in the organization of agricultural research in West Germany since World War II has been the assumption by the Federal Ministry of Food, Agriculture, and Forestry of primary responsibility for agricultural research. Only a few of the state experiment stations remain viable independent institutions. The research program of the federal ministry is organized around a series of 13 federal research centers. Of these the Federal Research Center for Agriculture (Bundesforschungsanstalt für Landwirtschaft Braunschweig-Völkenrode)—with its 14 research institutes (Soil Biochemistry, Soil Biology, Grassland and Fodder Crops, Crop Science and Seed Technology, Animal Nutrition, Animal Husbandry and Behavior, Poultry, Agricultural Engineering Science, Agricultural Machinery, Farm Buildings, Farm Mechanization, Farm Management, Agricultural Markets, and Agrarian Structure) —has the strongest scientific orientation. The programs of the other institutes have a more specific commodity orientation, and their research programs are more directly linked to the program needs of the ministry.

At the state level, agricultural research and teaching regained its traditional place in the universities. The states provide professional and technical staff positions, laboratory equipment, and financial support. Additional support for research activities is provided by project grants from the Federal Ministry of Agriculture or through other national funding agencies such as the German Research Association, the German equivalent of the U.S. National Science Foundation. Since the late 1970s, an effort has been made to achieve greater coordination in research programming and planning through the Senate of Federal Research Centers and the Research Planning Group within the ministry.

THE FEDERAL-STATE SYSTEM OF AGRICULTURAL RESEARCH IN THE UNITED STATES

The U.S. agricultural research system originally drew directly on the German system for its model of institutional organization and in its training of young scientists and science "entrepreneurs" who were responsible for the establishment of agricultural research stations.[10]

A belief that the application of science to solutions of practical problems represented a sure foundation for human progress has been a persistent theme in the intellectual and economic history of the United States. The nation "was born of the first effort in history to marry scientific and political ideas." However, the institutionalization of public responsibility for advances in science and technology as an instrument of national economic growth developed slowly.

In the 19th century, progress in agricultural science and technology was in the United States, as in Great Britain, primarily the product of innovative farmers and inventors and of the increasingly dynamic industrial sector. Progress in mechanical technology was impressive throughout the last half of the 19th century. Advances in tillage and harvesting machinery, induced by the westward march of the frontier and the associated shortage of labor relative to land, resulted in rapid growth of labor productivity. The advances in mechanical technology were not accompanied by parallel advances in biological technology. Nor were the advances in labor productivity accompanied by comparable advances in land productivity.

By the end of the 19th century, the earlier consistency between the generation of mechanical technology by the industrial sector and the growth requirements of American agriculture were breaking down. The rate of growth in labor productivity was beginning to decline. The rate of growth in agricultural output fell below the rate of growth in demand. Prices of agricultural products were beginning to rise relative to the general price level. U.S. agriculture appeared to be entering a period of diminishing returns. These changes set the stage for a dramatic increase in investment in public-sector agricultural research and for advances in biological technology.

Institutionalization of research capacity

The institutionalization of public-sector responsibility for research in the agricultural sciences and technology in the United States can be dated from the 1860s. The act of May 15, 1862, of the federal Congress, "donating public lands to the several states and territories which may provide colleges for the benefit of American agriculture and the mechanic arts" and establishing the Department of Agriculture with the principal function of acquiring "useful information on the subjects connected with agriculture," became the first federal legal authority under which a nationwide agricultural research system was to develop.

The institutional pattern that emerged for the organization of agricultural research drew heavily, as stated earlier, on the German experience. A number of the leaders who established state experiment stations had studied in Germany and in other European centers of graduate study.[11] Institutionalizing agricultural research in the United States created a dual federal-state system. The federal system developed more rapidly than the state system. Yet, it was not until the closing years of the 19th century that either the state or the federal system acquired any significant capacity to provide the scientific knowledge needed to deal with the urgent problems of agricultural development.

The emergence, toward the end of the century, of a viable pattern of organization for agricultural research in the U.S. Department of Agriculture involved breaking away from a discipline-oriented structure and organizing scientific bureaus focusing on a particular set of problems or commodities. A. Hunter Dupree cited the Bureau of Animal Industry, established in 1884, as an example: "The Bureau of Animal Industry thus had most of the attributes of the new scientific agency at its birth—an organic act, a set of problems, outside groups pressing for its interests, and extensive regulatory powers."[12]

The capacity of the new land-grant universities to provide new scientific and technical knowledge for agricultural development was even more limited than the Department of Agriculture's. The first state experiment station, the Connecticut State Agricultural Experiment Station, was not established until 1877. Prior to the passage of the Hatch Act in 1887, which provided federal funding for the support of land-grant-college experiment stations, only a few states were providing any significant financial support for agricultural research at the state level. And it was not until the early 1920s that it was possible to claim with some degree of confidence that a national agricultural research and extension system had been effectively institutionalized at both the federal and the state levels. It took 50 to 70 years of persistent effort to organize a productive agricultural research and advisory (extension) system in the United States. This seems an exceptionally long period when gauged by the impatient efforts of modern institution builders in the national and international aid agencies.

By the early 1900s, the scientific bureaus of the USDA included the Bureau of Plant Industry, the Bureau of Entomology, the Bureau of Soils, the Bureau of Biological Survey, and the Weather Bureau. The Office of Experiment Stations maintained liaison with state agricultural experiment stations and handled the transmission of federal funds appropriated for use by the state stations. The structure of the USDA's research remained relatively unchanged, with the Bureau of Agricultural Economics added in 1922, until the early 1950s. In 1953 a major reorganization broke up several of the bureaus and consolidated most of the USDA's agricultural research under the Agricultural Research Service (ARS). The organization of the state agricultural experiment stations remained relatively unchanged during this period. Research priorities and the allocation of resources along disciplinary and departmental lines experienced almost no change between the mid-1920s and the early 1960s.

Perhaps the most significant institutional development during the period was the emergence of coordinated federal-state commodity research programs. The coordinated national corn-improvement research program, established in 1926, was the first major attempt to organize both federal and state research capacities in an integrated effort. The initial effort was focused on the Corn Belt of the Midwest. In 1945 the coordinated corn program was extended to the South. Similar programs were also established for the improvement of wheat and other major crops during the late 1920s. The overlap of federal support and coordinating services made it possible to give more concentrated attention to specific problems of crop improvement of common importance to several states than would have been possible if researchers in each state had worked in isolation. This involvement with the state experiment stations gave the USDA's research program greater access to basic science capacity in fields such as genetics, entomology, and physiology than could have been assembled within the federal research system.

A second major institutional development of the period was the establishment of a national fertilizer research and development center administered by the Tennessee Valley Authority, at Muscle Shoals, Alabama.[13] The National Fertilizer Development Center is the only public-sector example in the United States of one facility that combines both agricultural and industrial research. In this unique research center, both basic research in soils and chemistry and applied research ranging from fertilizer trials, engineering design, and operation of pilot plants is performed. This was an evolutionary development in which the National Fertilizer Development Center gradually assumed research responsibilities formerly assigned to the USDA. Much of the TVA's agronomy research, in support of its fertilizer research and development program was carried out in co-operation with the state agricultural experiment stations.

Since the mid-1960s, the U.S. federal-state agricultural research system had undergone a series of internal and external reviews. It has been criticized and defended from a variety of scientific, popu-list, and ideological perspectives. A basic thrust of much of this effort has been to achieve more effective planning and coordination among the several components of the federal-state system. Attempts have also been made to exercise greater federal control over resource allocation in the state system. And there have been pressures to give greater weight in the research agenda to rural development and to environmental and consumer concerns relative to commodity pro-duction.

In April 1965 the Senate Committee on Appropriations recom-mended that the secretary of agriculture establish a Research Re-view Committee that would make recommendations for the co-ordination of the agricultural research system and establish agri-cultural research priorities as a basis for projecting resource require-ments for agricultural research over the next decade. One of the first efforts of the task force (which involved scientists and admin-istrators from the USDA, the state exeriment stations, and private universities and industries) was to attempt to construct an orderly system of classifying research. A three-way classification by activity, by commodity or resource, and by field of science was established. The establishment of a goals structure was also attempted. The effort was regarded as quite unsatisfactory by many of the partici-pants on the task force because of the short time available to develop the goals and relate them to the activities. The task force's work was also complicated by a simultaneous effort by the USDA to install a program, planning, and budgetary (PPB) accounting and evalua-tion system for the several USDA missions.

The committee's report, *A National Program of Research for Agriculture*, was completed and published in October 1966.[14] Its most useful product was the documentation of the existing alloca-tion of research resources and its recommendations and projec-tions of research resources needed in 1972 and 1977 to meet na-tional goals. (See figure 4.1.) In retrospect, it is clear that the task force underestimated the rapidly changing agricultural research agenda that was being established by a new set of agricultural re-search constituencies, particularly the environmental and consumer

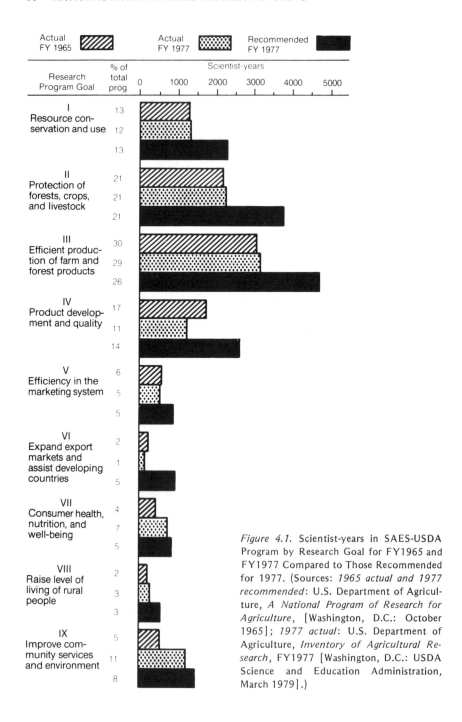

Figure 4.1. Scientist-years in SAES-USDA Program by Research Goal for FY1965 and FY1977 Compared to Those Recommended for 1977. (Sources: *1965 actual and 1977 recommended*: U.S. Department of Agriculture, *A National Program of Research for Agriculture*, [Washington, D.C.: October 1965]; *1977 actual*: U.S. Department of Agriculture, *Inventory of Agricultural Research*, FY1977 [Washington, D.C.: USDA Science and Education Administration, March 1979].)

constituencies and it overestimated the willingness of the executive branch and the Congress to expand the budget for agricultural research. Between the mid-1960s and the late 1970s, the scientist-years in U.S. public-sector agricultural research remained in the 11,000 to 12,000 range. During this same period, the total number of agricultural scientists in public- and private-sector employment rose from less than 50,000 to over 80,000.

Two elements entered into the changing perspective on the U.S. agricultural research system. One was a set of populist criticisms of the narrow focus of agricultural research. The two seminal publications in the public questioning of the goals of agricultural research were Rachel Carson's *Silent Spring* (1972) and Jim Hightower's *Hard Tomatoes, Hard Times* (1973).[15] Although both books are frustrating mélanges of errors and half-truths along with valid criticisms, they did serve to dramatize a major limitation of the U.S. agricultural research system—that it was too narrowly focused on plants, animals, and soils and inadequately oriented toward the problems of rural communities and consumers. The pressure to reexamine research priorities that was generated by the populist critics was reinforced at the state level by pressure from farm leaders, state legislative committees, and state budget officers for the state experiment stations to identify research priorities more carefully and to achieve more effective collaboration among researchers, among states in the same region, and between state and federal research systems.

A second criticism emerged from within the agricultural science community itself. A committee of the National Research Council chaired by Dean Glenn S. Pound of the University of Wisconsin severely criticized the quality of the research being conducted by the USDA and the state agricultural experiment stations.[16] The report indicated that much agricultural research is outmoded, pedestrian, and inefficient. It called for new initiative in reshaping administrative philosophies and organizations, in establishing goals and missions, in training and managing scientists, and in allocating research resources. It was particularly critical of what it viewed as inadequate interaction with the basic disciplines that underlie agricultural research and of the limited role of peer evaluation in project formulation and review.

Critics of the report argued that the report erred in applying criteria appropriate to basic research to the mission-oriented research of the federal-state system. They insisted that the appropriate criteria should include not only the professional quality of research proposals and reports but the impact of the new knowledge on technology generation and on technical change and productivity in agriculture.

It is difficult to evaluate the effect of the extensive period of internal and external analysis and evaluation that the federal-state agricultural research sectors have experienced since the mid-1960s. Responsibility for the research and advisory (extension) services of the USDA and for coordination with the state experiment stations and extension services have been consolidated under a director for science and education. The effect of a 1972 reorganization was to facilitate greater central direction of the USDA's research efforts. A 1978 reorganization mandated by the Agricultural Act of 1977 established the federal-state Joint Council on Agricultural Research, which has the responsibility for maintaining more effective coordination of the planning and conducting of federal and state research, and the Users Advisory Board, which monitors the responsiveness of the research system to national priorities.[17]

One of the more interesting new developments, also mandated by the 1977 legislation, has been the establishment of a federally funded competitive research grant program administered by the USDA and open to all scientists, including those from universities that had not been eligible for support under the formula funding arrangements used to provide institutional support to the state agricultural experiment stations. (See chapter 9.)

The responses to both the internal planning efforts and the external pressures that have been brought to bear on the federal-state research system have occurred in a very different economic and funding environment than that envisaged at the time the new planning and evaluation efforts were set in motion. Surplus capacity in American agriculture during the late 1960s generated complacency in the Office of Budget and Management and in the Office of the Secretary of Agriculture concerning the returns from increased research investment. Rather than expanding substantially, real federal support for agricultural research remained essentially unchanged between 1965 and 1980. The number of scientist-years available to the federal research system declined; the number available to the state system expanded slightly.

The future

A major issue facing the U.S. agricultural research system in the future is the appropriate division of responsibility for agricultural research among the federal, state, and private systems. The effectiveness of the state system has, to a greater extent than many state research administrators would care to admit, depended upon regional and interregional coordination and linkages provided through national program leadership. This dependence reinforces the seriousness of the erosion of program leadership and scientific capacity at the USDA. Although private-sector research in support of agricultural production and the processing and marketing of agricultural products has been expanding, its capacity is not well documented or understood. (See chapter 8.)

There is also concern about the declining share of support for the state system through the traditional Hatch Act formula method of providing federal support for research at the state experiment stations. The formula funding mechanism has represented partial compensation to the states for the spillover of benefits from the state performing the research to other states. It has also involved an implicit assumption that national priorities are, to a major degree, an aggregation of state and local priorities.

The critics who have attempted to broaden the agricultural research agenda have argued that the aggregation of state and local priorities gives too much weight to the concerns of organized agricultural producers and too little weight to the concerns of unorganized hired workers, small farmers, and consumers—that it has led to a neglect of rural development environmental quality, and nutritional and distributional objectives. The shift toward greater reliance on project relative to institutional support for agricultural research represents one attempt to induce greater responsiveness of the state system to national priorities. A question that remains unanswered is whether declining federal support for the state system will further erode the federal capacity for program leadership and coordination. In addition, there is a serious question of how far federal formula funding support can decline without reducing state incentives to support agricultural research.

ESTABLISHING AN AGRICULTURAL RESEARCH SYSTEM IN JAPAN

The Japanese agricultural research system, like that of the United States, drew directly on Germany's experience for its inspiration.[18] In the United States, the German model had its primary impact on the organization of state agricultural experiment stations. In Japan, the central government played a stronger role, relative to the prefectural governments, in establishing the agricultural research infrastructure. The role of the central government in agricultural research was a direct consequence of Japan's goal of utilizing industrial development to catch up with the Western powers. This required an agricultural surplus to support development in the nonagricultural sectors. The development of agricultural technology, through investment in research, was considered an essential means of increasing agricultural productivity and generating an agricultural surplus.

The initial stages of agricultural research

The institutionalization of agricultural research in Japan proceeded in three stages. The first stage involved an attempt to transfer Western technology from Great Britain and the United States to Japan. During the early Meiji period (1868-1911), Japanese leaders who traveled to the West were impressed by the performance of large-scale agricultural machinery. A demonstration farm, the Western Farm Machinery Exhibition Yard, was opened in Japan in 1871; machine operations were also demonstrated at the Naito Shinjaku Agricultural Station, established in 1878; and in 1879 the Mita Farm Machinery Manufacturing Plant was established to produce machines modeled after the imported machines.

The government also invited instructors from Great Britain to the newly opened Komaba Agricultural School (founded in 1877 and redesignated the University of Tokyo College of Agriculture in 1890). Instructors from the United States were brought to the Sapporo Agricultural School in Hokkaido in 1875. The curriculum at the new agricultural colleges was based on the idea of technology transfer that had led to the importation of British and American technology.

It soon became clear that, except on the northern island of Hokkaido, the factor endowments of Japanese agriculture, with an average arable land area per worker of less than 0.5 hectares, were simply incompatible with machinery of the Anglo-American type. (See chapter 2.) A Japanese farmer characterized the attempts to introduce Western technology as trying to dress a camel in the hide of an elephant.

The Meiji government quickly recognized the failure of the attempt to develop a mechanized agriculture modeled on the British and American pattern. In 1881, when the contracts of the British agricultural instructors at the Komaba School were completed, the instructors were replaced by a German agricultural chemist and a German soil scientist. The training and experience of the German agricultural scientists were more consistent with the needs of an agricultural system of small farms and intensive crop production. The curriculum of the agricultural colleges and the thrust of agricultural research in Japan were reorganized in line with the tradition established by Liebig in Germany. It was perhaps easier for Japan to make this transition in the 1880s than it is for a presently developing country in the 1980s, since 19th-century Japan was not encumbered by the vested interests of the several bilateral and multilateral aid agencies.

The second stage of Japanese agricultural development involved the rationalization and extension of indigenous technology. The Ministry of Agriclture and Commerce (founded in 1881) established an itinerant instructor system in 1885. Experienced farmers as well as new graduates of the Komaba School were employed as instructors. The Experimental Farm for Staple Cereals and Vegetables was set up in 1886. In 1893, a national agricultural experiment station was established at Nishigahara, with an experimental farm at Konosu. Six branch stations under the direct management of the national station were also established. The initial research consisted largely of field tests and experiments comparing the performance of different varieties and the effects of different practices.

The experiments did provide support for the growth of agricultural productivity during the latter years of the Meiji period. There was a substantial backlog of indigenous technological potential that could be further tested, developed, and refined at the new experiment stations. There was also a strong propensity to innovate among many leading farmers, comparable in many respects to the "improving" landlords of an earlier period in Great Britain. Rice production practices such as the use of salt water in seed selection, the improved preparation and management of nursery beds, the use of checkrow planting for weed control, and the selection of new varieties were developed by farming landlords (gōnō) and veteran farmers (rōnō). As the fertilizer industry developed, there was a strong inducement, on the part of both farmers and experiment-station workers, under conditions of land shortage to select varieties and develop practices that resulted in a favorable response to fertilizer inputs.

The prefectural agricultural research system in Japan initially developed somewhat more slowly than the national system. With the passage of a law of state subsidy for the prefectural agricultural experiment stations in 1899, this began to change. Since 1910, between two-thirds and three-fourths of the expenditures for agricultural research have been at the prefectural level.

The third stage involved the development of scientific research capacity. The potential for productivity growth from indigenous technology was largely exhausted soon after the turn of the century.

In 1904 the National Agricultural Experiment Station initiated a crop-breeding project. It was almost two decades, however, before new varieties of major significance were being produced by the National Experiment Station.

A major step in the move toward more effective research was the organization in the mid-1920s of a nationally coordinated crop-breeding program. The new system, termed the Assigned Experiment System, was established first for wheat in 1926 and then for rice in 1927. It was later expanded to include other crops and livestock. Research direction was highly centralized. Research planning—selection of parents and the initial crosses through the second (F_2) generation—was located at the Konosu Experimental Farm of the National Agricultural Experiment Station. Promising lines were then distributed to the branch stations of the National Experiement Station to test for adaptation to regional agroclimatic conditions. The varieties selected at the branch stations were transmitted to assigned prefectural stations to be tested for acceptability in specific localities. Researchers at the branch stations and the assigned prefectural stations were given almost no discretion in modifying the plans and designs established in Konosu. The varieties developed by this system were called Norin varieties (the abbreviation of the name of the Ministry of Agriculture and Forestry).

The new system of rice breeding was outstandingly successful. The first crossing of the parents of Norin No. 1 was made in the Rikuu branch of the National Experiment Station in 1922. The fourth (F_4) generation was sent to Niigata, the regional headquarters of the Assigned Experiment System in 1927. Three years later, the variety Norin No. 1 was selected for seed multiplication. It was released in 1931. It was a short-stature, early-maturing, fertilizer-responsive, high-yielding variety. Within a few years, it was the dominant rice variety grown in north-central Japan. By the late 1930s, Norin varieties had largely replaced the older selections throughout Japan.

A similar pattern was followed in wheat research. The Norin No. 10 wheat variety became an important genetic source of dwarfing for the new fertilizer-responsive wheat varieties that have been developed by the International Center for Improvement of Maize and Wheat (CIMMYT) and by a number of national research stations in the tropics during the 1950s and 1960s.

Toward scientific maturity

After World War II, the agricultural research system was reorganized to give the branch and prefectural stations greater autonomy.

The National Agricultural Experiment Station at Nishigahara was relieved of its coordinating role and assigned responsibility for basic research related to agriculture. The former branch stations and the Konosu Experimental Farm were designated as regional agricultural experiment stations.

In 1956 a further reorganization was undertaken. At that time, each research station or institute was managed by the major administrative branches of the Ministry of Agriculture and Forestry—agricultural experiment stations by the Agricultural Production Division, animal experiment stations by the Animal Industry Division, and forestry research stations by the Forestry Division. With the 1956 reorganization, all national research stations and institutes were placed under the umbrella of the Agriculture, Forestry, and Fisheries Research Council. The council is responsible for personnel and budget, for the development of major research programs, for cooperative research between the ministry and the provincial stations and the universities, and for international research collaboration.

It is of interest to contrast the new system for rice improvement with the pre-World War II system. At present, there are three national centers for rice breeding: (1) the Hokkaido National Agricultural Experiment Station for low-temperature environments, (2) the Hokuriku National Agricultural Experiment Station for more temperate environments, and (3) the Kyushu National Agricultural Experiment Station for semitropical environments. These three national centers, plus three other national stations and eight assigned prefectural stations, cooperate in a national rice improvement program. The regional centers and stations carry out selection of parents, initial crossings, and F_2 selection. The prefectural stations undertake further selection and screening for disease and insect resistance and adaptation to soil and water conditions and other local environmental conditions. The release of new Norin varieties is under the direction of a committee made up of officials of the Agriculture, Forestry, and Fisheries Council and of breeders from the several participating national and provincial centers and stations.

In the late 1970s, the National Institute for Agricultural Sciences, along with numerous other central agricultural research institutes, was moved from Nishigahara to the new science city complex at Tsukuba, 60 kilometers north of Tokyo.

The commitment of resources to agricultural research in Japan during the late 1970s can perhaps be best measured in terms of scientist-years devoted to research. In 1977, there were 3,400 researchers (plus about 3,000 technical and clerical workers) located at the 30 different research and experimental stations operated by the

Ministry of Agriculture and Forestry. There were, in addition, approximately 7,000 researchers and over 11,000 technical and clerical employees located at 379 stations, managed by Japan's 47 prefectures.[19] Approximately the same total number of researchers work in the U.S. federal-state system. This fact implies, of course, that there is a much higher level of research intensity per hectare or per unit of production in Japan than in the United States.

There is also a substantial private-sector agricultural research capacity in Japan. There are more than 400 private and semiprivate research organizations in Japan in addition to those supported by the government. These private research organizations give attention to food processing, livestock, pasture, fisheries, and forestry, as well as agricultural economics and policy. In contrast to the United States, however, where the private sector accounts for about two-thirds of agricultural research expenditures, the private sector in Japan apparently accounts for well under a tenth.

There are other interesting comparisons between the Japanese and the U.S. agricultural research systems. Both have established a dual national and state (prefectural) research system. In Japan, the prefectural system, which at the beginning was closely coordinated by the national system, has gradually acquired greater autonomy. Since World War II, the Japanese system has become less centralized and the U.S. system has become more centralized. In the United States, the state system, which was almost completely autonomous during its first half-century, has gradually become subject to greater central coordination. In Japan, prefectural experiment stations are not linked to universities, as they are in the United States. The universities in Japan are under the jurisdiction of the National Ministry of Education, and they are primarily teaching, rather than research, institutions. Although agricultural research at the universities has been poorly funded, a good deal of rather fundamental research related to agriculture is conducted by faculty researchers and as a joint product of graduate education.

Distortion of research priorities

In the discussion of the theory of induced innovation in chapter 2, it was pointed out that the distortion of factor (input) prices or product prices could induce distortion in the allocation of resources to research and bias the direction of productivity growth. The case of rice in Japan is an extreme illustration of this point. Between the early 1960s and the early 1970s, the price support program for rice resulted in a progressive widening of the gap between domestic and

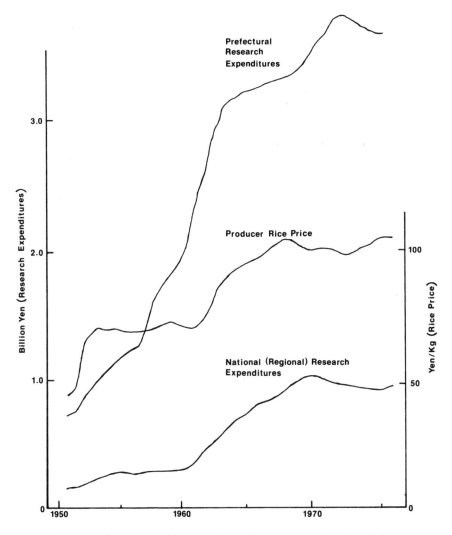

Figure 4.2. Comparison of Research Expenditures with the Producer Price of Rice, 1950-1976. (All figures are in real terms, rural price index, 1960 = 100.) (Used with permission from Keijiro Otsuka, *Role of Demand in a Public Good Market: Rice Research in Japan* [New Haven, Conn.: Yale University Economic Growth Center, 1979], p. 12.)

international prices for rice. The effect has been the expansion of rice production beyond the level required for domestic consumption; extremely high subsidy costs, equivalent to 50 percent of the agricultural budget and 5 percent of the national budget; and the allocation of excessive financial and scientific resources to rice research. (See figure 4.2.)

This distortion in the allocation of resources to rice research was significant even at the national level, where one might have expected more effective resolution of the inconsistency between price policy and the allocation of resources to research. The distortion has been even more severe at the prefectural level. There, the political pressure from organized farmer constituencies to expand rice research designed to increase yields and lower costs of production was consistent with prefectural government growth objectives.

This is a further illustration of the effect of the post-World War II reforms that increased the responsiveness of the Japanese agricultural research system to producer interests. The system by which resources are allocated to rice research is less effective at resolving conflicts of interest between consumers (or taxpayers) and producers. In the future, I would anticipate considerable political pressure for more effective coordination of prefectural and national agricultural research planning in order to balance producer and consumer interests more effectively.

BUILDING AGRICULTURAL RESEARCH INSTITUTIONS IN INDIA

The Indian agricultural research system developed primarily in response to the commercial and military interests of the colonial government. During the first decade and a half after independence, its evolution was strongly influenced by American aid programs. Its administration during both the colonial period and the postcolonial period has been complicated by the division of responsibilities of a state-federal system of government in which agriculture has been designated as a state concern but in which primary fiscal capacity is in the hands of the central government.

The colonial origins

The Agricultural College and Research Station at Coimbatore started as a model farm in 1868.[20] It was one of the two major centers for research and development on surgarcane in the British colonial system. It set the stage for modern sugarcane breeding in the 1920s by developing trihybrid canes that combined environmental adaptation with resistance to disease. The Coimbatore varieties have since become an important genetic resource for many national sugarcane improvement programs.

The Indian Veterinary Research Institute at Izatnagar traces its origins to the establishment in 1889 of the Bacteriological Research Laboratory in veterinary science in Poona in response to veterinary problems encountered by the Indian army. A camel- and ox-breeding farm had been established even earlier, around 1820, at Karnal. The Indian Agricultural Research Institute was established in 1905 at Pusa in Bihar, partly with American philanthropic assistance. After its facilities were destroyed by an earthquake in 1935, the institute was moved to New Delhi. It now serves as the central research institute and graduate training center of the Indian Council of Agricultural Research. A number of other research institutes were also established under the Ministry of Food and Agriculture during the colonial period. These included the Central Rice Research Institute (in Cuttack), the Central Potato Research Institute (in Simla), the Central Inland Fisheries Research Institute (in Barrackpore), the Central Marine Fisheries Research Institute (in Mandapam), and the National Dairy Research Institute (in Bangalore and Karnal).

The constitutional changes in 1919 made agriculture a provincial subject. The Imperial Council on Agricultural Research (ICAR) was established to coordinate provincial agricultural research. The primary source of support for agricultural research was provincial revenues from land taxes and irrigation fees and central government grants-in-aid channeled through the ICAR. An effort was also made to develop a series of central commodity committees to promote research, development, extension, and marketing of export crops. These committees had the authority to levy special cesses (taxes) on marketings to support their activity. However, only the Indian Central Cotton Committee provided any substantial support for agricultural research prior to the mid-1930s. The wheat, jute, sugarcane, and oilseed committees and trade associations played an important role in lobbying for public support of commodity research but were not an important source of financial support. Research on food crops, other than rice and wheat, was relatively neglected during the interwar period.

Research organization after independence

Following India's independence from Great Britain, a number of expert teams were organized to review the Indian agricultural research system and to make recommendations to the government. These included (1) the Indo-American Team on Agricultural Research and Education (1955); (2) the Second Joint Indo-American Team on Agricultural Education, Research, and Extension (1959); (3) the

Agricultural Administration (Nalagarth) Committee (1960); (4) the Committee for Agricultural Universities Legislation (1962); and (5) the Agricultural Research Review Team (1963). The First Indo-American Team concluded that the research institutes were, in general, ineffective in producing scientific or administrative leadership for the large number of central and state research efforts. The institutes were organized as subordinate offices of the Ministry of Food and Agriculture and were individually responsible to different sections or administrative heads in the ministry. Not even the research supported directly by the ministry was subject to adequate internal coordination.

One major result of the report of the two joint Indo-American teams was the reinforcement of the effort to strengthen graduate training at the Indian Agricultural Research Institute and to establish a major research-oriented agricultural university in each state. The need for an agricultural university system had been recognized in the 1949 report of the University Education Commission headed by Dr. S. Radhakrishnan. The first agricultural university came into existence at Pantnagar, Uttar Pradesh, in 1960. The university was established on a large land grant (at Pantnagar) and received very substantial assistance from the Indian Council of Agricultural Research and the U.S. Agency for International Development (USAID). The USAID contracted with the University of Illinois to provide technical and scientific support. By the time of the 1963 agricultural research review, agricultural universities had also been established at Udaipur in Rajasthan, at Bhubaneswar in Orissa, and at Ludhiana in the Punjab.

Little else of significance had been accomplished as a result of the several review reports until after the report of the 1963 Agricultural Research Review Team. This team, composed of eminent scientists from India, the United Kingdom, and the United States, was appointed during a period of increasing concern about India's capacity to meet the food requirements of its population. The team was appointed on October 31, 1963, and submitted its report on March 19, 1964. Within a year's time, the government of India had decided to reorganize its agricultural research system.

The first step was the reorganization of the Indian Council of Agricultural Research (ICAR) into a central agency with the authority for coordinating, directing, and promoting agricultural research in the whole country. All of the central research institutes that had been under the direct administrative control of the Department of Agriculture or the Department of Food were transferred to the

ICAR. The commodity committees were also abolished, and their research was put directly under the ICAR. The ICAR also became the source of central funding for the research activities of the agricultural universities. After some delay, the Agricultural Research Service, separate from the civil services administration of the Public Services Commission, was established. It was hoped that this would create an opportunity to develop personnel policies and pay scales consistent with the needs of a scientific organization.

The all-India coordinated research programs

A major administrative device for coordinating agricultural research in India, under the reorganized ICAR has been the All-India Coordinated Research Programs.[21] The concept was developed during the late 1950s and the early 1960s as a result of collaborative experience with the Rockefeller Foundation in the All-India Coordinated Maize Improvement Project. Maize is a minor crop in India. However, the successful experience in developing a coordinated maize research program involving the central Agricultural Research Institute, the state research stations, and the agricultural universities led to the establishment of other coordinated research programs.

The All-India Coordinated Rice Improvement Project represents a useful example of how the coordinated programs are organized. The Central Rice Research Institute, established in 1946 at Cuttack, was the first major federal rice research effort. As part of the third five-year plan, regional rice research centers were established by the Center government in several states. The All-India Rice Research Program was initiated in 1965 with the support of the USAID. The National Coordinating Centre was established at Rajendranagar in Hyderabad. The program incorporated research efforts at more than 100 stations located throughout the country. The Rockefeller Foundation lent the program a senior scientist to serve as joint coordinator. Four scientists from the International Rice Research Institute, funded by a contract between the USAID and the ICAR, were also attached to the program.

The objectives of the coordinated rice-breeding program included breeding varieties suited to different maturity regions (90 to 170 days), resistant to pests and diseases, and adapted to consumer preferences. Breeding was conducted at 24 centers in 7 different agroclimatic zones. Testing for local adaptability was conducted at 108 stations throughout the country.

The Indian agricultural universities

The establishment of the agricultural university system,[22] which began with the Uttar Pradesh Agricultural University (now the G. B. Pant University of Agriculture and Technology), was called in the 1978 report of the Review Committee on Agricultural Universities "one of the most significant landmarks in the history of agricultural education in India."[23] It was also criticized by a long-term observer of Indian rural development as an example of "thoughtless transfer of inapplicable experience from one civilization to another."[24]

The agenda that the leaders of the movement to establish an agricultural university system set for themselves in the early 1960s was monumental. They visualized the establishment in each state of an agricultural university with statewide responsibility for teaching, research, and extension. The process of establishing an agricultural university in each state is, except for a few of the smaller states, essentially complete. By 1978, 21 agricultural universities, some with several campuses, had been put in place.

The 1978 Review Committee on Agricultural Universities judged the progress of the educational programs at the new agricultural universities to be satisfactory. It was, however, concerned about the quality of undergraduate education at some of the multicampus institutions, and it cautioned against the too rapid proliferation of doctoral programs. The proposed transfer of state research responsibilities to the agricultural universities was initially confronted with serious problems in a number of states. In states such as Punjab, where the state governments immediately transferred statewide research functions and facilities to the new universities, strong state research programs have developed. In a number of states, however, bureaucratic infighting has delayed the transfer of facilities and support. The government of Rajasthan took 10 years to transfer research to Udaipur University and still retains responsibility for irrigation research. The government of Madhya Pradesh established a new rice research station at Raipar as an autonomous institution that remained unrelated to the agricultural university for a number of years. And even in Uttar Pradesh the state government has insisted that its obligations for support of research at the F. B. Pant University were satisfied by its initial land grant.

The Indian council of agricultural research

The Indian Council of Agricultural Research is, in addition to the state governments, the major source of research support for the agricultural universities. The ICAR provides on the average about

36 percent of the support for the agricultural university research. In a number of states, it accounts for more than half of the total support. The Council on Agricultural Research also plays a major role in coordinating state and Center research activities through the All-India Coordinated Research Programs. In addition, the council is also directly responsible for the research conducted at the Central Research Institutes. About 3,500 scientists are employed directly by the council or by the institutes directly responsible to the council.

In the past, it has been argued that the effectiveness of Indian agricultural research has been hampered by a heavy-handed administrative bureaucracy and by a civil service system that failed to distinguish between scientific and administrative capacity. Since the mid-1970s, a number of steps have been taken to improve both the management of research and the research environment. The role of the director-general of the ICAR has been enhanced through upgrading the position first to joint secretary and later to secretary, to the government of India, with administrative responsibility for the newly established Department of Agricultural Research and Education in the Ministry of Agriculture and Education. In 1975 the ICAR was granted authority to restructure its personnel policies. The scientific staff is now organized into the Agricultural Research Service, which is able to grant more appropriate recognition to scientific accomplishment.[25]

A retrospective and prospective view

Since independence, the Indian agricultural research system has gone through a period of consolidation in which the disparate institutions inherited from the past were molded into a national research system. The result is clearly one of the half-dozen most significant national agricultural research systems in the world in terms of resources employed and level of scientific activity. If estimates of the rate of return to research investment have any meaning, the Indian agricultural research system has been highly productive.[26]

The Indian system has also gone through a period of intense institutional innovation as it has put in place, in a matter of only 20 years, an agricultural university system patterned on the land-grant universities of the United States. Many of the problems that this innovation confronted are due to the imposition of these new institutions on the states by the Center. In the United States, the land-grant universities were, by and large, initially developed by the states. Later, the state experiment stations, supported by funds from the federal government, were established at the universities.

With these two accomplishments well on the way to completion, the Indian agricultural research system is now shifting toward a stronger emphasis on the quality of its staff and its program. The new administrative reforms were undertaken with this objective. Given the complexity of Indian agriculture and the size of the research system, this task will not be easy. Additional means will have to be sought to encourage a flowering of the intellectual vigor, the scientific capacity, and the research entrepreneurship of which India has such an abundant potential supply.

REFORMING THE BRAZILIAN AGRICULTURAL RESEARCH SYSTEM

Until the mid-1970s, Brazil had pursued a policy of increasing agricultural production through expansion of cultivated area.[27] As a result, Brazil entered the decade of the 1970s with an extremely weak national agricultural research system. In spite of a tradition of agricultural experiment station research in Brazil that extends back to the latter quarter of the 19th century, the research budget of the National Ministry of Agriculture was only about U.S.$14 million in 1973.

Among Brazilian states, only the highly developed state of Sao Paulo had established the agricultural research capacity needed to provide its farmers with a continuous stream of productive agricultural technology. Prior to the mid-1970s, the agricultural research budget and the scientist-years devoted to agricultural research by the state of Sao Paulo frequently exceeded those of the federal government.

In the decade between 1960 and 1970, pressure for higher land productivity began to develop. Expanded agricultural exports were wanted to finance industrial development, which in turn stimulated the rapid growth of the domestic demand for food in urban areas. As the frontier was pushed farther from the centers of population, transportation costs rose, leading to greater incentives to intensify agricultural production in the older areas. Attempts to stimulate more rapid growth in agricultural production through subsidized credit and heavy emphasis on extension were only marginally successful.

Inadequate agricultural research capacity

The agricultural research system that had evolved over the previous decades was clearly inadequate to serve as an effective source of the new knowledge and new technology needed to support productivity growth in Brazilian agriculture. In early 1972, Minister of Agriculture Dr. Luis Fernando Cirne Lima called a meeting of all state secretaries of agriculture and agricultural experiment station directors to discuss the requirements for modernizing the agricultural research system. A special committee was then formed to recommend reforms in the agricultural research system.

The committee was highly critical of the existing research system. It pointed out (1) that the basic national needs in respect to agriculture were unknown to most research personnel; (2) that there was little interaction between research personnel and farmers; (3) that the existing administrative structure inhibited the training and promotion of well-qualified personnel; (4) that inadequate internal communication among units and individual researchers was evidenced by the large numbers of parallel projects on unimportant products; (5) that the lack of suitable programming and evaluation mechanisms permitted researchers to undertake individual activities of doubtful value; (6) that of 1,902 individuals considered to be formal researchers, only 10 percent could be considered professionals with some kind of graduate training in research; (7) that the salary policy did not permit the government to compete in the professional labor market—that there were no means to hire and promote qualified personnel quickly or to demote unqualified persons; (8) that there were inadequate mechanisms for obtaining and managing financial resources that came solely from the federal government; and (9) that most existing research facilities were underutilized.

The committee also found a number of positive strengths in the federal research systems: (1) a geographically dispersed network of research units that covered practically the entire nation, (2) an infrastructure of equipment and facilities that was considered reasonably adequate for most of the units, (3) 16 technical journals for professional communication of scientific work, (4) several small but well-qualified groups of researchers whose talents would be better used if they were released from administrative overburdening, and (5) a recognition on the part of the researchers of the need for integrated research policy and planning.

The committee recommended that a public corporation would be the most effective institutional device to mobilize an effective agricultural research effort. This recommendation was supported by the U.S. Agency for International Development and the World Bank. In December of 1972, the Brazilian Congress created the Brazilian Public Corporation for Agricultural Research (EMBRAPA) that was to coordinate research in agriculture and animal husbandry. By April of 1973, responsibility for federal agricultural research and coordination of state research had been transferred to EMBRAPA.

The EMBRAPA system

The new EMBRAPA management was dominated by a group of exceptionally able young technocrats, many of whom had training in economics. They were granted the heady task of completely reorganizing the Brazilian agricultural research system. Once EMBRAPA was established as a separate entity, they proceeded in this task with great self-assurance.

A new pattern of organization that drew its inspiration from the single-commodity research model that characterized the initial units in the international agricultural research system was adopted. (See chapter 5.) National research centers were established for the major agricultural commodities—wheat, sugarcane, corn, beans, soybeans, rice, coffee, rubber, livestock, and dairy. Three centers that focused on key resource problems were also established—for the problem soils of the cerrado, the semiarid northeast, and the tropical conditions of the Amazon regions. Each national center was organized around a major central station plus a series of regional stations, depending on the several agroclimatic regions in which the several commodities were produced.

EMBRAPA visualizes the national centers as primarily applied research institutions. Their mission is to produce technology packages that will have an immediate impact on agricultural production. Explicit feedback mechanisms from producers and the extension services designed to focus research effort on priority production problems have been established. The commodity- and resource-problem focus of the research centers is viewed as an effective device for achieving multidisciplinary cooperation among researchers and for designing technology packages that take advantage of the interactions among the package elements (among crop variety, pest and disease control, fertilization, plant population, and weed control, for example).

In addition to the reorganization of the federal research system, a major effort has been made to strengthen the professional capacity of the EMBRAPA staff. The old system was overcrowded with the bureaucrats who were not trained for research. Only about 3,500 of the roughly 6,700 employees of the old system were transferred to EMBRAPA. Within the first few years, over 500 fellowships were granted for graduate study at Brazilian and overseas universities.

Salary levels were established at international levels, and an active recruiting effort was undertaken to attract foreign scientists into the Brazilian system. An attempt was made to strengthen the connections among Brazilian and foreign universities and with the international agricultural research centers.

By 1979 the EMBRAPA research budget was approaching U.S. $200 million. In terms of financial resources, this was roughly double the budget of the CGIAR-sponsored international agricultural research system. (See chapter 5.)

Some continuing problems

The EMBRAPA management thinks of their system of research management and planning as a concentrated, or focused, model rather than a decentralized, or diffuse, model characteristic of the research systems in other federal systems, such as the one in the United States.

The adoption of a highly centralized system of management was facilitated by the centralization of power that took place in 1964 with the military coup. In Brazil the states had traditionally exercised a great deal of economic and political power. The individual states were large (Minas Gerais, a medium-size state, is larger than France), the transportation network was weak, and there were great differences in the level of economic and political development among states. The new regime was determined to assert the primacy of federal power. The EMBRAPA system for organizing agricultural research was consistent with this objective. A reorganization of the extension service designed to create an effective federal service had already been completed.

In the original EMBRAPA plan, the state agricultural research systems were scheduled to be brought under the control of the federal system, to become EMBRAPINHAS. The strong state systems, in Sao Paulo, for example, refused to be brought under a system administered from Brasília. The attempted merger has led to some reforms in the state system but not to its complete integration into the federal system.

The EMBRAPA plan also saw the basic research in support of agriculture as a function of the universities. But the relationship has been complicated because EMBRAPA is associated with the Ministry of Agriculture and the universities are associated with the Ministry of Education. Furthermore, EMBRAPA's recruitment of personnel from the universities has weakened the universities' capacity to provide advanced training for the next generation of agricultural scientists.

It is too early to assess the effectiveness of the new EMBRAPA system in generating new knowledge and new technology. It is also too early to know whether EMBRAPA has established its credentials with clientele groups that will be capable of protecting it from the entrenched interests in the Ministry of Agriculture and the states when centrifugal political forces begin to reverse the centripetal forces that have dominated Brazilian policy since 1964. It is clear, however, that Brazil has been making the most serious effort of any country in Latin America to establish an agricultural research system capable of supporting the transition of Brazilian agriculture from a resource-based to a technology-based industry.

AGRICULTURAL RESEARCH
INSTITUTES IN MALAYSIA

The cases that have been reported so far have dealt with agricultural research in the older developed countries and with two large developing countries (India and Brazil). The smaller developing countries are, in most cases, smaller than the leading states or provinces in many of the large countries. They face particularly difficult problems in allocating scientific and technical efforts among commodities and resource problems.

In this final case study, I focus on Malaysia, one of the smaller and newer of the developing countries.[28] The reason for selecting Malaysia is that, during much of the colonial period and the period since the nation's independence, agricultural research was heavily concentrated on a single commodity at a single research institute, the Rubber Research Institute of Malaysia (RRIM). During the last decade, Malaysia has attempted to develop the research capacity to sustain productivity growth in other domestic food-crop and export commodities.

Establishment of the rubber industry in Peninsular Malaysia

Prior to 1900, the world supply of rubber came almost exclusively from wild rubber trees and primarily from the Amazon basin of Brazil. Efforts were made by the India Office in London during the 1870s to explore the possibility of propagating rubber in Asia. In 1872, the curator of the Pharmaceutical Society Museum in London was commissioned to obtain information about the prospects of introducing rubber to the East. The India Office, working with the Royal Botanical Gardens at Kew, collected rubber seeds from Brazil. The first consignment was obtained in 1873, and additional consignments were obtained in 1875 and 1876.

Enough seedlings were obtained, in spite of poor germination, to dispatch 1,000 seedlings to the Botanic Gardens at Heneratgoda in Ceylon. Fifty plants were also sent from Kew to the Singapore Botanic Gardens, but they failed to survive. A second consignment of 22 plants was shipped to Singapore in 1877, and these were successfully established. The transfer of rubber production to Peninsular Malaysia (formerly Malaya) started a sequence of events that led to rubber's dominant role in both the plantation and the small-holder sectors of the Malaysian agricultural economy.

In 1929, rubber represented more than 75 percent of the total value of the country's agricultural output. In the mid-1970s, rubber still occupied over half of the cultivated area, employed about one-third of the country's working population, and accounted for one-fifth of the gross national product. Research leading to higher-yielding clones, more efficient cultural and marketing practices, and improved processing and transportation has been an important factor in enabling Malaysian rubber to compete with synthetic rubber since World War II.

The establishment of an effective rubber research capacity in Malaysia proceeded slowly and with a number of interruptions. The first effort to bring science to bear on the growing of rubber was made by H. N. Ridley, who became director of the Singapore Botanic Gardens in 1888. His experimentation with tapping methods led to the replacement of the incision method (hitting the tree with an axe or a knife) with the excision method (paring off successive thin slivers of bark). This method resulted in improved yields and prolonged tapping life of the trees. Ridley's work also led to the discovery of "wound response" (i.e., increased yield of rubber in response to tapping in the *Hevea* species). He also conducted research on the means of coagulating latex for shipment and on the growth and diseases of trees.

As a result of the research at the Singapore Botanical Garden and the decline in the coffee industry in Malaya, substantial interest in rubber as a commercial crop developed during the last years of the 19th century. As planters began to take up rubber growing, a substantial body of knowledge was available to support the growth of the rubber industry. Incredible as it may seem in retrospect, Ridley's effort to develop rubber production technology at the Singapore Botanic Gardens was strongly opposed by his superiors in the colonial government.

In an attempt to bring more resources to bear on rubber research, the United Planters' Association urged the establishment of a department of agriculture. The Malayan Department of Agriculture was established in 1905 but was relatively ineffective at mobilizing resources for rubber research. A number of larger estates began to conduct their own research. The Lanadron estate hired a plant pathologist to work on disease control in 1909. The Rubber Growers' Association initiated a cooperative research program financed by participating companies and private estates in 1909. Within a few years, the association had assembled a staff of six scientists at its laboratory at the Pataling estate (outside of Kuala Lumpur) and had opened a branch station at Ipoh. Other research stations were established by the Malay Peninsula Agricultural Association and Societe Financiene des Cooutchoucs (a Franco-Belgian company).

Organization of the Rubber Research Institute

A proposal to merge official and private work under a central body was discussed as early as 1917; but it was not until 1925 that legislation was finally enacted to set up the Rubber Research Institute of Malaya (RRIM). The research on rubber in the Department of Agriculture was transferred to the new institute, and most of the private research stations were closed down. By 1927 the new institute had a senior research staff of over 20 and a budget of just under a quarter of a million dollars.

Initially, the research effort focused on two areas: chemical investigations into the preparation and the properties of rubber and pathological investigations into the diseases of the crop. Interest in crop improvement and soil management developed rapidly, and the Rubber Research Institute was organized around four research programs: chemistry, pathology, soils, and botany. The institute was initially funded by a cess (tax) of 10 cents a picul (133.33 pounds) on all rubber produced in and exported from the country. During the Great Depression, the research program came under considerable financial pressure. Adjustments were made in the cess, and the budget fluctuated with changes in the world rubber market. The rubber-breeding program was discontinued while the first breeder was on home leave in 1931 and was not resumed until 1937. It was again suspended during the Japanese occupation (1942-1945).

The major problem in varietal improvement was solved by Cramer, a Dutch scientist working at Buitenzorg (now Bogor), Indonesia, during the period 1910 to 1916. Cramer selected the seed from high-yielding trees and propagated the high-yielding selections by budding (cloning). Suitable methods of budding that were commercially feasible were developed in 1916. This methodology was transferred to Malaya during the 1920s. When varietal improvement work was resumed in 1938, it was realized that most of the progress possible from selection of superior seedlings had been made and a program of systematic breeding was initiated to create improved populations for subsequent selection and cloning.

Systematic *Hevea* breeding involves alternate clonal selection and generative breeding. Mother trees selected from the highly variable basic populations form a group referred to as primary clones. The primary clones, which provided the first clonal materials for commercial plantings, are also used as parents for hand-pollinated crosses to produce improved seeding materials. Then, improved seedling families can be used for large-scale planting or for further cloning. Elite secondary clones are selected and used for intercrossing. The process can be continued almost indefinitely. Each full breeding cycle takes about 15 years. Another 15 years is required for the tests necessary to determine the performance characteristics needed before large-scale planting can be recommended. In spite of the extremely long breeding cycle, dramatic improvements have been achieved. (See figure 4.3.)

The task of restaffing the RRIM after World War II began in 1945. In order to decide whether the institute was equipped to help the natural rubber industry withstand competition from the synthetic rubber industry, a major review of the research program and organization was undertaken. A decision was made to strengthen the production research by the RRIM in Malaysia and the utilization research by the Malaysian Rubber Producers' Research Association in the United Kingdom.

In 1954 the Rubber Research Institute initiated a new review of the program and its funding. The review (Blackman) committee's report recommended a more integrated organization of all the production and utilization research, marketing, and promoting activities supported by the Malaysian Rubber Fund. It also recommended an increase in the cess to support the work of the fund. The report also recommended a strengthening of the work of the Botanical Division, the establishment of the Estate Advisory Service to be separate from the researcn program, and a substantial strengthening of the Smallholders Advisory Service. Most of the Blackman report's recommendations were accepted and implemented. The Malaysian Rubber Fund Board (MRFB) was reconstituted, with representation from industry and government. The Coordinating Advisory Board, composed of eminent scientists from all over the world, was set up to advise the board on rubber chemistry and technology and on agricultural and biological research.

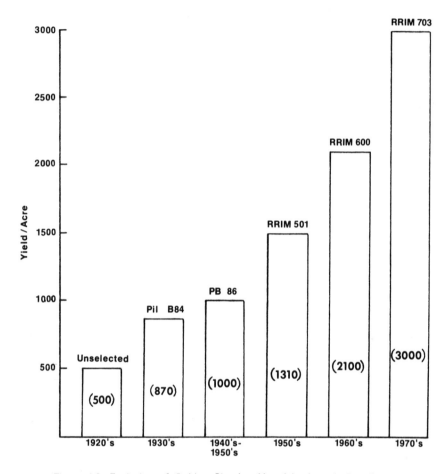

Figure 4.3. Evolution of Rubber Planting Materials through Breeding and Selection. (Used with permission from B. C. Sekhar, "Scientific and Technological Development in the NR Industry," *Malaysian Rubber Review*, 1 [July 1976], Table 1, p. 26.)

In 1972 the Malaysian Parliament passed a series of five bills designed to further simplify the governing strucutre, to encourage modernization of the small-holder subsector, to modernize marketing methods, and to encourage domestic rubber production. The Malaysian Rubber Fund Board was renamed the Malaysian Rubber Research and Development Board (MRRDB) and given the authority for the planning and formulation of research strategies and policies. At present, the MRRDB has three major programs: production research centerd in Malaysia, consumption and end-use research centered in the United Kingdom, and technical advisory offices servicing rubber users throughout the world.

The transition to national leadership in the management and operation of the Rubber Research Institute during the postindependence period was not easy. Research output lagged until the late 1960s, when the return of young scientists from training abroad began to take up the slack left by departing expatriates. Fortunately, the RRIM has had both the resources and the foresight to engage in an aggressive program of training and retraining. By 1980, half of its 220-member research staff had received training at the doctoral level and the rest at the master's level. The RRIM's management has continued to pursue a decentralized management structure, with primary program and budget responsibility in the hands of scientists.

One of the more notable developments during the postindependence period has been the redirection of research toward problems of small-holder rubber production and the strengthening of small-holder advisory services. During the colonial period, small-holder rubber production had, at times, been actively discouraged.

Diversifying agricultural research in Malaysia

When the Rubber Research Institute was established in 1925, responsibility for other crops remained with the Division of Agriculture. Pineapple research was later organized under the Malaysian Pineapple Industry Board along lines similar to the organization of rubber research. After independence, there was increasing concern with the relative neglect of research on other crops.

In 1967, the Economic Committee of the Cabinet proposed the establishment of the autonomous Malaysian Agricultural Research and Development Institute (MARDI), which incorporated the research being conducted in the Ministry of Agriculture and a number of other government agencies. Only rubber was excepted. After considerable hesitation, an act establishing the MARDI was passed by the Malaysian Parliament in 1969.

The new institute is an autonomous body with linkages to the government through the Ministry of Agriculture. It is governed by an 11-member board, which includes representatives from the government and the private sector. It is administered by a director-general and three deputy director-generals. Its program is organized around eight divisions: Annual Crop Production, Perennial Crop Production, Animal Production, Agricultural Product Utilization, Fundamental Research, Development Research, Research Services (which includes statistics, economics, and sociology), and Central Services (which includes publication, library, and administrative services). In addition to the headquarters station at Serdang, research is conducted at 23 other stations in Peninsular Malaysia and in each state in West Malaysia.

A shortage of trained and experienced personnel was recognized as a major limiting factor in the development of the MARDI program. Substantial funds were made available from government sources and from a World Bank loan to finance graduate fellowships for the training of potential staff members in universities abroad. Cooperative relationships were established with the International Rice Research Institute (IRRI), the Australian Commonwealth Scientific and Industrial Research Organization (CSIRO), the Canadian International Development Research Centre (IDRC), the CIMMYT-based Asian Maize and Sorghum Improvement Centre in Thailand, other international centers, and a number of bilateral and multilateral technical assistance programs.

Future directions

By the middle of the 1970s, Malaysia was well on its way to having, in addition to one of the world's major single-commodity research institutes, a broadly based agricultural research center capable of conducting both the commodity-oriented and the resource-base-oriented research necessary to sustain agricultural development. It was also rapidly building a capacity for higher education in agriculture through the upgrading of the former School of Agriculture at Serdang, adjacent to the MARDI headquarters station, first into a college of agriculture (1962) and later into an agricultural university.

There is continuing disagreement in Malaysia over the effectiveness of singe-commodity research institutes compared to multi-commodity research institutions. In September 1979, the Oil Palm Research Institute of Malaysia (OPRIM) was established under the Ministry of Primary Industries. Thus, the public-sector research in support of the two commodities that account for a large share of Malaysia's export earnings are organized in single-commodity institutes. In addition, several of the larger rubber and oil-palm plantation companies continue to maintain their own research programs. Research on the minor export crops and the domestic food crops remains the province of the MARDI. This restriction of the MARDI mandate to the less important commodities in the future could raise some rather difficult problems regarding the mobilization of political support for the MARDI's research budget.

SOME PERSPECTIVES ON THE ORGANIZATION OF AGRICULTURAL RESEARCH

The historical reviews of the several agricultural research systems presented in this chapter serve to illustrate and to reinforce a number of principles on the organization and development of national agricultural research systems. The historical case studies presented suggest that there are basically four models for the organization of agricultural research.

- The *integrated* research, extension, and education model. The U.S. land-grant-university system is one of the more clear-cut examples of this model.
- The *autonomous* or semi-autonomous publicly or privately supported research institute. This was the model pioneered in Great Britain. It was an important method of organizing research and export commodities during the colonial period. It remains an important model for the support of research on export commodities and other commodities produced under plantation or other large-scale forms of production.
- The *ministry* of agriculture model. This has been the most common pattern for research on domestic food crops in the smaller countries. It is usually an important component of the integrated federal-state (or national-provincial) systems in large countries.
- The agricultural *research council* model. This is a rather recent development. It has often been used, as in India, as a method of achieving more effective coordination of a system in which two or perhaps all three of the other models have developed alongside each other during an earlier period.

During the last decade, a common feature of all four systems has been an increased concern for more effective planning and management. (See chapter 11.)

Stages in the development of national systems

In the developed countries of Western Europe and North America and in Japan, agricultural research has typically progressed through a series of relatively distinct stages.

The first stage in the development of the older national systems was based on the innovative activities of individual farmers and inventors. In the United States, Thomas Jefferson's experiments with soil fertility, George Washington's efforts to introduce new crops, and the invention of new land-preparation and harvesting equipment by John Deere and Cyrus McCormick come immediately to mind. In the United Kingdom, Arthur Young dramatized the efforts of progressive landowners in the development of improved animal breeds and farming practices. In Germany, Johann Heinrich Von Thünen's investigations into estate management established the foundations for modern farm management and location economics. In Japan, veteran farmers (rōnō) and farming landlords (gōnō) were the sources of new technology, such as improved methods for seed selection, and they helped diffuse the best traditional farming practices.

Many other examples could be cited. The distinguishing feature of these early-stage developments was their highly personal character. Although there are records of active correspondence among many of the leaders in the agricultural improvement movement, the leaders' efforts proceeded with little of the institutional support that characterizes modern research. Futhermore, the early leaders were relatively uninformed about scientific principles and experimental methods. However, the accumulation of agronomic practice and craft experience did permit the gradual advancement of agricultural technology.

The second stage was characterized by the organization of agricultural experiment stations staffed by research workers with professional specializations in agricultural science or technology. Typically, the first experiment stations in a country were established by private estate owners or agricultural associations. J. B. Boussingault established an experiment station on his estate in Alsace in 1834. In 1842, Scottish farmers established the Agricultural Chemistry Association, which provided a chemical laboratory and a system for farm tests and experiments. The Rothamsted experiment station was established by Sir John Bennet Lawes on his family estate in 1843. In Germany, the impetus for the establishment of agricultural experiment stations was strengthened with the publication of Justus von Liebig's treatise on the chemical basis for plant nutrition.

The distinguishing feature of this second phase was the research laboratory or the experiment station. The laboratory and the station represented the physical and institutional infrastructure necessary to attract professionally trained scientists to agricultural research and to search for "methods of applying science to agriculture." It provided the framework both for assembling the resources needed to support agricultural research and for organizing the diverse scientific and technical skills necessary to advance knowledge.

Quite different patterns emerged during the second stage for the support of biological, chemical, and mechanical research. Support for research leading to advances in biological technology in most countries has been supported primarily by public revenues (or by organized producer groups granted the right to tax themselves); advances in chemical technology have been produced primarily in the laboratories of chemical companies; and advances in mechanical technology have continued to be based primarily on farmers' inventions, but engineering research performed by machinery-manufacturing companies has contributed to the commercial success of new machinery.

The third stage is characterized by the evolution of integrated national agricultural research systems. The characteristic feature of the third stage is the establishment of an agricultural research planning capacity that is capable of relating research priorities to the allocation of professional and financial resources. The third stage may evolve out of efforts by private scientific bodies or by quasi-public advisory councils such as a national academy of science to draw attention to gaps in the research effort—for example, between research in crop production and in animal or human nutrition or between private-sector research in plant breeding and public-sector research in genetics.

As the level of public-sector research support expands, the national agency that is primarily responsible for the funding of agricultural research typically develops substantial "in-house" planning and resource allocation capacity. The development of a national research planning capacity often results in an attempt to rationalize the organization of individual commodity research institutes or experiment stations, to reexamine the location of branch stations, and to attempt to achieve more effective integration of national, state or provincial, and private agricultural research.

The newer national systems have drawn heavily on the experience of the older national systems. But the pattern of evolution in the newer systems has been affected in some countries by their having evolved under colonial auspices or their having been influenced by particular bilateral or multilateral assistance programs.

Stress in the development of research systems

The modernization of agricultural research systems imposes substantial stress on administrators and personnel. Four issues that tend to be subject to almost continuous debate in every national system are discussed in this section.

Stress over linkages among research, education, and extension. Are education and research complementary or competitive activities? Is the dissemination of research results most effectively accomplished in association with, or separate from, the performance of research? Both questions were hotly debated in the United States during the early years after the establishment of state agricultural experiment stations within the colleges of agriculture at state land-grant colleges and universities.

Over time, a consensus seems to have emerged in the United States that research is highly complementary to graduate education but less so to undergraduate teaching. Similarly, systematic contact with the technical and economic problems of farmers is generally regarded as important in the focusing of research effort, but the effects of making the research staff responsible for participation in extension programs is still being debated. Members of the USDA's research staff working at or near college campuses are often encouraged to engage in a limited amount of graduate training and are expected to take some responsibility for the dissemination of research results.

Different perspectives have emerged in many other national systems. In Australia, for example, members of the research staff of the Commonwealth Scientific and Industrial Research Organization (CSIRO) often have little contact with graduate students even when a CSIRO laboratory is located on a college or university campus. In systems in which responsibility for agricultural extension is located in a ministry of agriculture and research is conducted by an autonomous research institute, there may be almost total separation between extension functions and research activities. In some research systems that are not linked administratively with extension programs, there is strong pressure to develop extension-type linkages with clientele groups in order to assure financial support.

Stress over centralization and decentralization. In countries with a dual state (provincial, prefectural) and national system, there typically is a continuous effort by the national system to achieve effective coordination of the combined efforts of the state and national systems. In systems that are highly centralized, there is a continuous struggle to assure consistency between regional and national priorities.[29] Such concerns underlie, for example, the reform efforts that have been directed at achieving greater centralization of planning and management in the Brazilian system. However, centralization may have costly consequences. In the Indian system, for example, reform efforts are being directed toward removing excessive bureaucratic restraints on research activity in a system with a strong central direction.

Conflict between basic and applied research. In every productive agricultural research institution, there is tension over the commitment of professional and financial resources to mission-oriented research or to more fundamental research. (See chapter 3.) The problem of achieving convergence between the professional objectives of the individual research scientists and the social objectives of the research institution is subject to continuous debate between the research staff and the administration. A partial solution to this problem is to link support for fundamental research to the anticipated demand for basic knowledge in mission-oriented applied research programs. Fundamental research on nutrient uptake may, for example, have a high payoff for applied research on varietal development.

The issues involving the appropriate mix of basic, supporting, and applied research are particularly difficult to handle for many of the smaller national research systems. The agricultural sectors in many of the smaller countries are no larger in economic terms than the areas served by the smaller states in India or the United States or even the prefectural stations in Japan. Developing substantial domestic capacity in the basic and supporting sciences would be very expensive relative to the value of commodity production. The development of regional research networks (such as the West African Rice Development Association) and the establishment of effective links with the international agricultural research system represent one way of dealing with the limitations imposed by the small size of many national systems. (See chapter 7.)

Conflict over who should pay for research. In countries organized along federal lines, the question of what share of the costs of research should be borne by the local governments and what share should be borne by the central governments is of continuous concern. Central funds are often accompanied by central control. Yet, the benefits of research done by one state or province spills over into other states or provinces. What part of the total agricultural research effort is appropriately within the public sector and what part is appropriately conducted in the private sector? Of that part that is conducted in the public sector, what part should be paid for by consumers in the form of general revenue measures and what part should be paid for by producers in the form of a marketing cess, or tax? The answers to these questions must be sought in three areas: in the division of benefits of research between producers and consumers, in the economic organization or structure of the agricultural sector, and in the fiscal and administrative capacity of governments to support agricultural research.

A major gap in our knowledge about the development of national agricultural research systems is the lack of a body of research on the mobilization of political resources for the support of agricultural research. Why do some societies find it so difficult to sustain support for an activity that pays such high growth dividends? Are organized producer groups capable of capturing substantial benefits from research essential for sustained support? How does the agrarian structure, the farm size distribution and tenure structure, affect the demand for agricultural research? The case studies in this chapter suggest the importance of these questions but not the answers.

NOTES

1. I am indebted to Ralph W. Cummings, Jose D. Drilon, P. V. Hariharasankaron, Albert M. Moseman, Jose Pastore, Teck Yew Pee, G. Edward Schuh, M. S. Swaminathan, W. D. Ting, Eduardo Trigo, Tilo L. V. Ulbricht, Adolf Weber, Floyd W. Williams, and Noboru Yamada for making comments on an early draft of this chapter. I have benefited in ways that go beyond what can be acknowledged in specific references to Isaac Arnon, *Organization and Administration of Agricultural Research* (Amsterdam: Elsevier, 1968); and Albert H. Moseman, *Building Agricultural Research Systems in the Developing Nations* (New York: Agricultural Development Council, 1970). Among the other very useful reports on which I have drawn are the following: Committee on African Agricultural Research Capabilities, *African Agricultural Research Capabilities* (Washington, D.C.: National Academy of Sciences, 1974); J. D. Drilon, ed., *Agricultural Research Management: Asia*, vols. 1 and 2 (College, Laguna, Philippines: Southeast Asian Regional Center for Graduate Study and Research in Agriculture, 1975, 1977).

2. A more complete review would include case studies of the evolution of agricultural research systems in Africa during the colonial and postcolonial periods. For an excellent study that became available just as this chapter was being completed, see: Francis Sulemanu Idachaba, *Agricultural Research Policy in Nigeria* (Washington, D.C.: International Food Policy Research Institute, August 1981, Research Report 17). A similar study for the Ivory Coast would be very useful.

3. This section draws very heavily on S. C. Salman, and A. A. Hanson, *The Principles and Practice of Agricultural Research* (London: Leonard Hill, 1964); Tilo L. V. Ulbricht, "Contract Agricultural Research and Its Effect on Management," in *Resource Allocation and Productivity in National and International Agricultural Research*, Thomas M. Arndt, Dana G. Dalrymple, and Vernon W. Ruttan, eds. (Minneapolis: University of Minnesota Presss, 1978), pp. 381-93.

4. *The Agricultural Research Service* (London: Agricultural Research Council, 1963), p. 5.

5. Ulbricht, "Contract Agricultural Research," p. 382.

6. In this section I draw on a memorandum written by Adolf Weber, "Some Prelimi-nary Notes on the Early History of the German Agricultural Research System" (Institut für Agrarpolitik und Marktlehre der Christian-Albrechts-Universität, Kiel, December 1979). For a useful history, see Werner Tornow, *Die Entwicklungslinien der Landwirtschaftlichen For-schung in Deutschland unter besonderer Berücksichtigung ihrer institutionellen Formen* (Itiltrup [Wesf.]: Landwirtschaftsverlag GmbH., 1955). I have also benefited from com-ments Theodore Dams and Gunther Weinschenck made on an earlier draft of this section.

7. David S. Landes, "Technological Change and Development in Western Europe, 1750-1914," in *The Cambridge Economic History of Europe VI, The Industrial Revolution and After*, Part 1, H. J. Habakkuk and M. Poston, eds. (Cambridge: Cambridge University Press, 1966), p. 571.

8. Salman and Hanson, *Principles and Practice*, p. 22.

9. H. C. Knoblauch et al., *State Agricultural Experiment Stations: A History of Re-search Policy and Proceedings* (Washington, D.C.: U.S. Department of Agriculture, Miscel-laneous Publications No. 904, May 1962), p. 16.

10. This section draws very heavily on material previously developed in Yujiro Hayami and Vernon W. Ruttan, *Agricultural Development: An International Perspective* (Baltimore: Johns Hopkins Press, 1971), pp. 138-53. I am indebted to Ned D. Bayley and Roland R. Robinson for their comments on an earlier draft of this section.

11. The results of foreign training were, however, not unlike those observed in many underdeveloped countries today. "European professors were puzzled by American students who, after beginning so well abroad, lapsed into mediocrity upon returning home. And one recalls cases in which Americans, inspired by European science, actively began to make basic contributions, but never went on to fulfillment of the potentialities so revealed." (Richard Harrison Schryock, "American Indifference to Basic Science during the Nineteenth Century," in *The Sociology of Science*, Bernard Barker and Walter Hirsch, eds. [New York: Free Press, 1962], pp. 104-5.) These comments obviously applied to men like John P. Norton, who studied under Liebig in Germany and went on to found the Connecticut Agricultural Experiment Station, and Evan Pugh, who worked under Sir Henry Gilbert at Rothamsted and went on to found the Pennsylvania station.

12. A. Hunter Dupree, *Science in the Federal Government: A History of Policies and Activities to 1940* (Cambridge, Mass.: Harvard University Press, 1957), p. 165.

13. Office of the General Manager, *The Federal Government's Role in Fertilizer Re-search and Development* (Knoxville, Tenn: Tennessee Valley Authority, June 1978 [mimeo-graphed]).

14. U.S. Department of Agriculture and Association of State Universities and Land Grant Colleges, *A National Program of Research for Agriculture* (Washington, D.C.: U.S. Government Printing Office, 1966). The work of the committee has been reviewed in several of the papers in Walter L. Fishel *Resource Allocation in Agricultural Research* (Minneapolis: University of Minnesota Press, 1971). By the late 1970s, a continuing but somewhat cum-bersome planning process had become institutionalized. See Regional and National Planning Committees and Interim National Research Planning Committee, *1979-84 Cycle for Pro-jecting and Analyzing Research Program Adjustments with Historical Trends and Compari-sons* (Washington, D.C.: Joint Council on Food and Agricultural Sciences, SEA/USDA, July 1980).

15. Rachel Carson, *Silent Spring* (New York: Houghton-Mifflin, 1962); Jim Hightower, *Hard Tomatoes, Hard Times* (Cambridge, Mass.: Schenkman, 1973).

16. National Research Council, *Report of the Committee on Research Advisory to the U.S. Department of Agriculture* (Springfield, Va.: National Technical Information Service, 1972). For a review of the findings of the committee, see the series of articles by Nicholas

Wade, "Agriculture: NAS Panel Charges Inept Management, Poor Research," *Science*, 179 (January 5, 1973), pp. 45-57; "Agriculture: Critics Find Basic Research Stunted and Wilting," *Science*, 180 (April 27, 1973), pp. 390-93; "Agriculture: Signs of Dead Wood in Forestry and Environmental Research," *Science*, 180 (May 4, 1973), pp. 474-77; "Agriculture: Social Sciences Oppressed and Poverty Stricken," *Science*, 180 (May 18, 1973), pp. 719-22; "Agriculture: Research Planning Paralyzed by Pork-Barrel Politics," *Science*, 180 (June 1, 1973), pp. 932-37.

17. For the most recent reports, see *Report of the National Agricultural Research and Extension Users Advisory Board* (Washington, D.C.: U.S. Department of Agriculture, Science and Education Administration, October 1980); Joint Council on Food and Agricultural Sciences, *1980 Annual Report to the Secretary of Agriculture* (Washington, D.C.: U.S. Department of Agriculture, Science and Education Administration, January 1981). For a review of earlier research coordination efforts at the USDA, see Vivian Wiser and Douglas Bowers, "Research and Its Coordination in USDA: A Historical Approach" (Washington, D.C.: Economics, Statistics, and Cooperative Service, May 1979, mimeographed report).

18. This section draws heavily on Hayami and Ruttan, *Agricultural Development*, pp. 153-66; Yujiro Hayami, *A Century of Agricultural Growth in Japan* (Minneapolis and Tokyo: University of Minnesota Press and University of Tokyo Press, 1975), pp. 44-86 and 140-46; Keijiro Otsuka, "Public Research and Price Distortion: Rice Sector in Japan" (University of Chicago, Ph.D. dissertation, 1979). I am indebted to Dr. Noboru Yamada, former director of the Tropical Agricultural Research Center, for his review of an earlier draft of this section and for information on recent developments in agricultural research organization in Japan.

19. Agriculture, Forestry and Fisheries Research Council, *Directory of Research Council and National Research Institutions for Agriculture, Forestry and Fisheries in Japan* (Tokyo: Ministry of Agriculture, Forestry and Fisheries, 1979).

20. M. S. Randhawa, *A History of the Indian Council of Agricultural Research, 1929-1979* (New Delhi: Indian Council of Agricultural Research, 1978). See also, K. P. A. Menon, "Building Agricultural Research Organizations—The Indian Experience," in *National Agricultural Research Systems in Asia*, Albert H. Moseman, ed. (New York: Agricultural Development Council, 1971), pp. 23-28; Rakesh Mohan, D. Jha, and Robert Evenson, "The Indian Agricultural Research System," *Economic and Political Weekly*, 8 (March 31, 1973), pp. A21-A26. This section has been reviewed by M. S. Swamanathon and Carl Pray.

21. S. V. S. Shastry, "The All-India Coordinated Rice Improvement Project," in *National Agricultural Research Systems in Asia*, Albert H. Moseman, ed. (New York: Agricultural Development Council, 1971), pp. 120-34.

22. For an excellent review of the development and status of the Indian agricultural university system, see *Report of the Review Committee on Agricultural Universities* (New Delhi: Indian Council of Agricultural Research, June 1978).

23. *Report of the Review Committee on Agricultural Universities*, p. 5.

24. Guy Hunter, *Modernizing Peasant Societies* (London: Oxford University Press, 1969), p. 185. Mr. Hunter was particularly critical of attaching extension responsibilities to new universities.

25. *ICAR Agricultural Research Service* (New Delhi: Indian Council of Agricultural Research, 1977).

26. R. E. Evenson and D. Jha, "The Contribution of Agricultural Research Systems to Agricultural Production in India," *Indian Journal of Agricultural Economics*, 28 (1973), pp. 212-30; A. S. Kahlon, P. N. Saxena, H. K. Bal, and Dayanath Jha, "Returns to Investment in Agricultural Research in India," in *Resource Allocation and Productivity in National Agricultural Research*, Thomas M. Arndt, Dana G. Dalrymple, and Vernon W. Ruttan, eds. Minneapolis: University of Minnesota Press, 1977), pp. 124-47.

27. This section draws heavily on Jose Pastore and Eliseu R. A. Alves, "Reforming the Brazilian Agricultural Research System," in *Resource Allocation and Productivity in National and International Agricultural Research*, Thomas M. Arndt, Dana G. Dalrymple, and Vernon W. Ruttan, eds. (Minneapolis: University of Minnesota Press, 1977), pp. 394-403. See also G. Edward Schuh, *The Agricultural Development of Brazil* (New York: Praeger, 1976), pp. 227-46.

28. This section draws primarily from Ani Bin Arope, "The Malaysian Agricultural Research and Development Institute," in *National Agricultural Research Systems in Asia*, Albert H. Moseman, ed. (New York: Agricultural Development Council, 1971), pp. 65-74; and Teck Yew Pee, "Social Returns from Rubber Research in Peninsular Malaysia" (Michigan State University, Ph.D. dissertation, 1977). I have also benefited from correspondence with Yusof Bin Hashim, deputy director of the Malaysian Agricultural Research and Development Institute, and with Ani Bin Arope, director of the Rubber Research Institute of Malaysia.

29. Idachaba pointed out in *Agricultural Research Policy in Nigeria* (p. 53) that "there are already grumblings about the present financing of all agricultural research by the federal government. . . . the Institute of Agricultural Research (Samaru) feels that the present budget cuts would not have occurred if the Northern states had remained its financial sponsors. There are many benefits to the local economy that the federal leadership is unable to fully appreciate."

Chapter 5

The International
Agricultural Research System

At the end of World War II, the Food and Agriculture Organization (FAO) of the United Nations was established to perform the functions of a global ministry of food and agriculture.[1] The FAO headquarters were established in Rome. Its responsibilities included technical assistance to individual countries; support of agricultural education; collection of statistics on land use, food consumption, and agricultural production and trade; and publication of periodicals, bulletins, and yearbooks on the world food and agriculture situation.[2]

The FAO has—through its technical assistance, educational, and regional networking activities—made significant contributions to the development of national research capacity in agriculture. But the FAO's Governing Council and Program Committee was reluctant to approve a substantial role in the sponsorship and conduct of research as part of the FAO's regular program.

This was due in part to an inadequate intellectual perception of the causes of lagging agricultural development in the poor countries at the time the FAO was established. Lagging agricultural development was interpreted primarily in terms of the failure to make effective use of available technology. The lack of more productive technology was not itself seen as a major barrier to agricultural development.

By the late 1950s, this view had changed. Technical assistance and community development programs, based explicity or implicitly on a technology-transfer model, failed to generate either rapid modernization of traditional farms and villages or rapid growth in agricultural production. (See chapter 2.)

116

Between the late 1960s and the late 1970s, a new international agricultural research system was established. The development of this new system represents a dramatic example of scientific entrepreneurship. The product of this effort has been to complete the vision of the agricultural leaders who established the FAO after World War II. The global ministry of agriculture has been complemented by a global agricultural research service.

In this chapter I attempt to place the evolution of the international agricultural research system within the context of other post-World War II agricultural development assistance. I also indicate some of the organizational and management issues that have confronted, or continue to confront, the management of the international agricultural research system. And I present some of my own perspectives on future policy and program directions.

THE INTERNATIONAL AGRICULTURAL RESEARCH INSTITUTE MODEL

Experience with development assistance in agriculture over the last several decades has resulted in three models for organizing professional resources to work on problems of agricultural and rural development. These can be described as the counterpart, the university contract, and the international institute models.

The counterpart model is perhaps the most familiar. It refers to a situation in which the individual scientist, or "expert," is employed by a technical assistance agency to function in an advisory role in close cooperation with counterpart scientists or professionals in national research, education, or program operations. The staff member from the external technical assistance agency is typically viewed as an expert who functions in an advisory role relative to his or her national counterpart.

It was gradually recognized, however, that the transferability of expertise from temperate- to tropical-region agriculture was severely limited. The scientific knowledge needed to improve both agricultural productivity and institutional performance could only be obtained by the development of location-specific research capacity on the part of both expatriate and indigenous scientists. As a result, by the mid-1960s, the use of foreign "experts" as advisers was increasingly regarded as an inadequate response to the technical and scientific assistance needs of the less-developed countries. It continues to be an important way of organizing technical assistance efforts in a number of national and multilateral and technical assistance programs. The World Bank, for example, often incorporates a substantial amount of technical assistance, organized in the counter-part model style, as part of its project-lending activity.

The university contract model typically has involved the establish-ment of a special relationship between a university in a developed country and a university in a less-developed country. At times, it has also been employed to link a consortium of institutions in developed countries and/or less-developed countries. Occasionally, the link has also involved a ministry-level research division or insti-tute. The university contract model has been employed where insti-tution building and training have represented the major objectives of technical assistance activity. Frequently, the institution-building objective has involved, either explicitly or implicitly, positive assump-tions about the relevance of the United States' land-grant experience to the solution of problems of technical or institutional innovation in the host country.

In the past, the university contract model has not provided an environment conducive to the long-term commitment of professional resources to the pursuit of scientific and technical research on prob-lems of agriculture in developing countries by staff members of institutions in developed countries. It has made an important contri-bution to the training of large numbers of agricultural scientists from developing countries at the postgraduate level. But the univer-sity contract model has not, with few exceptions, been an effective instrument for research leading to the discovery of new knowledge or the invention of new technology needed to expand productive capacity in developing countries. In the United States, a new effort has been initiated (under Title XII of the International Development and Food Assistance Act of 1975) to establish a more adequate financial and administrative environment for the participation of the U.S. colleges of agriculture in an effort to strengthen national research programs in the less-developed countries. Experience thus far seems to indicate that the lessons of the 1960s—that the com-parative advantage of institutions in developed countries lies more in the training of scientists from less-developed countries than in tech-nology-oriented research for the less-developed countries—will have to be relearned.

Between the mid-1960s and the early 1970s, the international research and training institute emerged, in the perception of the international aid agencies, as the most effective way to organize scientific capacity to generate technical change in agriculture for the developing countries. The international institute model draws on two historical traditions. One is the experience of the great tropical research institutes that played an important role in increasing the production of a number of export commodities, including rubber, tea, sisal, and sugarcane. The Rubber Research Institute of Malaya, discussed in chapter 4, was an important example of the colonial research institute. The sugar and pineapple research institutes in Hawaii were examples of private research institutes supported by industrial organizations.

The international institute model draws more directly on the experience of the Rockefeller Foundation in Mexico and the Ford Foundation and the Rockefeller Foundation in the Philippines in support of the research that led to dramatic increases in food crop production, particularly wheat and rice. The complex of international agricultural research institutes listed in table 5-1 evolved directly from two institutions, the International Rice Research Institute (IRRI) and the International Center for Improvement of Maize and Wheat (CIMMYT), established by the Rockefeller Foundation and the Ford Foundation in 1959 and 1963.[3] The Rockefeller Foundation's agricultural sciences program was initiated in 1943 with the establishment of the Office of Special Studies (Oficina de Estudies Especiales) in cooperation with the Mexican Ministry of Agriculture. Field research programs were first initiated with wheat and corn. The program was later expanded to include field beans, potatoes, sorghum, vegetable crops, and animal production. A common pattern of staffing was followed for each commodity program. A U.S. specialist was brought in as each commodity program was initiated. Each specialist assembled a staff of young Mexican college graduates who were trained in research methods and practices through participation in the research program. Some of the most promising participants were later selected for further training at the postgraduate level.

The wheat program was successful almost from the beginning. Stem rust was diagnosed as a major factor limiting wheat production. Resistant wheat varieties were being distributed to farmers by the fall of 1948. By 1956 the impact on productivity was sufficient to make Mexico independent of the need for imported wheat for a number of years.

The success of the wheat program in Mexico led to conversations between the Ford Foundation and the Rockefeller Foundation about the possibility of collaborating on a program in Asia. After a good deal of investigation, an agreement was reached between the two foundations and the government of the Philippines to establish the International Rice Research Institute in the Philippines. The

Table 5.1. The International Agricultural Research Institutes

Center	Location	Research	Coverage	Date of Initiation	Core Budget for 1980 ($000)
IRRI (International Rice Research Institute)	Los Banos, Philippines	Rice under irrigation, multiple cropping systems; upland rice	Worldwide, special emphasis on Asia	1959	16,119
CIMMYT (International Center for the Improvement of Maize and Wheat)	El Batan, Mexico	Wheat (also triticale, barley): maize (also high-altitude sorghum)	Worldwide	1963	17,035
IITA (International Institute of Tropical Agriculture)	Ibadan, Nigeria	Farming systems: cereals (rice and maize as regional relay stations for IRRI and CIMMYT); grain legume (cow-peas, soybeans, lima beans); root and tuber crops (cassava, sweet potatoes, yams)	Worldwide in lowland tropics special emphasis on Africa	1967	15,106
CIAT (International Centre for Tropical Agriculture)	Palmira, Colombia	Beef; cassava; field beans; swine (minor); maize and rice (regional relay stations to CIMMYI and IRRI)	Worldwide in lowland tropics, special emphasis on Latin America	1968	14,998
WARDA (West African Rice Development Association)	Monrovia, Liberia	Regional co-operative effort in adaptive rice research among 13 nations with IITA and IRRI support.	West Africa	1971	2,768
CIP (International Potato Centre)	Lima, Peru	Potatoes (for both tropical and temperate regions)	Worldwide, including linkages with developed countries	1972	8,048
ICRISAT (International Crops Research Institute for the Semi-Arid Tropics)	Hyderabad, India	Sorghum; pearl millet; pigeon peas; chickpeas; farming systems; groundnuts	Worldwide, special emphasis on dry semiarid tropics, non-irrigated farming. Special relay stations in Africa under negotiation	1972	12,326

Table 5.1 – *Continued*

Center	Location	Research	Coverage	Date of Initiation	Core Budget for 1980 ($000)
IBPGR (International Board for Plant Genetic Resources)	FAO. Rome, Italy	Conservation of plant genetic material with special reference to crops of economic importance	Worldwide	1973	3,124
ILRAD (International Laboratory for Research on Animal Diseases)	Nairobi, Africa	Trypanosoiasis; theileriasis	Mainly Africa	1974	10,443
ILCA (International Livestock Center for Africa)	Addis Ababa, Ethiopia	Livestock production system	Major ecological regions in tropical zones of Africa	1974	8,986
ICARDA (International Centre for Agricultural Research in Dry Areas	Lebanon Syra	Crop and mixed farming systems research, with focus on sheep, barley, wheat, broad beans, and lentils	West Asia & North Africa, emphasis on the semiarid winter precipitation zone	1976	11,825
IFPRI (International Food Policy Research Institute)	Washington, D.C. United States	Food policy	Worldwide	1975	2,400
ISNAR (International Service for National Agricultural Research)	The Hague, Netherlands	Strengthening the capacity of national agricultural research programs	Worldwide	1979	1,199

Sources: J. G. Crawford, "Development of the International Agricultural Research System," in *Resource Allocation and Productivity in National and International Agricultural Research*, Thomas M. Arndt, Dana G. Dalrymple, and Vernon W. Ruttan, eds. (Minneapolis: University of Minnesota Press, 1977), pp. 282-83. Budget data for 1980 were obtained from the Secretariat for the Consultative Group on International Agricultural Research, World Bank, Washington, D.C.

agreement was signed in 1959. By 1962 the institute's facilities had been constructed, a staff was being recruited, and the program was under way.

The establishment of the IRRI as an autonomous research institute with its own governing body and its own staff was a sharp departure from the collaborative pattern employed by the Rockefeller Foundation in Mexico. A major reason for the success of the Rockefeller Foundation's program in Mexico was its economical use of the scarce professional manpower available in Mexico at the time the program was initiated. The shortage of professional manpower and indigenous educational resources was conducive to the development of an internship system that intimately linked professional education with investigation. However, in situations in which the scientific and technical problems were more complex than in the early wheat work, the patterns worked out during the early years of the Mexican program were not as effective. In more complex situations, gathering together a team of highly skilled senior scientists from several disciplines appeared to be a more appropriate staffing strategy. This was the pattern followed in the establishment of the IRRI. It was also the approach adopted in 1963 when the Rockefeller Foundation's program staff in Mexico was reorganized into the new International Center for Improvement of Maize and Wheat in order to facilitate a stronger international collaborate research effort in wheat and maize.

During the late 1960s, the Ford Foundation and the Rockefeller Foundation again collaborated in the establishment of the International Institute of Tropical Agriculture (IITA) in Ibadan, Nigeria, and the International Center for Tropical Agriculture (CIAT) in Palmira (near Cali), Colombia. With the establishment of these two new centers, it became apparent that the financial requirements of the system would soon exceed the capacity of the two foundations. Consultations were held between the Ford and Rockefeller foundations, the World Bank, the Food and Agriculture Organization, and the United Nations Development Program (UNDP) in 1969. Following several informal meetings in January 1971, a formal meeting was held in May of 1971 to organize the Consultative Group on International Agricultural Research (CGIAR). The initial membership included the World Bank, the FAO, and the UNDP as sponsors, plus 9 national governments, 2 regional banks, and 3 foundations. Membership had grown to 39 institutions by 1980.

The leadership for the Consultative Group is now centered at the World Bank, which provides a chairperson and a secretariat for the group. To provide technical guidance for its work, the Consultative Group established the Technical Advisory Committee (TAC). The TAC consists of a chairperson and 12 scientist members. The FAO provides the TAC secretariat. Technical matters such as new institute initiatives and program changes at existing institutes are referred to the TAC for technical review before action by the CGIAR. The TAC develops draft policy statements for the CGIAR's consideration on priorities within the system and has the authority to initiate investigations and to suggest initiatives and program changes to the CG on its own initiative. Since 1976, the TAC has been charged with the responsibility of organizing comprehensive quinquennial reviews of the programs of the several international research centers and with the periodic analysis of programs that have common elements in the several centers, such as cropping systems or mechanization research.

The international research system has grown rapidly. Expenditures rose from only $1.1 million in 1965 to almost $120 million in 1980. (See figure 5.1.)

The rapid growth has been due to several factors. One is the perception of research as a high-payoff and an efficient source of agricultural growth. The intellectual basis for this perception was provided in T. W. Schultz's iconoclastic book, *Transforming Traditional Agriculture.*[4] The empirical support was provided by the rapid diffusion of both the new wheat varieties produced in Mexico, first through the collaborative effort of the Ministry of Agriculture and the Rockefeller Foundation and then by CIMMYT, and of the new rice varieties produced in the Philippines by the IRRI.

A second factor was the efforts of both the multilateral and the bilateral Assistance agencies to find a viable alternative to the counterpart and university contract models. Although the university contract model was viewed as an effective device for building professional capacity, there was a tendency to give it low marks in terms of its ability to bring that capacity to bear on the generation of more productive agricultural technology.

On this point it may not be inappropriate to quote my own perspective of a decade ago.

The United States is characterized by a highly developed institutional infrastructure linking the university to other private and public institutions involved in technical, social and economic change. In societies where such an infrastructure has developed, research and education within the framework of the traditional academic disciplines and professions have represented an effective link in a larger system devoted to the production, application and dissemination of new knowledge. . . . The American pattern of academic and professional organization, when transplanted into societies where the presumed institutional infrastructure does not exist, rarely performs as an effective instrument of technical, social, or cultural change.[5]

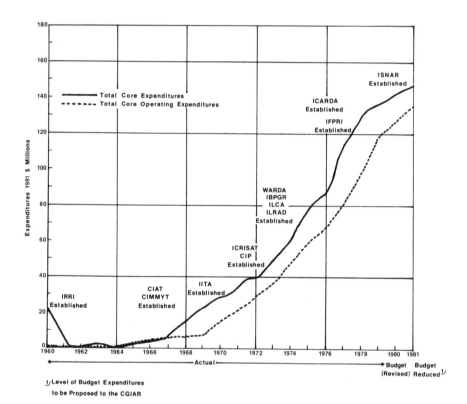

1/Level of Budget Expenditures
to be Proposed to the CGIAR

Figure 5.1. International Agricultural Research Centers' Annual Core Expenditures and Core Operating Expenditures, 1960-1980 (in terms of constant 1981 dollars). (From CGIAR Secretariat 1980 [Washington, D.C.: World Bank].)

SOME ORGANIZATION AND MANAGEMENT ISSUES

In this section, I focus on several areas of concern with respect to the future of the international agricultural research system. These include (1) relations with less-developed countries, research institutions, (2) relations with developed countries' research institutions, and (3) financial and program viability. Finally, I attempt to deal with some issues that the institutes are facing in the area of program development as they evolve toward becoming more comprehensive agricultural development institutes rather than simply being commodity research institutes. My comments in this section draw on my experience as an IRRI staff member, a member of several institute review teams, and a member of the CGIAR's Technical Advisory Committee. I also draw on the 1976 report of the review team that the Consultative Group on International Research commissioned to review the work of the international agricultural research institutes.[6]

Relations with less-developed countries' research systems

By the late 1960s, many of the bilateral and multilateral aid agencies were recognizing serious shortcomings in the results of their efforts to support the development of national agricultural research systems. Most national systems in the less-developed countries were unprepared to absorb effectively large amounts of financial, material, and professional assistance. The capacity for scientific management and entrepreneurship of the newly trained scientific community was often underdeveloped. Many systems were plagued by cyclical sequences of development followed by erosion of capacity as budgetary priorities responded to change in political regimes.

Impatient staff members at aid agencies were often unaware of the history of their own national institutions. They had forgotten that the national agricultural research systems of the United Kingdom, Germany, the United States, and Japan had taken decades, not years, to acquire the research and training capacity needed to generate the new knowledge and technology needed to sustain agricultural development. (See chapter 4.) Furthermore, the political support available to many national and international aid agencies was often so fragile that support for institution building was difficult to sustain unless a short-term payoff could be visualized.

In addition to a sense of frustration with efforts to strengthen national research systems, there was a growing conviction of urgency about the problem of meeting food requirements in the poor countries. The initial success of the IRRI's rice program and the CIMMYT's wheat program combined to create a conviction that the international agricultural research institute, which could operate independently of the vagaries of the local political environment and could draw on the global agricultural science community for its staff, represented an effective instrument for the management of research resources and for the generation of new agricultural technology.

It was recognized, of course, that effective liaison with national systems would be essential for individual countries to make productive use of the knowledge and the technology generated at the international institutes. Efforts were made to establish collaborative research networks linking the international institutes with regional organizations supported by the FAO and bilateral aid agencies and with national research and training institutions. The linkages represented in the international wheat network are illustrated in figure 5.2.[7]

By the mid-1970s, it had become increasingly clear that the productivity of the international agricultural research system was severely constrained by the limited capacity of many national systems and that the adaptation and dissemination of the knowledge and technology generated at the international institutes was dependent on the development of effective national systems. Robert E. Evenson's work has demonstrated that the ability to screen, borrow, and adapt scientific knowledge and technology requires essentially the same capacity as is required to invent new technology.[8] Capacity in the basic and supporting biological sciences is at least as important as capacity in applied science.

The outreach programs of the international institutes, even when working through networks such as the international wheat research network, the inter-Asian corn program, and others, did not have the capacity to take on the role of strengthening national systems. The regional commodity networks played an important role in enabling the institutes to conduct research and to test materials and methods under diverse ecological conditions, but they could not assume a larger role without diverting effort from the institute's research programs. By the mid-1970s, only a few national systems— those of India, Brazil, and the Philippines, for example—had developed the managerial and professional capacity to effectively absorb, transmit, and adapt the knowledge and technology that were becoming available to them through the international institutes, from developed countries' research systems, and from the stronger developing countries' institutions.

The bilateral and multilateral assistance agencies had no alternative, therefore, but to place the strengthening of national research systems high on their assistance agendas. Several steps have been taken by the CGIAR's system to respond to this need. The first direct effort was taken in 1973 when the TAC endorsed support for the coordinated trials program of the West African Rice Development Association (WARDA).[9]

The WARDA had been established in 1968, with the support of

International Winter × Spring Wheat Research Network, 1979

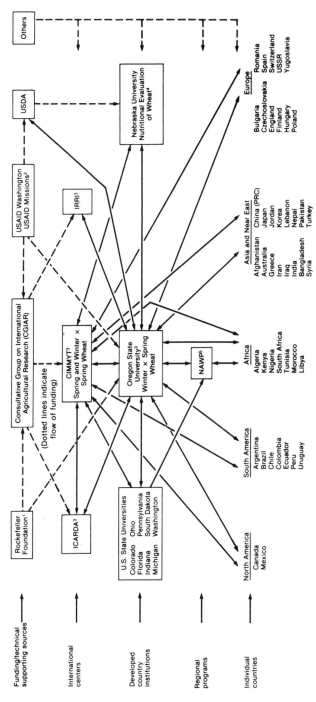

Notes: [1]The Rockefeller Foundation provides funds to Oregon State University for graduate training. [2]USAID missions contribute through bilateral support of individual country programs. [3]The links with individual countries take the form of information and materials exchange, the conduct of seminars and workshops, cooperative research projects, advisory services, and training. [4]USAID funding of nutritional work expired in late 1979.

[5]North African Wheat Program.   Source: Adapted from figure provided by Department of Crop Science, Oregon State University, Corvallis.

Figure 5.2. International Winter and Spring Wheat Research Network, 1979. (From Dana G. Dalrymple, *Development and Spread of Semi-dwarf Varieties of Wheat and Rice in the United States: An International Perspective* [Washington, D.C.: USDA, Agricultural Economics Report No. 455, June 1980], p. 133.)

the FAO and several bilateral donors, as a cooperative rice research and development program among 11 West African countries. The region is characterized by a great diversity of rice production systems—upland rice, irrigated rice, mangrove swamp rice, and deep-flooded and floating rice. The WARDA has a limited professional capacity in its own staff, and it works in a region where national capacity is even weaker. Some of these difficulties are overcome through cooperative programs with the IITA, the IRRI, and the Institute pour la Recherche dans Agronomie Tropicale (IRAT). The effective use of available professional capacity has been enhanced by deciding to operate as a decentralized research and development program in collaboration with national programs. The WARDA'S facilities and staff are largely located at national research centers.

When I visited a number of the WARDA's and national facilities in 1978, in connection with the TAC's quinquennial review mission, I was impressed by the importance of the WARDA's linkage role. The effectiveness of researchers at the most remote stations, at Mopti in Mali, for example, was greatly enhanced through their access to the world collection of germ plasm and the communication network that made the methods and the results of rice research throughout the world available to them. Their participation in the WARDA's training activities and research conferences gave them a sense of sharing in a scientific effort that extended beyond the constrained environment of their own stations.

A second step in the effort to strengthen national research systems has involved the development of a new CGIAR-sponsored institution charged with supporting the development of national research systems. The Rockefeller Foundation played an important entrepreneurial role in this development. After a series of consultations with the leaders of national research systems, the International Agricultural Development Service (IADS) was established, with initial funding from the Rockefeller Foundation, to provide contract research management and development services to national research systems.

The initiative of the Rockefeller Foundation influenced the CGIAR to intensify its own deliberations. In 1977 the CGIAR organized a task force to explore the possibility of establishing an international service for the strengthening of national agricultural research within the CGIAR's systems.[10] These deliberations led to the establishment of the International Service for National Agricultural Research (ISNAR) in 1979. There had been some expectations that, in establishing the ISNAR, the CGIAR might absorb the IADS, much as it had incorporated the IRRI, CIMMYT, IITA, and CIAT under its umbrella in 1971. By 1979, however, the CGIAR had become somewhat sensitive about absorbing activities initiated prior to the CGIAR/TAC assessment and evaluation process.[11] Some European donors were also sensitive about the fact that staffing patterns at the institutes had not drawn effectively on European professional capacity. The FAO, one of the CGIAR's sponsors, expressed strong concern that the new service was infringing on an area of traditional FAO responsibility.

By 1980 the IADS had acquired a substantial portfolio of projects. The ISNAR had appointed a director-general and was beginning to assemble a staff. But it was premature to attempt to evaluate the effectiveness of either effort. It is clear, however, that the strengthening of national research systems is only partially, and perhaps only marginally, amenable to the efforts of the assistance agencies. Unless national governments take the necessary steps to create viable national research systems—particularly those measures to establish hospitable and productive professional environments for indigenous professional capacity—the contributions of the bilateral and multilateral assistance agencies will have limited impact.

The international institutes have not always been the most effective instruments for reinforcing the emerging professional capacity in the national systems. Relationships have sometimes been competitive rather than collaborative. In efforts to establish their credentials and to provide the donors with evidence of prouductivity, the international institute staff and management have, in some cases, overemphasized their own contributions and failed to give appropriate credit to the work at national institutes. For example, the diffusion of the rice variety C-4, developed at the University of the Philippines College of Agriculture, did not receive the same support by the international system as did the diffusion of the varieties IR-5 and IR-8, developed at the IRRI. The International Institute for Tropical Agriculture initially viewed the West African Rice Development Association as a competitor for donor resources rather than as a complementary resource.

Relations with developed countries' research institutions

The initial years of the new international institutes were characterized by a tendency for keeping relationships at arm's length among the institutes and the developed countries' universities and research institutions. The new institutes tended to view, and not without some justification, efforts by developed countries' scientists and their graduate students to establish collaborative relationships as efforts to exploit new sources for the funding of personal research interests that were only marginally complementary to the commodity-centered mission orientation of the new centers.

This relationship has changed over time. As the institutes have identified problems in which lack of knowledge in areas such as physiology, pathology, and other fundamental or supporting areas of science have become constraints on their ability to expand yield frontiers, they have taken steps to institutionalize their relationships with developed countries' institutions. Examples include the relationship among the CIMMYT and several Canadian institutions for work on triticale. The International Potato Center (CIP) has used contract linkages with institutions of developed countries for work on fundamental problems related to the CIP's mission more extensively than any of the other international centers. At the time of the 1977 quinquennial review mission, the CIP identified 12 such contracts with developed countries' institutions and 7 with less-developed countries' institutions. In a number of cases, the CIP's contracts induced additional effort and expenditure on CIP-related problems by the developed country's contracting institution.[12]

There are clear dangers in the growing relationships among the international centers and institutes engaged in supporting and fundamental research in the developed countries. If the less-developed countries are to establish a viable base for self-sustained scientific effort leading to productivity growth in agriculture, it is important that they establish a capacity to work on the fundamental problems that are of particular significance in tropical environments. In conversations with aid agency and university representatives from developed countries, exception is often taken to this view. The point is sometimes made that the scientific capacity already exists to conduct the relevant basic research and supporting research in the developed countries. What is the justification for duplicating the expensive facilities and scientific capacity? My response to this question is that the creativity of the scientist is strongly conditioned by the institutional and physical environment in which he or she works. Scientific imagination is as surely conditioned by the scientist's environment as the growth of the rice plant is conditioned by its environment.

In my view, a systematic effort to establish centers of fundamental research in the tropics, such as the International Center for Insect Physiology and Ecology (ICIPE) in Nairobi, should be encouraged. The international crop and livestock improvement centers should be encouraged to establish linkages with these institutions wherever they represent viable alternatives to institutions in developed countries. If such institutions cannot be funded within the CGIAR's system, the developed countries' donors should seek another mechanism to assure the development of centers of research and training in the supporting sciences that are related to the technology-oriented research of the international agricultural research system and of the national research systems.

Viability of the international system

The first four institutes (the IRRI, the CIMMYT, the CIAT, and the IITA) were initially funded jointly by the Ford Foundation and the Rockefeller Foundation. The Consultative Group on International Agricultural Research was organized to provide a broader resource base for the expanding system. Funding requirements have risen rapidly. (See figure 5.1.) And funding problems have become more complex. The CGIAR provides a forum in which the international agricultural institutes set forth their program and funding needs and the donors make known their funding commitments. The commitments of the individual donors are typically, however, made to particular institutes rather than to the system as a whole. Although there is an exceptionally good record of funding continuity, institute directors (more recently director-generals) have had to devote an increasing percentage of their time to donor relationships in order to assure the continuity and growth of funding. The fact that the funding is typically made available on an annual rather than a long-term basis puts particularly severe pressure on the institutes' management to cultivate relationships with donors and to respond to donors' priorities.

During the early years of the system, the participation of the institute's director in scientific leadership and personnel recruitment was an important factor in establishing the quality and productivity of the research program. By the mid-1970s, the institute's management had become more heavily involved in matters of financing and public relations, and the institute's administration had become more highly bureaucratized. The external relationship responsibilities of the institute's director-generals are increasingly insulating them from program leadership. Most directors no longer have the professional capacity or the inclination to be vigorous leaders of the scientific programs of the institutes.

As the capacities of the less-developed countries' national research systems improve, the relative contributions of the international agricultural research institute system to the generation of knowledge and technology will decline. One possible outcome of this process is that the institutes will lose the distinct leadership roles that they now possess. My own model for the future of the institutes is for an expansion in their role as centers for the conservation and diffusion of genetic resources and of scientific and technical information, relative to their role as producers of new knowledge and new technology. If they are careful to select staff members for their leadership capacities as well as for their scientific and professional competence, they will also be able to continue to play a strategic role in establishing research priorities.

There is, however, what might be considered a natural history of research institutes. A new institute that is able to bring together a team of leading scientists tends to go through a period of high productivity that often lasts a decade or longer. There is a tendency, after this initial period of creativity, for an institute to settle down to filling the gaps in the scientific literature and to fine tuning incremental changes in technology.

The international institutes will need to guard against this tendency. There are numerous technical and institutional constraints that dampen the ability of farmers to achieve present yield potentials economically. The world will continue to need a system of international institutes that will play a strategic role in the area of crop and livestock improvement. The political fragmentation that characterizes much of the world in which the international centers operate assures that relatively few nations will become self-sufficient in their capacity for either training or research in the agricultural sciences.

If the international institutes develop a capacity to link the national systems into a carefully articulated international system, they will assure their own continued viability. If they become viewed as being competitive with national research systems, they could fade away into mediocrity.

FUTURE POLICY AND PROGRAM DIRECTIONS

There are continuous pressures for the broadening of program responsibilities at the international agricultural research centers. They are subject to pressures to become the primary centers for conservation of genetic resources, to expand their programs into research on cropping and farming systems, and to develop research programs in pre- and postharvest mechanical technology. There are also pressures on them to place greater emphasis on small-farm technology and to become broad-based agricultural or rural development institutes rather than simple crop or livestock improvement centers.

Genetic resources

Each institute has found it necessary to develop at least a working collection of genetic material for use in its crop improvement, technology development, and supporting science programs. Some institutes (such as the IRRI) from the beginning have set out to establish a world collection of genetic materials. The International Board for Plant Genetic Resources was established under the CGIAR's auspices in 1974 to encourage and assist in the collection, preservation, and exchange of plant genetic materials.

Upon its inception, the International Board for Plant Genetic Resources faced a situation in which the motives and aims of the several available collections were very different. Many useful collections, kept by individual plant breeders, included only the cultivars that had proven to be most useful to a particular breeder's program and did not represent the full range of heritable variation. Such collections were often dispersed or lost because of a lack of funding or the retirement of the leader of a breeding program.

At the other end of the spectrum were great conservation collections containing a substantial part of the genetic variation of a wide range of species, such as those of the National Seed Storage Laboratory at Fort Collins in the United States and the N. I. Vavilov Institute in the Soviet Union. Some centers have functioned more as museums than as centers of plant breeding. Among the international institutes, the IITA in Nigeria, the IRRI in the Philippines, and the International Crop Research Institute for the Semi-Arid Tropics (ICRISAT) in India had assembled large genetic resource collections that are intimately related to the institutes' breeding programs. Most collections consist exclusively of seed materials. The problems of storing genetic material for crops that are normally propagated with vegetative materials (such as cassavas, potatoes, and yams) remained unresolved.[13]

Decisions with respect to the appropriate locations for the facilities and programs for collection, maintenance, and use of genetic resources are urgent. Issues of the most effective way of linking the various collections, establishing exploration and collection priorities, and making materials available to international and national breeding programs are not easily resolved. Albert Moseman, in a personal correspondence (May 1977), has argued convincingly that, if genetic resource centers are to make productive contributions to the development of biological technology, they must be institutionalized in locations where strong plant breeding and genetics research programs are being conducted.

In my judgment, a strong argument can be made for having each of the major international crop institutes evolve into a world center for the conservation and evaluation of genetic materials. Even if contributions to knowledge and technology at the international centers decline relative to the contributions of less-developed countries' research institutions, the national agricultural research systems will still need a reliable source for materials and information on genetic resources and potentials.

The budgetary requirements for a serious commitment to genetic conservation and use are likely to be larger than many of the donors to the international system realize or the director-generals of the international institutes are willing to admit. Financial resources committed to the support of the genetic materials "archives" will face continuous pressure for diversion to use in technology development research unless the genetic material collections are seen as contributing in a very direct way to technology development. The value of the world collections to technology development will be most apparent if the collections are organized and managed as part of genetic resources utilization programs.

Farming and cropping systems research

The initial programs of the international centers focused primarily on crop improvement and advances in the technology of crop production. Some institutes were given responsibility for research on a single crop. Even those institutes that had responsibilities for more than one crop often functioned as a set of loosely linked single-commodity institutes held together by the economies of scale in administration and related institutional infrastructure (motor pool and library, for example).

Farming or cropping system research was included in the mandates of several of the newer institutes (the CIAT, the IITA, and the ICRISAT, for example) at the time the institutes were established. Several of the other institutes have also established farming or cropping system programs. Although the CIAT has discontinued its farming system program, it has continued to work on cropping systems within the framework of its individual commodity research program. At several of the institutes, the cropping or farming system programs have evolved as an important instrument to achieve more effective integration of disciplinary research.

The initiation of farming and cropping systems research has clearly been one of the exciting contributions of the international agricultural research system. In most national programs, research efforts have been fragmented along disciplinary lines. Even scientists from the two integrating fields of agronomy and farm management are often confronted with institutional barriers to effective collaboration. The farming and cropping systems initiatives at the international institutes have, therefore, generated a great deal of intellectual excitement among scientists and agricultural program managers who were looking for a way to integrate the knowledge emerging from disciplinary and commodity research programs with farming practice.

The issues surrounding the appropriate role of farming and cropping systems programs within the international institute have, however, been difficult to resolve.[14] Even the terminology that is appropriate to categorize the activities has been the subject of dispute. The issue of the appropriate role of farming and cropping systems research at the international agricultural research institutes must be evaluated, not on the basic merits of the systems approach, but in terms of the specific role that the international institute can reasonably be expected to play in the development and dissemination of knowledge, materials, and technology.

A modest program of farming or cropping system research is clearly essential even if a research institute takes as its primary goal the limited objective of developing improved plant materials—and other technical components, such as plant protection technologies—for a single commodity. Research directors and program leaders must have accurate information about the technical and environmental constraints influencing the productivity of the farming systems that incorporate the commodities on which the institutes are conducting research. Research designed to produce knowledge about the cropping and farming systems employed by farmers in the major agroclimatic regions in which the institutes work represents an important part of the knowledge needed to focus research resources.

At some institutes, the cropping systems programs have broader objectives. One stated objective has been to develop the methodology of farming and cropping systems research and to transfer such methodology to national systems. Another stated objective is to develop more productive cropping systems that can be extended to farmers.

The first objective has some merit. There are areas in which radical departures from traditional practice may be required for the evolution of viable intensive farming systems. The development of "prototype" farming systems for such environments may, in some cases, be an appropriate objective of an international institute. The maize-based zero tillage system at the IITA is sometimes cited as an example of a "prototype" system that is approaching the stage at which further development and testing by national or provincial systems is warranted.

The second objective is not, in my judgment, a credible objective for cropping systems research at the international institutes. The fine tuning of cropping systems is highly location specific. It is more appropriately a function of the local (state and prefectural) stations of national systems than of the international institutes.[15] While modest productivity gains may be made as a result of cropping systems research at the international centers, it is unlikely that these gains will approximate those that can be realized from the resources devoted to the development of improved components for cropping systems.

PRE- AND POSTHARVEST
MECHANICAL TECHNOLOGY

The allocation of resources for the development of pre- and post-harvest mechanical technology has been an issue that has continuously faced the management of the individual institutes and the CGIAR and its Technical Advisory Committee.

A small-machinery development program was initiated at the IRRI in the mid-1960s with support from the U.S. Agency for International Development. Limited work on small machinery has also been initiated at the ICRISAT and the IITA. The program at the IRRI has included research and development work on irrigation, land preparation, and weed and pest control as well as on harvesting, threshing, and drying equipment. The IRRI's program has had imaginative engineering leadership and has developed an effective outreach program with manufacturers of agricultural machinery in the Philippines. After a decade and a half of effort, however, the machinery development program still must measure the impact of its machine designs in terms of thousands of units manufactured and distributed in contrast to the work on genetic improvement, which measures its output in terms of millions of hectares planted to new varieties or millions of metric tons of additional rice production.

The impact of the machinery development programs at the IITA and the ICRISAT are even less impressive.

There has also been considerable discussion of the role of research on postharvest technology at the international institutes. The establishment of a postharvest research and training institute within the CGIAR's system has also been suggested. Although neither of the options has been implemented, several of the CGIAR's donors committed themselves in 1976 to supporting an information and advisory service on postharvest research to be organized under the auspices of the Southeast Asian Regional Center for Graduate Study and Research in Agriculture (SEARCA) in the Philippines. It is my own impression that the potential gains achievable in postharvest technology, though potentially large, have thus far been achieved at a relatively high cost or else they have generated relatively low returns.[16]

In my judgment, the evidence with respect to the assumed complementarity between tillage equipment and cropping intensity that has provided a rationale for the small-machinery development programs is less than compelling. Neither am I convinced that the private sector lacks incentives for the development of pre- and postharvest technology when economic conditions are such as to encourage a demand for such equipment. The extremely rapid diffusion of the power tiller and mechanical tillage equipment in Taiwan since the mid-1960s and rice-milling equipment on Java since the early 1970s would seem to suggest that in most areas demand constraints have been more serious than supply constraints in the diffusion of mechanical technology.

The reasons for the greater effectiveness of public-sector (or philanthropic) research activity in biological and chemical technology compared to that in mechanical and engineering technology is fairly straightforward. The invention of new biological and chemical technology for agriculture typically draws on high-level scientific skills in genetics, physiology, pathology, chemistry, statistics, and related disciplines. The design of new machine technology in agriculture depends to a much greater extent on mechanical skill and ingenuity and on farming experience and practice. Furthermore, the relevant skills in mechanical technology are pervasive in developed countries and are widespread even in developing countries except for the most backward countries of the "fifth world."

In my judgment, work on equipment design at the international agricultural research institutes should be limited to work on prototype equipment needed for radically new production technologies or farming systems that have not previously been available to farmers. It would be inadvisable for the institutes that are members of the CGIAR system to expand their machinery design or postharvest technology programs without more careful analysis of the technical and economic potential of such programs than has been available in the past.

SMALL-FARM TECHNOLOGY

The issues of differential access of large and small farms to the new varieties and crop production technologies generated by the international agricultural research institutes have been the subject of interdisciplinary aggression and ideological conflict almost from the time the new wheat and rice varieties appeared on the scene. The view was expressed that the new institutes represented "big science for big farmers." The small farmer was viewed as an actual or potential victim rather than as a beneficiary of the new technology.

Much of this criticism was based on the crudest sort of casual empiricism. Some of it appeared to reflect little more than interdisciplinary pique or aggression. The green revolution had shifted attention from the early architects of development—planners, economists, and social scientists—to geneticists, plant breeders, and agronomists. Critics, who welcomed the green revolution's technology in Cuba and China, viewed the introduction of the high-yielding varieties in nonsocialist economies as raising the cost of radical change by channeling new income to the middle- and upper-level peasantry.[17] The reaction to these criticisms by many sponsors of the CGIAR's system was to call for a redirection of scientific and technical effort at the institutes. Sometimes this was cast in terms of producing labor-intensive technologies. Sometimes it was cast in terms of technology for small farmers.

What has not been generally appreciated in this argument is the fact that in labor-intensive, low-wage economies simply focusing technical effort in the direction that would have the highest payoff will bias the generation of technology in the direction desired by the critics. The message I tried to convey in chapter 2 is that the generation of a technology designed to use relatively scarce and expensive inputs—a capital-intensive technology in a labor-surplus economy, for example—would be less profitable and diffuse more slowly than a technology designed to absorb more abundant and less expensive labor inputs.[18]

This is precisely what most of the institutes were doing. The emphasis on new biological technology rather than mechanical technology resulted in the development of technologies that were essentially scale-neutral. The decision at the CIAT to deemphasize the beef program and to place increased emphasis on beans was a decision consistent with the factor endowments in the poorer countries and regions of Latin America, and it was also consistent with developing technologies for small farmers.

When the hard data began to come in and replace the casual observations that had fueled the early arguments, it was apparent that the green revolution technology was being adopted at least as rapidly by small farmers as by large farmers in the areas where the new technology had been adapted to the agroclimatic conditions. This has been shown rather dramatically in a series of studies conducted by the International Rice Research Institute and cooperating national institutions in 36 villages in 6 countries in Asia. On the average, small farmers have adopted the new high-yielding technology faster than large farmers. (See figure 5.3.) One reason this has occurred is that the technology was suited to those crops produced by small farmers and those regions characterized by small farms.

Once the rhetoric of ideological debate is left behind, two issues

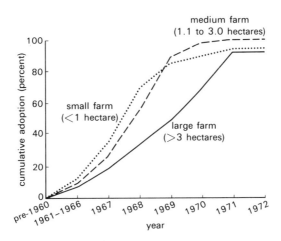

Figure 5.3. Cumulative Percentage of Farms in Three Size Classes Adapting Modern Varieties in 30 Villages in Asia. (Adapted from Randolph Barker and Robert W. Herdt, "Equity Implications of Technology Changes," in *Interpretive Analysis of Selected Papers from Changes in Rice Farming in Selected Areas of Asia* [Los Banos, Laguna, Philippines: International Rice Research Institute, 1978], p. 91.)

remain that are much more difficult to deal with. The first is whether technologies can be invented that confer a larger advantage on small farms than on large farms. A small-farm bias may be introduced through selection of the commodities mix on which to conduct research. Research on beans in Latin America and dairy production in India are examples. The farming systems research program of the Centro Agronomico Tropical de Investigacion y Enseñanza (CATIE) in Central America is another example. Once the geographic and commodity focus of a research effort has been made, however, I find it very difficult to find many examples of technology that can be biased strongly toward small farmers. It seems rather clear that institutional changes, in land-tenure relations for example, have a potentially much more powerful influence on the distribution of gains from technology than attempts to go beyond the objective of designing technology that will be profitable for the farm structure and resource endowments that actually exist.

The second major issue is whether productive technologies can be developed for resource-poor regions—those areas where the soils are poor, the slopes are steep, or the rainfall is low. Can productive "low-input" technologies be developed? Or must the ultimate solution be the reduction of cropping pressure in marginal areas combined with more intensive use of more favorable areas? All of the technical answers to this question are not yet in. And the institutional answers will depend on the pace of development in the total economies of which the more fragile environments are components.

Agricultural development centers

There is also continuous concern that the institutes should develop an analytical capacity and a flow of research results designed to influence agricultural and rural development policies. This pressure has risen with the perception that institutional constraints—such as (1) the intervention into input and product markets; (2) the performance of extension, credit, and market institutions; and (3) the capacity for investment and management of physical infrastructure, such as irrigation systems—represent an increasingly serious constraint on the realization of the productivity gains that the new biological technology makes possible.

A decade and a half ago, it had become quite clear that low returns were being realized from investments directed toward institutional reform and institutional innovation. The response was a redirection of institution-building efforts to focus on creating a capacity to produce a continuous stream of more productive biological technology. This effort has been highly successful.

It seems clear that it is now time again to redirect the capacity for institutional innovation toward enabling rural people in the developing countries to realize the potential gains from the higher productivity potential—higher output per hectare and per worker—that is becoming available. This shift in the demand for institutional innovation is creating a derived demand for new knowledge in the rural social sciences. How should the international crop and livestock research institutes respond to this emerging perception of the growing demand for the social science knowledge that is needed as an input into institutional innovation?

Clearly, a strong social science research capacity can represent a high-productivity use of institute or center resources even if the primary objective of the center remains simply the invention or design of improved components for cropping systems. The economics staff of the International Rice Research Institute, for example, is able to provide the research director and the program leaders with relatively precise information regarding the value of the potential productivity gains that can be realized by alternative efforts to release yield constraints under farm conditions. The staff is also able to identify with some precision the extent to which productivity gains depend on investment or management decisions that lie outside of the scope of the IRRI's research program and that are amenable only to policy intervention or institutional innovations over which the IRRI's has little influence.[19]

In 1973 and 1974, the CGIAR and its Technical Advisory Committee initated a series of consultations about the role of social science research at the international institutes and in the international system. One result was a modest strengthening of social science research capacity at several of the institutes. Another result was the recommendation that the International Food Policy Research Institute (IFPRI) be established. The institute would be devoted to research on agricultural production and price- and trade-policy issues. Initially, the recommendation was not accepted by the CGIAR. Some donors felt that an institution designed to analyze the burden of agricultural and food policy on agricultural development in the less-developed countries would not be able to avoid turning its attention to the effects of the developed countries' policies, particularly the European Common Market's policy, on the agricultural export potential of the less-developed countries. The IFPRI was, however, established in 1975 in Washington, D.C., by a group of CGIAR donors (the Ford Foundation, the International Development Research Center, and the Rockefeller Foundation), and in 1979 it was finally incorporated into the international agricultural research system.

My present perspective is that each institute needs a social science research staff of at least the size and capacity that have now been established at the IRRI and the ICRISAT if the institute's director and program leaders are to be in a position to effectively allocate the research resources available to them. However, even at the IRRI and the ICRISAT there would be a substantial payoff to increasing the staff's capacity in anthropology and sociology relative to economics. Anthropologists, in particular, have demonstrated a capacity to understand the dynamics of technology choice and impact at the household and village level that is highly complementary to both agronomic and economic research. The internal dynamics of a staff's interrelationships and professional balance will make it difficult for directors to allocate a very significant share of institute resources to social science research on more macroeconomic agricultural and rural development issues. The international system should seek ways to strengthen graduate training capacity in the rural social sciences at the university level and research capacity at the central research or ministry level within less-developed countries' national training and research institutions. Chapter 12 is devoted to a more detailed consideration of the role of the social sciences in agricultural research.

A PERSPECTIVE ON THE FUTURE

A major burden on the future development of the new international system is the spectacular contribution to productivity growth in rice and wheat resulting from research conducted at the IRRI and the CIMMYT. Both institutes were able to build on a large backlog of relevant knowledge and on a major gap between the yields being obtained on the same crops under tropical and temperate-region conditions.

Even the most conservative estimates of the result of the research conducted at the IRRI suggest that, by crop year 1974-1975, the supply of rice in all developing countries was approximately 12 percent higher than it would have been if the same total resources had been devoted to the production of rice using only the traditional rice varieties available prior to the mid-1960s.[20] The impact of the CIMMYT's research on wheat production has probably been even greater.

One result of the very high rates of return to the initial investments in research at the IRRI and the CIMMYT has been to create expectations about the performance of the other institutes in the CGIAR's system that are unlikely to be realized. Other centers are working on crops and farming systems for which the research base at the time their programs were established was much less adequate. It seems reasonable to anticipate much more modest returns on investment in the CGIAR system in the future than in the past. It seems likely that in the future the institute's accomplishments will be described more in terms of incremental gains rather than in terms of revolutionary breakthroughs. The returns from such research are, however, likely to be significantly higher than the returns that can be realized on most of the other investments in agricultural or rural development that are available to the CGIAR's donors.

A second result of the early success in crop technology is the credibility generated by institutes in areas in which they have little capacity. This has led to an increase in efforts by donor agencies to channel bilateral technical assistance activities through one or another of the international institutes or to involve the institutes in the implementation of technical assistance programs. The financial pressures operating on the director-generals of individual institutes are difficult for them to resist. It will be necessary for the CGIAR's system to develop greater capacity to avoid taking on projects that divert the limited capacity for scientific management to low-productivity activities. In the future, the system will have to develop greater institutional capacity than in the past to discontinue unproductive activities and to avoid adding activities that are marginal for its mission.

In my judgment, the international agricultural research system should be viewed as a permanent feature of the global agricultural development infrastructure. Its significance as a source of new knowledge and new technology is not limited to its immediate accomplishments in exploiting the potential for low-cost productivity gains that were available during the third quarter of the 20th century.

I do have several major concerns about the future of the international agricultural research system. One is that the managerial structure of the system might, in the future, become overly bureaucratic. It is possible that its informal funding arrangements and its decentralized managerial structure may protect the system from the evolution of a burdensome administrative structure either at the level of the CGIAR or the TAC secretariat. One student of science policy, when confronted with the organizational and financial structure of the system, insisted that "it can't exist." An excessive impetus for rationalization of management and planning could dampen incentives for scientific entrepreneurship and productivity.

A second concern is with the evolution of a steering mechanism that will assure a continuing focus on the most significant research problems. As each new institute was established, priorities were set through an extended series of consultations at both the scientific and policy levels. As a research institute matures, the problem of redirection of effort becomes more difficult. It was noted in the last chapter that in the American and Japanese systems the tension between the national and state (prefectural) systems and the involvement of clientele organizations had acted to induce an efficient portfolio of projects. In the international system, the clientele—the developing countries—is only beginning to become involved in a significant way in the financial support of the system. This limited financial involvement weakens the weight given to less-developed countries' perspectives in research policy discussions at the individual institute governing boards and at the CGIAR level.

Finally, I continue to be concerned about how slowly national research systems in the developing world are acquiring the capacity to make effective use of the support that has become available as a result of the organization of the international system. The manpower devoted to agricultural research in the developing countries is small, and most research institutions are poorly managed.[21] Consistent political support for the funding of agricultural research has been difficult to institutionalize. The effectiveness of the international system depends on the development of strong national systems.

NOTES

1. This chapter is a revised and expanded version of Vernon W. Ruttan, "The Role of the International Agricultural Research Institute as a Source of Agricultural Development," *Agricultural Administration*, 5 (1978), pp. 298-308; the original paper was also published in *Transforming Knowledge into Food in a Worldwide Context*, William F. Hueg, Jr., and Craig A. Gannar, eds. (Minneapolis: Miller Publishing Company, 1978), pp. 124-38. Many of the issues discussed in this chapter are also discussed in Per Pinstrup-Anderson, *The Role of Research and Technology in Economic Development* (New York: Longmans, in press), ms. pp. 111-47; and in Sterling Wortman and Ralph W. Cummings, *To Feed This World: The Challenge and the Strategy* (Baltimore: Johns Hopkins University Press, 1978), pp. 113-43. I have benefited from comments made on an earlier draft by Per Pinstrup-Anderson, Randolph Barker, Hans Binswanger, Nyle Brady, John Coulter, Ralph W. Cummings, Dana Dalrymple, James Goering, Robert W. Herdt, Richard DeMuth, Haldore Hanson, William Gamble, Max Langham, Kenneth Mentz, John Nickel, Peter Oram, Albert Moseman, Ralph Retzlaff, and Grant Scobie.

2. For a history of the establishment and early development of the FAO, see Gove Hambidge, *The Story of FAO* (New York: Van Nostrand, 1955).

3. For a more complete background on the establishment of the IRRI and the CIMMYT, see E. C. Stakman, Richard Bradfield, and Paul C. Mangelsdorf, *Campaigns against Hunger* (Cambridge, Mass.: Harvard University Press, 1967). For more recent discussion of the organization of the international agricultural research system, see J. G. Crawford, "Development of the International Agricultural Research System," in *Resource Allocation and Productivity in National and International Agricultural Research*, Thomas M. Arndt, Dana G. Dalrymple, and Vernon W. Ruttan, eds. (Minneapolis: University of Minnesota Press, 1977), pp. 281-94.

4. Theodore W. Schultz, *Transforming Traditional Agriculture* (New Haven, Conn. and London: Yale University Press, 1964).

5. Vernon W. Ruttan, "The International Institute Approach," in *Agents of Change: Professionals in Developing Economies*, Guy Benveniste and Warren F. Ilchman, eds. (New York: Praeger, 1969), p. 226.

6. Alex McCalla, Ewert Aberg, James McWilliams, and Arthur Mosher, *CGIAR Review Committee, Final Report* (Washington, D.C.: CGIAR Secretariat, October 1976). A second review of the institute system was initiated in 1980 under the direction of Dr. Michael H. Arnold. For additional information on the institute programs, the reader is referred to the series of quinquennial reviews commissioned by the CGIAR and the TAC and the annual integrated reports prepared by the CGIAR's secretariat.

7. For greater detail, see the material prepared by Dana Dalrymple, "Global Agricultural Research Organization," in National Research Council, *World Food and Nutrition Study: Supporting Papers*, vol. 5 (Washington, D.C.: National Academy of Sciences, 1977), pp. 44-133.

8. Robert E. Evenson, "Cycles in Research Productivity in Sugarcane, Wheat and Rice," in *Resource Allocation and Productivity in National and International Agricultural Research*, Thomas M. Arndt, Dana G. Dalrymple, and Vernon W. Ruttan, eds. (Minneapolis: University of Minnesota Press, 1977), pp. 209-236.

9. For a more complete review of the WARDA's program, see *Report of the TAC Quinquennial Review Mission to the West African Rice Development Association* (Rome: TAC Secretariat, FAO, 1979).

10. *Report of the Task Force on International Assistance for Strengthening National Agricultural Research* (Washington, D.C.: CGIAR Secretariat, World Bank, August 1978).

11. After a good deal of deliberation, the CGIAR had earlier decided not to provide funding, under the CGIAR's auspices, for the Asian Vegetable Research and Development Center (AVRDC), the International Fertilizer Development Center (IFDC), or the International Center for Insect Physiology and Ecology (ICIPE). The reasons were somewhat different in each case. In the case of the AVRDC, the uncertain political status of Taiwan was an issue. In the case of the IFDC, the reluctance to open the door to the direct support of institutes in developed countries was of concern. In the case of the ICIPE, a reluctance to fund what some donors viewed as basic research seemed to be involved.

12. *Report of the TAC Quinquennial Review Mission to the International Potato Centre* (Rome: TAC Secretariat, FAO, 1977), pp. 85-86.

13. International Board for Plant Genetic Resources, *The Conservation of Crop Genetic Resources* (Rome: IBPGR/FAO, 1975). See also *Annual Report, 1979* (Rome: IBPGR/FAO, May 1980).

14. *The Review of Farming Systems Research of the International Agricultural Research Centers: CIAT, IITA, ICRISAT, and IRRI* (Rome: TAC Secretariat, FAO, April 1978). See also Hans P. Binswanger, B. A. Krantz, and S. M. Virmani, *The Role of The International*

Crops Research Institute for the Semi-arid Tropics in Farming Systems Research (Hyderabad, India: ICRISAT, April 1976).

15. This view is generally consistent with, but somewhat less optimistic than, the perspective presented in E. H. Gilbert, D. W. Norman, and F. E. Winch, *Farming Systems Research: A Critical Appraisal* (East Lansing, Mich.: Michigan State University, Rural Development Paper No. 6, May 1980).

16. For a more positive view, see National Research Council, *Post-harvest Food Losses in Developing Countries* (Washington, D.C.: National Academy of Sciences, 1978).

17. For more detailed discussion of the arguments, see Hans P. Binswanger and Vernon W. Ruttan, *Induced Innovation: Technology, Institutions and Development* (Baltimore: Johns Hopkins University Press, 1978), pp. 381-408.

18. See Grant M. Scobie, "Investment in International Agricultural Research: Some Economic Dimensions" (Washington, D.C.: World Bank Staff, Working Paper No. 361), p. 25.

19. See, for example, International Rice Research Institute, *Farm Level Constraints to High Rice Yields in Asia* (Los Banos, Laguna, Philippines: IRRI, 1979). See also John H. Sanders, "After the Dwarf Varieties: Crop Program Strategies and the Role of the Economist" (Cali, Colombia: International Center for Tropical Agriculture, September 20, 1980), mimeographed report).

20. Robert E. Evenson, Piedad M. Flores, and Yujiro Hayami, "Costs and Returns to Rice Research," *Economic Consequences of New Rice Technology* (Los Banos, Laguna, Philippines: IRRI, 1978), pp. 243-65. The authors pointed out that there are several ways of estimating the advantage of the high-yielding varieties. They estimate that under farm conditions the yield advantage of the new high-yielding varieties over traditional varieties was in the order of 6 percent in South Asia and 20 percent in Southeast Asia during the early 1970s. The cost advantage of producing a ton of rice using high-yield varieties was about 25 percent in South Asia and 15 percent in Southeast Asia (p. 13). See also the results presented in Evenson's Table 11-3.

21. A report by the World Bank, *Agricultural Research Systems-Sector Policy Paper* (Washington, D.C.: 1980), noted that in the entire developing world the number of agricultural research staff members (about 23,000) is hardly more than the number in Japan alone.

Chapter 6

Reviewing
Agricultural Research Programs

It has been repeatedly documented that agricultural research represents one of the more productive investments available to both developed and developing societies.[1] (See chapter 10.) In most countries, regardless of their stage of development or form of economic organization, a relatively high percentage of agricultural research is supported and carried out by public-sector agricultural research institutes and experiment stations. Indeed, the formation of the publicly supported agricultural research institute in Saxony in 1852 can be regarded as one of the great institutional innovations of the 19th century.

Increasing attention has been devoted to the practice of establishing research priorities and to the methodology of research resource allocation in the literature on agricultural science policy during the last several decades. (This literature is reviewed in considerable detail in chapter 11.) Much less attention has been given to the issue of managing agricultural research institutions and to monitoring the performance of agricultural research programs. Traditionally, the external review has represented an important device employed by funding agencies and national and international research administrators to monitor the performance of agricultural research institutes and stations. For example, the U.S. Department of Agriculture conducts subject-matter reviews of programs at the state agricultural experiment stations every three or four years. Comprehensive reviews are conducted less frequently, usually at the request of a station director. Research institutes in the United Kingdom are on a six-year review schedule, and the Technical Advisory Committee (TAC) of the Consultative Group on International Agricultural Research (CGIAR) schedules external reviews of each of the international agricultural research institutions funded through the CGIAR every five years.

My personal experience as a member of research review teams and as both a staff member and an administrative officer at institutions subject to external review has led me to the conclusion that such reviews are usually regarded as highly unsatisfactory by members of the review teams and by the institute's and station's management and staff. Members of the review teams are often concerned with the difficulty of obtaining an in-depth discussion of research strategies and methodologies. Research administrators often consider individual team members to be lobbyists for their disciplinary interests and the reviewers' reports to be punitive. Institute and station staff members often complain that external reviews tend to whitewash administrative deficiencies and to exacerbate interpersonal and interdisciplinary tensions. A more fundamental criticism involves the tendency for review teams to focus on the details of scientific, agronomic, or engineering research methodology rather than on strategic considerations in the design of the institute or station research program. For example, review teams often focus on the quality of the research carried out by the individual bench or field researcher and fail to address problems of research strategy.

In spite of the pervasiveness of the use of the external review as a monitoring device and the widespread dissatisfaction with its effectiveness, I have not been able to locate any serious discussion of the external review process in the literature that is comparable, for example, with the recent discussions of the peer review process in research project selection[2] or with the extensive literature on the evaluation of social programs.[3] The material that is available tends to be in the form of mimeographed cookbook or how-to materials designed to brief members of review panels with respect to their responsibilities.

In this chapter, I attempt to clarify what can be reasonably expected from an external review, and I suggest some guidelines for the conduct of external reviews. In the section on the research review process, I draw heavily on the experience of the system of international agricultural research centers supported by the CGIAR and of the state agricultural experiment stations in the United States. The functions of the international institutes are similar to, though perhaps more complex than, the functions of many national, state (provincial), and university-related agricultural research institutes and experiment stations.

THE RESEARCH REVIEW PROCESS

The questions that must be asked in evaluating an agricultural research program can be classified under separate headings having to do with the supply of, and demand for, new knowledge and technology.

The potential supply of new knowledge and technology is determined by what it is feasible to accomplish and by what the institute or station is willing or competent to do. Only a limited number of scientists working at the leading edge of their disciplinary fields or with exceptional understanding of the constraints that limit the production of a particular commodity or the use of a particular resource can respond effectively to the question of feasibility. There are a number of subsidiary questions. They include: How long will it take to accomplish a particular goal? What will it take in the nature of scientific and technical manpower? What alternative strategies might be followed to achieve the same objective (such as protection from a particular pest or disease)? What is the trade-off between a more concentrated effort over a limited time and a less concentrated effort over a longer time (will 4 scientist-years for 5 years accomplish as much as 2 scientist years for 10 years)? How risky is a new research initiative, and can the risk be hedged?

The institute's willingness to supply new knowledge is determined, within its resource constraints, by its objectives. What weight is given to the advancement of knowledge as compared to the development of technology? Do the objectives include the production of information about the institutional constraints on technology choice by farmers? Do the objectives include attempts to understand the effects of agricultural policy on agricultural production and on the incomes of farm people? Is the ultimate test of success the development and adoption of new crop varieties or new machines or equipment? The institute's capacity to supply new knowledge and technology will depend on the quality of the program's management, scientific staff, and research design.

The social demand for new knowledge or technology determines what research is worth doing and the desirable amount of each type of research. The demand for research is derived from a number of subsidiary questions. How will the new knowledge or technology contribute to national or regional development priorities? Who will receive the benefits from the research—other scientists, large or small producers, landless laborers, or urban consumers? Will any segment of society be harmed or made worse off as a result of the research?

There is a tendency for the typical research program evaluation to focus primarily on questions dealing with the supply of knowledge and technology. Within this general area, major attention typically is given to the scientific, engineering, or agronomic quality of the program. But the fact that a research activity is conducted with skill and imagination is not adequate evidence that it is worth doing. The objectives of a research review should not be simply to assess the quality of the program at the time of the review, but rather they should be to engage in a dialogue with the management and staff that will not only contribute to improvements in research performance but will encourage the institute's or experiment station's management and staff to be more responsive to the changing demand for knowledge and technology.

It is essential that the review team and the institute's management arrive at an agreement on the criteria against which each major activity in the institute's program is to be evaluated either prior to initiating the review or at the start of the review process. Joseph S. Wholey referred to this as "evaluability assessment." If there is ambiguity or incomplete agreement at this level, there is little prospect for reaching agreement on the review team's evaluation of the research program. And there is little prospect that there can be an agreement on research, training, and information or outreach program objectives unless the institute itself is prepared to make these objectives explicit enough for effective communication.

Among the reasons that review teams often fail to produce a useful evaluation report is that the research institute or station being evaluated has made an inadequate effort to specify the objectives of its own program or program components clearly. My perspective falls within the goal-based, rather than the goal-free, tradition of evaluation. In his case for a goal-free evaluation, M. Scriven argued that, rather than concentrating on the program's goals, the appropriate approach is to evaluate the actual impact of the program, regardless of the program's goals.[4] I emphasize goals because of the importance of goal setting as a management device within the research institute. I agree with Scriven that the full range of a program's impacts should be identified and that the research goals or strategies should themselves be the subject of evaluation.

The review should adopt a positive and forward-looking approach rather than a backward-looking and negative approach to the institute's program. Few members of the review team will have as much depth of knowledge or research experience in any area as the research institution's own staff. But it is difficult for a review team to focus its attention effectively on research program objectives and strategy in the short time available to it unless the institute's or station's management and staff have developed rather clear-cut research program objectives. The management of the institution being reviewed should be prepared to specify what it would regard as acceptable research program achievement, in the form of new technology and new knowledge, over a planning period of five to eight years. These research output projections should be accompanied by careful identification of the resources and modification of the intermediate processes and activities required to realize the output projections. (See figure 3.4.) This is a management tool that every research institute should develop for internal purposes. It would be of great value for the continuing review process.

THE RESEARCH PROGRAM

The fact that a research institute has been highly productive (or cost-effective) when evaluated from a total social accounting perspective does not mean that each activity in the institute's research portfolio has made or can be expected to make an effective contribution to the total program. An important step in the review process is, therefore, orienting the review to a separate evaluation of the components of the research program. This requires identifying those activities that are expected to produce a reasonably direct economic payoff in terms of new knowledge, new materials, more productive technologies, and better institutional performance. Supporting services and supporting research activities then can be evaluated in terms of their contribution to the research output of the institute. It is important that the program's objectives be specified in rather broad terms so that changes in research resource allocation can be made in response to changes in the scientific or technical environment (such as advances in basic knowledge) or in the socioeconomic environment (such as a change in relative product prices or input costs) without substantial revision of the institute's ideology or goals.

It is useful to illustrate the points made in this section by reference to the organization of the International Rice Research Institute (IRRI).[5] In the mid-1970s, the IRRI's program activity was organized around nine interdisciplinary research and training programs. The program areas and the percentages of core budget allocated to each program were as follows:

	1974	1980
Genetic evaluation and utilization	38	40
Control and management of rice pests	14	10
Irrigation-water management	3	5
Soil and crop management	18	12
Environment and its effects on rice	4	2
Constraints on rice production	6	3
Rice-based cropping system	13	19
Machinery development	4	7
Consequences of new technology	—	2

This pattern of organization represents a more relevant focus for the efforts of a review team than the IRRI disciplinary units (Agricultural Economics, Water Management, Agricultural Engineering, Soil Chemistry, Soil Microbiology, Agronomy, Chemistry, Plant Physiology, Entomology, Plant Breeding, Rice Production Training and Research, Multiple Cropping, Statistics, Experimental Farm and Phytotron Operations, and Chemistry Analytical Laboratory) which cut across the interdisciplinary research and training programs.

At the IRRI the interdisciplinary research programs are expected to result in changes in agricultural production technology, improved materials, and/or more effective rural institutions. The plant breeding and/or genetics evaluation unit is expected to produce new breeding materials or varieties that can be evaluated in terms of actual or potential yield changes under farm conditions. The machinery development program is expected to produce prototype machines that, after further development, can reduce the costs of production, contribute to increased production or reduce losses during harvesting or storage. Some programs or disciplinary departments are directed primarily toward the production of knowledge. Work under the program on constraints on rice yields is an example. Knowledge of the economic significance of constraints on rice yields becomes an input into the IRRI's strategy. The program may also have a direct impact on economic policy because, through its operation, it finds out things about farmers' behavior that could influence governmental policy in putting forward schemes to help farmers. In other programs there are dual objectives. For example, the program on the control and management of rice pests produces information that then becomes an input into genetics evaluation and utilization. It also produces knowledge about the effectiveness of insecticides or cultural practices that can be transmitted directly to farmers (through extension services and sales representatives).

Evaluation of the supporting disciplinary research and service departments (such as statistics) poses a more difficult problem than the interdisciplinary research and training programs that produce outputs in the form of new materials and new technologies. Work in these fields is valued within the institute primarily in terms of their contribution to the output of materials and technology. (See figure 3.4.) The problem of evaluating these activities is similar to evaluating the purchase of a new electron microscope, a new computer, or a phytotron. What is its contribution to the institute's effectiveness in achieving the primary objectives against which its major programs should be evaluated? As when considering a new piece of equipment, it is relevant to ask whether a supporting program is too large or too small in relation to the value of its contribution.

It is also relevant to ask whether the research results could be obtained in a more cost-effective manner by contracting out rather than developing in-house capacity. A number of the international agricultural research centers have, for example, entered into contractual or collaborative relationships with other research institutes in developed or developing countries for the conduct of basic investigations related to the institute's mission. The International Potato Center has developed a much more self-conscious policy of contracting out for research requiring basic or supporting science capacity than the other international agricultural research centers.

THE TRAINING PROGRAM

Training is regarded as an essential component of the program of many agricultural experiment stations and research institutes. The training efforts reflect a consensus that the transfer of agricultural technology involves the transfer of the human and institutional capacity to invent, to adapt, and to diffuse new technology and new knowledge. The diffusion of the prototype technology and the knowledge developed at the international institutes depends on the capacity of national programs to adapt the prototype technology to local environments and to initiate institutional innovations that make it possible for farmers to utilize the improved technologies and the new knowledge. Similarly, effective diffusion of the knowledge and materials developed by national research institutes, provincial experiment stations, and agribusiness laboratories depends on the development of a cadre of extension workers or industrial representatives whose background includes training in research as well as in communication and marketing.

There is also a philosophy, particularly at academically based agricultural research institutions, that a graduate training program is essential to the continued viability of the institute's staff itself.

Participation of the research staff in training activities is one way of achieving the intellectual renewal, or recycling, that prevents a mature institute from developing into a geriatric society. There is also the principle at some institutes that trainees represent a source of imaginative and committed research assistants that work as a useful complement to a permanent cadre of research technicians.

The training programs that have been organized at the international agricultural research institutes have widely different objectives and styles of organizations.[6] The IRRI has operated both a research program and a production training program. The research program is built around the conduct of a research project. Research training at the IRRI may also involve simultaneous study toward an advanced degree at the University of the Philippines or at a home-country university. The objective is to provide well-trained research workers who will return to their national ministries or university research programs. The IRRI's rice production training program was designed to provide extension specialists and extension training program staff with a high level of knowledge about rice production problems, advanced rice production technology, and research communication skills needed to be able (1) to conduct applied rice research trials and to modify the technology to suit local conditions and (2) to organize and staff in-service rice production training programs for extension personnel in their home countries.

This is in contrast to the training program of the International Center for Improvement of Maize and Wheat (CIMMYT), which has been directed more specifically to the development of the technical skills needed to staff national research and production programs that make use of the breeding materials supplied by the center. B. E. Swanson's comparative study indicated that the objective of the CIMMYT's training program is to upgrade the technical skills of research personnel in order to build strong, independently functioning national programs that can cooperate effectively with other national programs and draw on the CIMMYT for new genetic resources and technical information. The program attempts, for example, to impart the research skills and the knowledge needed to run a wheat improvement program. It is also designed to strengthen the trainees' ability to develop improvements in wheat production technology, and it attempts to foster attitudinal changes leading to a strong field-research motivation.

There are also substantial differences among the several institutes in their relationships to national training institutions. The IRRI's research training program has been closely related to the graduate program at the University of the Philippines College of Agriculture (UPCA). The IRRI's senior staff members have held appointments in the UPCA graduate school, and many of the IRRI's scholars have simultaneously pursued graduate training at UPCA, with their IRRI Research projects providing the data for their theses. In contrast, other institutes—the International Institute for Tropical Agriculture (IITA) during its initial years, for example—have tried to keep their relationships with universities in the host countries at arm's length in the belief that this would avoid the diversion of resources from research. Some institutes were located with little attention to the availability of local professional resources that might complement their training programs.

The evaluation of training programs is, if anything, even more difficult than the evaluation of research programs. The impact of a training program on production is often less direct than the impact of research. The study by Swanson referred to above is one of the few successful examples of training evaluation in the field of agriculture.

A first step in the evaluation of a training program should be to arrive at a consensus with the institute's management as to the objectives the training program is attempting to achieve. To what extent are the objectives and the context of the training program based on an objective analysis of the professional requirements of the commodity research and extension programs of the political or administrative units served by the institutes? To what extent does it reflect the particular research interests of individual institute staff members? To what extent is it viewed as a substitute for the employment of technicians on the institute's regular staff? Finally, what is the optimum size of the training program? Over what range is it complementary to, and over what range is it competitive with, the research program of the institute?

THE OUTREACH PROGRAM

The social return, or payoff, to investment in agricultural research is not realized until the new knowledge or technology produced by the institute is embodied in new materials or practices and put into use at the farm level. Delays in the embodiment and diffusion processes lower the social return on investment in research. These considerations typically lead to activities designed to increase the relevance of the institute's research and to speed up the embodiment and diffusion processes. At the very minimum, most research institutes engage in research and testing activities that are carried on at off-campus locations in order to expand the range of observations needed to solve a scientific or production problem or to subject materials to environmental conditions that are not available at the central station. Such off-campus activities are typically funded as part of the institution's core budget and should be evaluated in the same manner as the on-campus research. The institutes also engage in other off-campus programs designed to make institute materials, methods, and professional capacity available to other research institutes and to extension organizations or to private firms that embody the new knowledge resulting from research in the products they market.

The international agricultural research institutes may engage in off-campus activities that benefit national programs under several circumstances. There may be a genuine opportunity to assist a country's commodity program in realizing a rapid return from the utilization of materials or methodologies that have been developed at the international centers. There may be a possibility of achieving a demonstration effect with one or more commodities of the payoff to investment in modern agricultural science that will lead to a self-sustaining, ongoing research effort by other organizations. Under some circumstances, no other technical assistance agencies or organizations that may be acceptable to the host country appear to have the program management or scientific capacity to assist in getting a national program moving. Finally, there may be pressure from one or more donors to the CGIAR's system who are impatient about obtaining a quick payoff for their investment in the international agricultural research system. The institute's management may also be concerned that it must demonstrate a quick payoff to its research program. Similar types of considerations enter into the role of national and provincial research institutes relative to their linkages with extension services, cooperatives, and private-sector agribusiness organizations.

The off-campus outreach and service programs of an agricultural research institution are, therefore, affected by a much wider range of motivations and objectives than the research and training programs. They are also subjected to much greater managerial stress. Because they are usually conducted in collaboration with other instituions, there is frequently less freedom in the choice of outreach program objectives and design, in personnel management, and in access to research and training facilities. The problems of staff recruitment are often more difficult. The professional rewards associated with being part of an innovative, scientific program are often not realized by staff members in the outreach program. In many cases, the outreach staff does not have an opportunity to become fully familiar with the research programs and staff at the central station.

One set of questions centers on the appropriate mix of outreach activities. These include (1) the disseminating of information, materials, and other research outputs; (2) the holding of seminars and workshops; and (3) the stationing of liaison staff in related research and extension institutions. What is the most cost-effective mix of outreach activities when evaluated against the institute's objective of rapid diffusion of knowledge and technology?

Clearly, each of these activities can be highly complementary to the central research program through the feedback of experience, knowledge, and materials from research conducted under a wide variety of environmental conditions. As in the case of training programs, it is useful to ask at what point the outreach activity program becomes competitive with, rather than complementary to, the research program. This point may come when it begins to divert a substantial share of the effort and attention of the institute's management away from the research effort of the center or station. It must be recognized by the review team that this issue poses unusually difficult choices for the management of a research instituion since it may involve trade-offs among the several outputs of the institution's program (See figure 3.1) and since it may have long-term effects on the financial viability of the research institution.

PERSPECTIVE

In contrast to the chapter on resource allocation (chapter 11), there has been little emphasis in this chapter on the quantification of research benefits and costs. Quantification of projected costs and benefits is not always feasible and when feasible the advantages may not be commensurate with the costs. When quantification is feasible, it is (as emphasized in chapter 11) most appropriately viewed as part of the continuing management information program of the research system rather than a specific response to the needs of a review team. Nevertheless, the research institute, or its management, should be pressed to go as far as possible toward quantification in stating the objectives against which the program of a research center designed to produce new materials, new technologies, more effective institutions, or greater human capacity should be evaluated.

Regardless of the level of quantification that is feasible or desirable, the objectives can and should be verbalized in reasonably precise terms.

In this chapter, it has been assumed that the research institutes, such as the international agricultural research centers that are members of the CGIAR system or the USDA's Meat Animal Research Center, are to be evaluated primarily in terms of their effectiveness as agricultural research institutes. No attempt has been made to develop the criteria that should enter into the evaluation if the research institutes are, as Arthur Mosher has suggested, in the process of developing into research-based agricultural or rural development institutions.[7] This is a direction in which the agricultural colleges of the state land-grant universities in the United States and the agricultural colleges and universities in some other countries have evolved. As a research system matures, however, it is not unreasonable to anticipate that the resources devoted to the maintenance and diffusion of genetic resources, the communication of research methodologies and results, and the training of commodity research and program personnel will expand relative to the resources devoted to research. The implications of such shifts in the allocation of research and related professional resources should be examined by the research review team and its implications brought to the attention of the institute management.

It is worth returning again to the question of what can reasonably be expected from the research review process. In the introductory section, I pointed to the disappointment frequently experienced by research administrators and staff and by members of review teams over the conduct and results of research reviews. Similar views have also been noted in the evaluation literature. It has been suggested by cynics that the review or evaluation process can most appropriately be reviewed as a ritual rather than as an analytical exericse.[8]

My own view is much more positive. Preparation for a review, when taken seriously, often creates an environment that enables a research center or department staff to engage in self-analysis and to resolve latent or overt conflict over objectives and performances. My major criticism, noted in the section on the research review process, is the failure of most research institutions and reviewing agencies (1) to take the steps necessary to acquaint the research review teams with the longer-term strategic issues that face the research unit and (2) to assure continuity in the research review process over time. When such continuity is lacking, the signals that should lead to a more intensive evaluation of the total program and of critical program components are often missed.

In his book, *Evaluation: Promise and Performance,* Joseph Wholey identified four sequential steps in effective program evaluation: (1) evaluability assessment, (2) rapid feedback assessment, (3) performance monitoring, and (4) intensive evaluation. Most research review efforts focus on the second step. I have argued in this chapter that this second step could be more effective if more attention were given to the first step. I have also argued that research reviews should be organized in a manner that permits the more effective accomplishment of the third step.

Achievement of the final step in evaluation involves a longer-term effort. Its implementation is highly dependent on the personal qualities of the research system or the research center's leadership and on the system of governance employed by the research system. I have, over time, become increasingly concerned that the hierarchical system of governance employed in most national agricultural research systems weakens responsiveness to new scientific and technical opportunities or to changes in the demand for knowledge and technology. These dangers are somewhat muted where strong state or provincial research systems are capable of engaging in effective dialogue with a national system. Traditional systems of governance have a strong propensity to protect the system from new sources of knowledge and to limit the capacity of the system to respond to new demands.

NOTES

1. This chapter represents a revision and an expansion on an earlier paper, Vernon W. Ruttan, "Reviewing Agricultural Research Programs," *Agricultural Administration,* 5 (1978), pp. 2-19. The initial draft of the paper was prepared when the author was a member of the Technical Advisory Committee of the Consultative Group on International Agricultural Research to assist in the formulation of procedures for the quinquennial review of CGIAR-sponsored institutes. I am indebted to Nyle C. Brady, Walter L. Fishel, Haldore Hanson, Lowell S. Hardin, Ralph H. Retzlaff, Roland R. Robinson, William J. Siffin, and members of the CGIAR/TAC for comments on an earlier draft.

2. T. Gustafson, "The Controversy over Peer Review," *Science,* 190 (December 12, 1975), pp. 1060-66; R. Bowers, "The Peer Review System on Trial," *American Scientist,* 63 (6) (November-December 1975), pp. 624-26.

3. Joseph S. Wholey, *Evaluation: Promise and Performance* (Washington, D.C.: Urban Institute, 1979); Robert S. Floden and Stephen S. Weiner, "Rationality to Ritual: The Multiple Roles of Evaluation in Governmental Processes," *Policy Sciences,* 9 (1978), pp. 9-18.

4. M. Scriven, "Pros and Cons about Goal-Free Evaluation," *Journal of Educational Evaluation,* 3 (December 1972), pp. 1-4.

5. For greater detail, see International Rice Research Institute, *IRRI Long Range Planning Committe Report* (Los Banos, Laguna, Philippines: IRRI, 1979). Data for 1980 were supplied by the IRRI.

6. B. E. Swanson, *Organizing Agricultural Technology Transfer: The Effects of Alternative Arrangements* (Bloomington, Ind.: International Development Center, 1975); B. E. Swanson, "Impact of the International System on Natural Research Capacity: The IRRI and CIMMYT Training Programs," in *Resource Allocation and Productivity in National and International Agricultural Research*, Thomas M. Arndt, Dana G. Dalrymple, and Vernon W. Ruttan, eds. (Minneapolis: University of Minnesota Press, 1977), pp. 336-66.

7. A. T. Mosher, "Unresolved Issues in Evolution of the International System," in *Resource Allocation and Productivity in National and International Agricultural Research*, Thomas M. Arndt, Dana G. Dalrymple, and Vernon W. Ruttan, eds. (Minneapolis: University of Minnesota Press, 1977), pp. 567-77.

8. "Evaluation may be seen as a ritual whose function is to calm the anxieties of the citizenry and to perpetuate an image of government rationality, efficiency and accountability. The very act of requiring and commissioning evaluations may create the impression that the government is seriously committed to the pursuit of publicly expressed goals, such as increasing student achievement or reducing malnutrition. Evaluations lend credence to this image even when programs are created primarily to appease interest groups" (Floden and Weiner, "Rationality to Ritual," p. 16).

Chapter 7

Location and Scale
in Agricultural Research

One hundred years ago there were few places in the world where grain yields were significantly higher than 1 metric ton per hectare.[1] Since that time, differences in output per hectare and output per worker in agriculture have widened. These differences have not been due to changes in physical resource endowments. They have been due primarily to advances in the technology of crop and animal production that have been products of institutional innovations that were only beginning to take shape a century ago—the agricultural experiment station and the research laboratory.

Little formal knowledge about the economics of location and scale is available to help in decision making either at the level of the individual experiment station or research center or at the level of the naor international agricultural research system. Issues of location and scale are frequently obscured by the thick clouds of political interest and the strong winds of professional controversy. They are issues on which the planners and the policymakers in national research bureaucracies and the political partisans in national legislative bodies have rarely asked for or received professional analysis. In this chapter I am not able to advance knowledge on these questions significantly. I do attempt, however, to help focus the issues in a manner that may lead to more formal analytical treatment and empirical investigation.

THE LOCATION OF AGRICULTURAL
RESEARCH FACILITIES

The literature on the location of agricultural research facilities is extremely limited. Little attempt has been made to distinguish between those elements of a location that contribute to a facility's efficiency in supplying knowledge and technology and those that operate on the demand side to enhance the impact of new knowledge and technology on production.

A supply-side argument for the location of agricultural research facilities has been made in two seminal articles. Peter Jennings argued that the optimum location of an international crop research center is at or near the geographic center of origin of each particular crop.[2] Jennings and James Cock developed this point, giving particular significance to the coevolution of biological constraints on a particular crop in its area of origin. They argued that in crop improvement work it is important to subject the improved varieties to the biological stress that exists in such locations in order to assure that they will exhibit minimal inherent weakness in their response to biological or biochemical stress. Jennings and Cock also argued that a technology that is productive in the center of origin can then be more successfully introduced where there is less biological stress on the new cultivar. If, on the other hand, when crop improvement research is conducted outside the center of origin, development of cultivars suited to the local environment may be very rapid but transferability may be limited. They also argued that incomplete technological packages developed at the center of origin may have a much more dramatic impact on crop yields outside the center than in or near the center of origin. They pointed to the much greater impact of the modern rice varieties on rice yields in Colombia than in Southeast Asia as an example. Although they did not explicitly mention considerations of soil quality, temperature, and related factors, I assume that Jennings and Cock would have viewed these as microlocation factors within the area of origin.

There is also a somewhat different supply-side argument from Theodore W. Schultz to the effect that "the comparative advantage of an agricultural experiment station associated with a major research-oriented university is clearly large."[3] Schultz's insistence on the existence of a symbiotic relationship between research and training is related to my own observation on the "natural history" of research institutes. A new institute is often able to bring together a team of leading scientists who tend to experience a period of high productivity, which often lasts for more than a decade. When the team is able to exploit existing knowledge not previously brought to bear on a problem, a period of explosive growth in technology development may occur, as was the case for wheat development in Mexico and rice development in the Philippines during the 1960s.

There is a tendency, however, after an initial period of creativity for an institute's research program to settle down to filling in the gaps in the literature. During this "plateau" phase, team efforts tend to give way to disciplinary fragmentation, with each specialist attempting to advance knowledge along the line that seems most likely to remove the constraints on technology development. One of the more effective protections against geriatric tendencies in a research institution is an environment that facilitates interaction with trainees or graduate students and with colleagues in related disciplines.

A somewhat similar argument was put forth by Albert H. Moseman with respect to the location of genetic resources centers.[4] Moseman insisted that, if genetic resource centers are to make productive contributions to the development of biological technology, they must be institutionalized in locations that are conducting strong plant breeding and genetic research programs. Financial commitments to the maintenance of genetics resource archives will face continuous erosion unless the archives are seen as contributing in a very direct way to biological technology development programs.

A demand-side argument is implicit in the early work of Zvi Griliches on the diffusion of hybrid corn.[5] Aside from the initial basic research, both public and private research and development efforts in the United States were observed to be most heavily concentrated in those areas in which the size and density of the hybrid seed markets were greatest. Griliches concluded that the location of a research effort by both the state agricultural experiment stations and the private seed companies could be explained by a model in which the cost of research, development, and marketing per unit of hybrid seed sold would be minimized.

The demand-side analysis was formalized and applied to the location of commodity research institutes in Nigeria in a recent study by Francis Sulemanu Idachaba. He developed a research station location suitability index based on intensity of crop production. When the three states nearest to a central commodity research institute account for at least one-third of the nation's output of the commodity, the location is regarded as suitable. Of the 14 commodity research institutes for which location suitability indexes were calculated, 6 fell below the minimum criteria for suitability.[6]

Both the supply-side and the demand-side location factors seem to support a conclusion that there should be a major internationally supported research institute located at the center of origin for each crop that is capable of playing a key role in the production of generalized genetic technology. The international center needs to have effective linkages for the transfer of knowledge and materials with environmentally specific and market- or demand-oriented systems of public-sector national research centers and with the private-sector genetic supply industry. (See chapter 5.)

One issue that this information does not deal with effectively is the issue of the location and intensity of research efforts in areas where neither the supply-side nor the demand-side criteria appear to be operative. The decision by the International Center for Tropical Agriculture (CIAT) to locate a forage improvement station in the Colombian llanos represents a case in point. The problem soils of the llanos are characteristic of an area of perhaps 300 million hectares of low-density forage and livestock production and human population in the South American tropics. What criteria can one employ in making a decision with respect to the location or intensity of research directed to improving potential productivity under such circumstances?

The actual decision-making process with respect to the location of agricultural research institutions typically bears little relationship to the answers that would be turned up by a simultaneous solution to the demand and supply relationship suggested above. Research programs at several of the new international centers have been burdened by poor location decisions. Within national systems, non-agricultural considerations associated with employment and income generation at politically sensitive locations in the district or the state of a strategic member of the legislature or of the administration in power, for example, are often of dominant importance.

A recent analysis written by Don Hadwiger indicated that the pork barrel approach to the location of agricultural research facilities resulted in 44 percent of all USDA research facility construction between 1958 and 1977 in states represented by members of the Subcommittee on Agriculture of the Senate Appropriations Committee. The district of the chairman of the Subcommittee on Agriculture of the House Appropriations Committee was also favored by substantial facilities investment. Hadwiger noted that this has forced "the federal Agricultural Research Service to operate a 'traveling circus' opening up new locations in current Senate constituencies, while closing some locations in states whose Senators are no longer members of the subcommittee."[7] It is clear, however, that a body of technical and economic location criteria have not yet been conclusively established, let alone accepted by planners and legislators, to enable research planners to counter the political criteria that have dominated decisions on research facility locations.

The result of the failure to give more formal attention to research location decisions has been costly. I have been continuously confronted with comments from research administrators and staff to the effect that "our work at the Northwest station should not be directly compared with the results achieved by farmers in this area since we are located in a heavy soils area that is outside the northern margin of soybean production," or that "the rice station is located on some of the most productive soils in the country; however, this limits the transferability of the results of plant nutrition research conducted at the station," or that "the Center was located so far from the nearest university or urban center that it has had problems recruiting and retaining the best young scientists and research team leaders."

In most developed countries with large, established agricultural research systems, a careful rationalization of the locations and functions of research institutes, centers, and stations would contribute significantly to the effectiveness of the resources devoted to research. The history of facilities reviews does not suggest, however, that rationalization efforts have been highly productive. In the United States, there have been five facilities reviews in the last 20 years.[8] Perhaps the only solid generalization one can make is that efforts to close down or relocate research facilities are usually accompanied by considerable political and psychological stress. Research directors rarely enhance their popularity with constituency groups by closing down a research facility!

Countries that are just now developing their research systems have an opportunity to avoid the location errors of the older, established systems. Yet, this opportunity is frequently lost, sometimes with the advice and support of donor agencies. The World Bank has encouraged a series of incredibly bad experiment station locations.

The Sukamandi Rice Research Station in Indonesia, located about 100 kilometers west of Jakarta on the north coast of Java, is one example. An initial decision was made that the Central Rice Research Institute should be moved from Bogor to a location in a major rice-growing area on the north coast of Java. A decision was then made to locate the facility at Sukamandi on land that the Indonesian government had expropriated from a British plantation.[9] When the plantation had been privately owned, the area had been devoted to cassava production. The station occupies over 400 hectares. When fully developed, it will have facilities comparable to some of the international agricultural research centers. The staffing plan is built around a cadre of 20 to 25 senior scientists trained at the doctoral level.

The location decision involved two costly errors. The first error was to locate a research station in an isolated area completely lacking in either physical or institutional infrastructure. The station has had to build its own power plant because of the lack of a dependable power supply. Housing units for approximately 150 scientific and technical staff members had to be constructed. It has been difficult to recruit and retain scientific and technical employees because of the location's lack of educational, health, and cultural facilities. Even though the station's development was started in 1974, it was not yet fully operational in 1980. And it appears unlikely that it will be adequately staffed in terms of numbers and quality to make effective use of the facility within the next decade.

The second major error was the choice of the specific site. The site is not representative of rice soils on the north coast area. Maximum yield trials that result in yields of above 8 metric tons per hectare at nearby locations result in yields in the 6 ton range at the station. The soils are more acid (in the 4.0 to 4.5 pH range) than the soils on which rice is typically grown. Heavy lime application and other land development will be necessary to make the station an effective site for rice research. As one station scientist commented, "It wasn't an accident that the plantation company was growing cassava here instead of rice!"

What were the alternatives to the Sukamandi site? Certainly, attention should have been given to a location in close proximity to one of the north coast cities, where the institutional infrastructure would have been an asset rather than an obstacle to attracting capable senior scientists and where more of the physical infrastructure was already in place.

It is clear that the Sukamandi location has reduced the rate of return on the donors' investment. It has imposed excessively expensive operation and maintenance burdens on the Indonesian agricultural research systems, and it has delayed the flow of benefits from new rice technology to Indonesian farms and consumers. The tragedy is that the lessons of Sukamandi are not being learned. Similar mistakes are being made in the location of the World Bank-financed Rubber Research Institute near Medan and in the USAID-financed research facility developments in Sumatra and eastern Indonesia.

SIZE AND PRODUCTIVITY IN
AGRICULTURAL RESEARCH

There is even less solid evidence on the relationship between scale (or size) and productivity in agricultural research than on the issue of location. There has been very little analytical or empirical research on the relationship between size and productivity in agricultural research. And what there is, even in the way of casual observation, often lacks precision as to whether the size-output relationship being referred to is with respect to the size of the individual research unit (team, laboratory, department), the individual research institution (center, institute, faculty), or the national or international research system. The view that small is better has often been advanced with considerable heat but with relatively little precision in concept or definition and with little empirical evidence.

The optimum scale of the research is affected by factors both external and internal to the research process. The optimum level of resources devoted to a commodity research program, as demonstrated rigorously by Binswanger, is positively related to the area planted to a commodity in a particular agroclimatic region.[10] Determining the optimum scale of a research unit or program involves, therefore, balancing the increasing returns associated with the area devoted to the commodity (or problem) on which the research is being conducted against the possible internal diseconomies of scale of the research process or system. In this section, I consider primarily the issue of internal factors that affect scale economies and size of the research unit or system.

Evidence has been assembled and summarized by Jacob Schmookler and by Morton Kamien and Nancy Schwartz that indicates that industrial research and development productivity, measured in terms of patents per engineering or scientific worker, is lower in the large laboratories of the largest firms than in the smaller firms in the same industry.[11] Similar evidence was presented by G. S. Pound and P. E. Waggoner for agricultural research.[12] There also are a number of case studies that suggest very high rates of return to individual public, philanthropic, and private research units, often with fewer than 20 scientific or technical staff members per unit.[13] Many of the smaller "freestanding" agricultural research units are, however, engaged primarily in "genetic engineering" or in technology screening, adaption and transfer activities that depend only minimally on in-house capacity in such supporting areas as physiology, pathology, chemistry, and even modern genetics.

Robert Evenson also noted that, during the early stages in the development of national research systems, experiment stations tend to be widely diffused, to utilize primarily technical and engineering skills, and to be characterized by a strong commodity orientation. He also pointed to a trend toward hierarchical organization and consolidation into a smaller number of larger units at later stages in the development of agricultural research systems. These centralizing trends are apparently motivated in part to take advantage of economics resulting from research activities in the basic and supporting sciences and to use economically the laboratory, field, communications, and logistical facilties. The urge for consolidation can easily be overdone, however. In the United States, for example, there is now rather strong evidence supporting the value of decentralization even within individual states. For a given level of expenditures, a state system that includes a strong network of branch stations gets more for its research dollar than a state system that is more concentrated.[14] What decentralization gives up in terms of lower costs seems to be more than compensated for by the relevance of the research and the more rapid diffusion of results.

There are, of course, limits to the gains from decentralization. The gains vary among commodities and are influenced by the diversity of agroclimatic conditions and the area devoted to the crop in each agroclimatic region. In Minnesota, for example, soybean breeding is conducted at four locations (Rosemount, Waseca, Crookston, and Morris) selected to reflect the major agroclimatic, or ecological, regions in which soybeans are grown. Breeding programs could, of course, be conducted in each county in which soybeans are grown. The agroclimatic variations and the area devoted to soybeans in each county are, however, not sufficient for the gains that might be realized to offset the additional costs.[15]

Scale considerations must also be disaggregated by commodity and discipline. Some disciplines require rather capital-intensive facilities. Certain kinds of research in plant physiology, for example, may require access to a phytotron. The high capital and operating costs of such a facility mean that it must be used intensively. Large animal research programs make up another area that appears to be characterized by substantial economics of scale.

In the United States, a substantial share of large animal research conducted by the U.S. Department of Agriculture has been concentrated at a single location—the Meat Animal Research Center (MARC) at Clay Center, Neb. The development of MARC began in 1966.[16] It is located on a 35,000-acre facility, which, when fully developed, will employ about 100 resident scientists and 400 support personnel. It is intended to be the only center in the United States with the resources required to investigate total life-cycle production systems for meat animals, including large-scale crossbreeding programs. The research program includes research on genetics and breeding, nutrition, reproduction, life-cycle production systems, facilities and equipment, and meat science and technology.

The operating budget for the MARC in 1980 was $5.1 million. This amounted to approximately 25 percent of the resources allocated to large animal production research (but only 8 percent of combined production and protection research) by the U.S. Department of Agriculture Science, and Education Administration's Agricultural Research Service (USDA/SEA/AR). Roughly 50 percent of the budget is allocated to beef cattle, 25 percent to swine, and 25 percent to sheep. Center scientists serve as national coordinators for both the U.S. Department of Agriculture's and the cooperating state agricultural experiment station's beef cattle-, sheep-, and swine-breeding efforts.

The MARC conducts a substantial amount of research in close cooperation with the University of Nebraska, but its location approximately 100 miles from the Lincoln campus imposes some limits on the effectiveness of the collaboration. It has established cooperative programs with 15 different universities in the United States and with the Research Division of Agriculture Canada. A program of predoctoral and postdoctoral studies and a visiting scientist program have been initiated. Program consultations and reviews are held with the National Cattlemen's Association, the National Pork Producers' Council, the National Wool Growers' Association, the USDA/SEA/AR national program staff, and internal and external scientific committees.

Yet, there has been criticism that the MARC has been slow (1) in establishing effective collaborative working relations with a number of the strong state stations and (2) in effectively institutionalizing a continuing professional consultation and advisory system. A 1975 advisory committee recommended the establishment of an extramural research program at the annual level of $3.0 million, or 30 scientist-years. Although this program has not yet been funded, the MARC does allocate about 15 percent of its research budget to the funding of cooperative extramural research.

It seems apparent that in beef-animal breeding there are substantial advantages to the concentration of a major research effort in a single location. This advantage is not as great in the cases of swine and sheep. There are several state systems that have swine-breeding programs at least as large as the Clay Center program.

The facilities that have been established at the Meat Animal Research Center for research on life-cycle production systems from conception to consumption are unique. It would be very expensive to attempt to duplicate such facilities elsewhere. Yet, the advantages of the Clay Center location for research on animal production systems appear to be less compelling than for large-animal breeding.

Livestock performance reflects the interaction between genetic traits and environment. Livestock improvement research thus involves the design of genotypes and of management practices adapted to differences in the environments under which animal production is carried out. This typically involves a trade-off between wide adaptability and optimum performance under specific environmental regimes.

Although the MARC is located near the geographic center of the United States' beef production and feeding, it would be expensive to attempt to duplicate at a single location the wide range of pasture, range, and feedlot conditions under which beef animals are produced. The advantages of the Clay Center location for production system research on swine and sheep are weaker than the advantages for research on beef cattle. For research on sheep, this disadvantage may be partially offset by the fact that, with few exceptions, support for sheep research at the state experiment stations had been relatively weak. Some observers have suggested concern that the MARC could become primarily a regional center for the central plains region rather than a truly national research center. It is hard to avoid the conclusion that, if the Clay Center facility is to exert a substantial national impact, it must evolve a more decentralized collaborative research style designed to offset the limitations of its location and to take greater advantage of the research capacity at other locations.

The desirability of a decentralized style of governance and collaboration was recognized in the planning for the National Dairy Forage Center. The MARC's headquarters are located on the campus of the University of Wisconsin at Madison. A major laboratory-greenhouse and dairy production facilitiy is located nearby at Prairie du Sac. Research at the Wisconsin facilities is complemented by the location of federal (USDA/SEA/AR) scientists at the state agricultural experiment stations in New York, Minnesota, Pennsylvania, Michigan, Ohio, Iowa, and Missouri, where they are engaged in collaborative dairy forage research with state agricultural experiment station scientists.

The record of research productivity at many of the older, large national USDA laboratories suggests that great caution should be exercised in concentrating a large share of the national research effort in a single facility. When economics of scale imply the desirability of such a concentration, efforts should be made to institutionalize the governance of such centers and the professional linkages among research scientists in a manner that will limit the drift toward intellectual conformity.

The new international agricultural research centers have approached the problems of governance, consultations, and collaboration with greater sophistication than is typical of most U.S. (or other national) research centers. (See chapter 5.) The U.S. federal research system (and many other national research systems) are organized hierarchically. Advisory and consultative committees have no real authority. The international system employs a highly decentralized style of governance. Each center is governed by a board of directors that is responsible for program strategy and has the authority to hire and fire the center's director.

STRUCTURAL RIGIDITY

It seems apparent that very little can be said about scale economics or optimum size of individual research units in the absence of a more adequate understanding of the effects of the size, structure, organization, and administration of the national and international research systems to which the individual units are related. The appropriate organization and structure of individual experiment stations or of national and international agricultural research systems is, however, an area in which formal knowledge, in contrast to folklore and insight arising out of administrative experience, is also in short supply. The need to take these system relationships into account represents a major challenge to future research on size and productivity in agricultural research.

Almost all university-based agricultural research and much of the work in national agricultural research systems is organized within discipline-based units. This is, however, a relatively recent development. During the period when research in agriculture was struggling to establish its legitimacy as a field of scientific inquiry, departments and faculties of agriculture were usually organized as multidisciplinary administrative units. Even the initial administrative decentralization typically involved organization in multidisciplinary units (field crops, horticulture, animal husbandry, agronomy, farm management). In the U.S. Department of Agriculture, the emergence of a viable pattern of organization toward the end of the 19th century involved the organization of scientific bureaus focusing on particular sets of problems or commodities. For example, "The Bureau of Animal Industry thus has most of the attributes of the new scientific agency at its birth—an organic act—a set of problems, outside groups pressing for its interests, and extensive regulatory powers."[17] In the

United States, both the USDA's national research system and the university-based state systems evolved a structure that includes some units that tend to be based primarily on a single discipline (agricultural economics, plant pathology, soil science) and others that retain a broader multidisciplinary base (agronomy, animal science, horticulture).

In the university-based state system, this structure had become established by the 1920s. Very little change in structure has occurred since that time, except for a tendency in the 1960s for mergers among separate animal science and crop science departments. Once this structure had become established, there was very little resource reallocation among the disciplinary and commodity groups.[18] The USDA's research system has gone through a number of reorganizations since the establishment of the research bureau pattern of organization. There is some evidence that since the mid-1960s there has been a modest shift away from the traditional commodity areas and toward some of the new agenda items—consumer health and nutrition, rural levels of living, community services, and quality of the environment. (See figure 4.1.)

In view of the stability of university disciplinary and departmental organization, it appears that any significant reallocation of research has been primarily within the disciplinary or commodity units. I have in mind, for example, the greater concern in recent years by soils units with nitrogen pollution, the concern by entomology units with integrated pest control, the renewed emphasis in agronomy departments on nitrogen fixation, and the shift of resources in agricultural economics departments to work in resource economics and on rural development problems. It is hard to escape the conclusion that the concept of parity is the dominant principle employed in research resource allocation. (See chapter 11.)

The aging process in the new international system does not yet appear to have imposed the same degree of structural rigidity that one observes in the U.S. system and perhaps other more mature national systems. The original institutes were organized as multidisciplinary crop research institutes (wheat, maize, rice). In recent years the amount of resources devoted to research on animals has risen relative to the amount of research on crops. Within the crops area, research on cropping and farming systems has risen relative to research on individual commodities. Within several of the international institutes, the IRRI, for example, a dual system of organization employing both disciplinary and problem orientation has been introduced in a deliberate effort to resist the hardening of arteries around particular disciplinary and problem sets.

At both the national and the international levels, system growth appears to have been a major factor supplying the lubrication necessary to avoid declining productivity in the older national systems. There are some indications that rates of return in at least some mature systems are beginning to decline. (See chapter 10.) As yet it is not known whether such apparent declining marginal productivity reflects an approaching exhaustion of technological potential or whether it is a function of the organizational and structural deficiencies of mature research systems.

THE SMALL COUNTRY PROBLEM

One of the most difficult issues related to size and productivity in agricultural research is the problem faced by the smaller countries in the development of their agricultural research systems.[19] Most of the smaller countries—those in the 2- to 10-million population range—do have the resources, or have access to donors' resources, that would permit them to develop, over a 10- or 20-year period, an agricultural research and training capacity of 250 to 500 postgraduate-level agricultural scientists and technicians capable of staffing the nation's public- and private-sector agricultural research, education, planning, and service institutions. A professional cadre in this range, trained at the masters and doctoral levels, represents an essential component of any serious effort to enhance agricultural productivity or to mount effective rural development programs. The 50 or so low-income countries with populations of less than 2 million must, however, think of research systems that will often be little larger than a strong branch station in a state such as Texas or Minnesota.[20]

But how can the government of a small country decide on the appropriate size and organization of its national agricultural research system? The case studies of national agricultural research programs and the generalizations drawn from the case studies (chapter 4) are most relevant for the development of national research systems in the middle-size and larger countries. Malaysia was the smallest country included in the case studies. Its level of economic development is well above the level that most smaller developing countries will be able to achieve in the next generation. Its educational system is relatively well developed, and its capacity to reproduce its own professional manpower has been substantially augmented during the 1960s and 1970s. For a country like Sierra Leone or Nepal, even the financial and professional agricultural research resources of a small American state or a Japanese prefecture are probably at least a generation in the future. The time required to achieve viable research systems for many of the smaller national systems must realistically be calculated in terms of a generation rather than the 5- or 10-year project cycles used by most development assistance agencies.

One major focus of the research effort in these smaller research systems must be the direct support of agricultural production and rural development programs. This means a primary focus in applied fields such as agronomy, plant breeding, animal production, crop production, farming systems, and agricultural planning and policy. Yet, the size of the smaller countries' agricultural sectors cannot justify the relatively large research facilities and staff needed to conduct research on each crop in each agroclimatic region. Even a research system with a research staff of 250 to 500 people will have only a few staff members in each specialized field. How can such a system avoid spreading its scientific and technical capacity too thin?

Even in larger countries with advanced agricultural research systems—the United States, the Soviet Union, Japan, India, and Brazil, for example—are not able to be entirely self-sufficient in agricultural science and technology. An effective national agricultural research system must have the capacity to borrow both knowledge and materials from the entire world. The problem of how to link effectively with an increasingly integrated, and interdependent, global agricultural research system is difficult for the state and provincial research units in the larger national systems. It is even more difficult for the national agricultural research systems in the smaller countries.

One approach to this problem has been to attempt to establish cooperative regional research programs. In chapter 5, reference was made to the West African Rice Development Association (WARDA) and the international crop research networks that are linked to the international agricultural research institutes. Other regional institutions not directly linked to the international (CGIAR) system include the Inter-American Institute of Tropical Agriculture (CATIE), the Caribbean Agricultural Research and Development Institute (CARDI), and the Southeast Asian Fisheries Development Center (SEAFDEC). It is hard, however, to find many outstanding success stories among these efforts. Program activities and cooperative efforts often appear stronger in the glossy pamphlets issued by the organizations than they do in practice.

Neither have the international crop research networks, centered around the international institutes, been without problems. When the institutes have had confident and effective leadership, they have often played an exceedingly useful role in creating opportunities for productive professional interaction and collaboration. But the institute research networks tend to be selective. At times they have found it hard to bend institute priorities to meet national priorities. Collaborative efforts tend to involve the strongest institutions and the leading scientists rather than those who have the greatest need.

A richer institutional infrastructure is needed to strengthen and to sustain the capacity of the smaller national agricultural research systems. In spite of ideological considerations, many small countries have found it advantageous to encourage the transfer and adaptation of technology by the private-sector genetic supply industry or by the multinational firms engaged in commodity production, processing, and trade. Firms engaged in the production of crops grown under plantation systems and independent growers producing under contract arrangements with processors have at times provided their own research and development facilities. In other cases, associations of producers have been willing to tax themselves to support commodity research stations. Although such arrangements have often been associated with discarded colonial systems of governance, their utility should be reexamined.

The establishment of the International Agricultural Development Advisory Service (IADS) by the Rockefeller Foundation and the International Service for National Agricultural Research (ISNAR) by the Consultative Group on International Agricultural Research may also lead to more effective capacity in the smaller countries.[21] In spite of its limitations, the WARDA model is among the more promising of the several regional systems. Initially, its single-commodity focus was a source of strength in that it permitted the WARDA to maintain a necessary sharp focus in its program. Over time, either the commodity focus should be expanded or other regional commodity research and development institutions will have to be established.

Research staffs in smaller countries must be as well trained as those in larger countries. A much larger share of the research effort than in a large country should be devoted to screening and adapting knowledge, materials, and technology to fit the requirements of local resource and cultural endowments. This focus on adaptive and applied research, however, does not imply that the smaller national systems can get by with a poorly trained or inadequately supplied research staff. Essentially, the same scientific and technical skills are required to transfer, screen, and adapt technology as to develop new technology for agriculture. The national system designed to transfer, screen, and adapt technology requires essentially the same skills as a system devoted to the invention of new technology.

The problem of institutional development is compounded when research and training capacity must be expanded at the same time. It may be useful to illustrate the problem by an example from a planning exercise conducted by the agricultural economics group of the University of Nairobi Faculty of Agriculture in the mid-1970s.[22] The planning exercise involved the development of a teaching-research department over a 10-year period. It was assumed that at the end of the 10 years the unit would consist of 12 staff members trained at the doctoral level and 4 at the masters level. During the planning period, deficiencies in local staff capacity would be filled by expatriates. It was also assumed, somewhat conservatively, that 2 staff members would have to be selected and sent abroad for training for each staff member who would eventually end up filling a staff position. The total cost of the development program for this one small department was projected at over U.S.$8.6 million in 1976 dollars. My own judgment, based on my experience with other institutional development programs, is that such an effort might easily take closer to 20 than 10 years.[23]

But would it not be more cost-effective to focus only on building research capacity rather than attempting to build domestic training capacity simultaneously? My first response is that there appears to be significant benefit to linking training, particularly at the graduate level, with research. If the two activities were developed separately, more resources would be needed to achieve the same output than if they were developed in an integrated manner.

But all professional activity in agriculture is not complementary with training. The planning cell in the ministry of agriculture in even a small country will require a staff of 3 to 5 senior staff members trained at the doctoral level and 10 to 12 junior staff members trained at the masters level who are fully committed to meeting the information and economic analysis needs of the ministry of agriculture and the central planning organization. A small country that does not have the training capacity to replace the professional capacity needed for research, planning, and regulatory responsibilities will find great difficulty in maintaining continuity in its agricultural research and development programs. At the beginning, when only limited professional capacity is available, there is a tendency for research staff members to move from research into administrative and policy positions after a short research or teaching career. The capacity to provide a continuous stream of replacements is essential to the continued viability of the research effort.

Considerable thought needs to be given to the pattern of governance and funding of institutions to serve the agricultural research needs of the small countries. There is always the danger that they will be more responsive to the needs of the suppliers of technology and assistance than to the needs of the recipient countries. Careful consideration needs to be given to the incentive effects of alternative patterns of domestic and donor funding of regional collaborative research programs. They should be governed by autonomous boards with representatives from both national research agencies and donors. A large share of the regional effort should be located at national stations with the research organized as collaborative projects between national and regional program staffs. The funds from external donors should be made available on some sort of a formula basis involving matching contributions from national governments and multilateral donors or a donors consortium and should be available for program support rather than being tied to specific projects. The commitment of national resources to the regional program is an important incentive for the national research program leaders to insist that the regional efforts should pay off in terms of their contributions to national programs.

The perspectives outlined in this section are highly tentative. Although they are drawn from considerable experience, they should be treated as hypotheses rather than as conclusions. Institutions such as the IADS and the ISNAR should devote a reasonable amount of analytical effort to attempts to understand the problem of developing and sustaining effective agricultural research in the smaller national research systems.

In spite of the limited knowledge that is available, there are a few generalizations about smaller agricultural research systems that can hardly be avoided. One is that the research investment per acre or per hectare will have to be higher in a small system than in a larger system in order to achieve an equal level of effectiveness. This is because the cost of developing, for example, a new millet variety that will be grown on a million acres is not likely to be substantially greater than one that will be grown on half a million acres.

A second generalization is that the cost of developing productive farming systems for a small country with great agroclimatic variations will be greater than for a small country that is more homogeneous. For example, the cost per hectare of developing an effective agricultural research system for Sri Lanka is likely to be much larger than developing one for Uruguay. The issue of guns versus butter in national budgets is also likely to cut more sharply in a small country than in a large country.

Finally, there is no way that a small country can avoid being dependent on others—on the international agricultural research system, on the research systems of large countries in the same region, on multinational firms—for much of its agricultural technology. Furthermore, a small nation with a strong research program but a limited agricultural or industrial base cannot capture as high a proportion of the benefits from its investment in basic research as can a large nation with a diversified economic base. Much of the benefit will spill over to other countries. If it has a weak agricultural research system, it will lack the knowledge needed to capture the benefits of research in other countries or to choose a technological path consistent with its own resource and cultural endowments. Even having a strong agricultural research system cannot assure autonomy. But small countries do need to develop sufficient agricultural science capacity to enable them to draw selectively on an interdependent global agricultural research system. They need to be able to choose what is useful to borrow from other national systems and from the international system.

NOTES

1. This chapter represents a revision and an expansion of Vernon W. Ruttan, "The Organization of Research to Improve Crops and Animals in Low Income Countries: Comment," in *Distortions in Agricultural Incentives*, Theodore W. Schultz, ed. (Bloomington, Ind.: Indiana University Press, 1978), pp. 246-53. I am indebted to Lowell Hardin, Keith Huston, S. C. Hsieh, and Albert H. Moseman for comments on an earlier draft of this chapter.

2. Peter R. Jennings, "Rice Breeding and World Food Production," *Science*, 186 (December 20, 1974), pp. 1085-88; see also Peter R. Jennings and James H. Cock, "Centers of Origin of Crops and Their Productivity," *Economic Botany*, 31 (January-March 1977), pp. 51-54.

3. Theodore W. Schultz, "The Allocation of Resources to Research," in *Resource Allocation in Agricultural Research*, Walter L. Fishel, ed. (Minneapolis: University of Minnesota Press, 1971), p. 114.

4. Albert H. Moseman, *Building Agricultural Research Systems in the Developing Nations* (New York: Agricultural Development Council, 1970). Also in a personal communication, May 3, 1977.

5. Zvi Griliches, "Hybrid Corn: An Exploration in the Economics of Technological Change," *Econometrica*, 25 (October 1975), pp. 501-22.

6. Francis Sulemanu Idachaba, *Agricultural Research Policy in Nigeria* (Washington,

D.C.: International Food Policy Research Institute, Research Report 17, August 1980, pp. 55-58.

7. Don F. Hadwiger, *The Politics of Agricultural Research* (Lincoln, Neb.: University of Nebraska Press, in press), pp. 7-9.

8. *Facilities for Food and Agricultural Research*, Report to the Secretary of Agriculture (Washington, D.C.: Joint Council for Food and Agricultural Sciences, March 1979).

9. It has not been possible for me to adequately document the steps in the decision process that led to the location of the headquarters of the national rice research program at Sukamandi. In addition to appraisals made by the government of Indonesia and the World Bank, there were also consultations with the International Rice Research Institute. The Sukamandi location was not mentioned in a preliminary study conducted for the World Bank in 1973. The plans for the development of the Sukamandi station were developed in considerable detail in a 1975 report prepared for the World Bank by W. P. Panton, J. A. M. Loup, A. H. Moseman, and D. Shoesmith. See *Appraisal of Agricultural Research and Extension Project: Indonesia* (Washington, D.C.: International Bank for Reconstruction and Development, Report 6460-IND, April 8, 1975).

10. For a more technical discussion, see Hans P. Binswanger, "The Microeconomics of Induced Technological Change," in Hans P. Binswanger, Vernon W. Ruttan, and others, *Induced Innovation: Technology, Institutions and Development* (Baltimore: Johns Hopkins University Press, 1978), pp. 91-127.

11. Jacob Schmookler, *Invention and Economic Growth* (Cambridge, Mass.: Harvard University Press, 1966); Morton I. Kamien and Nancy L. Schwartz, "Market Structure and Innovation: A Survey," *Journal of Economic Literature*, 13 (March 1975), pp. 1-37.

12. G. S. Pound and P. E. Waggoner, "Comparative Efficiency, as Measured by Publication Performance of USDA and SAES Entomologists and Plant Pathologists," in *Report of the Committee on Research Advisory to the U.S. Department of Agriculture*, G. S. Pound, Chairman (Washington, D.C.: National Academy of Sciences, 1972), pp. 145-70. A more recent study by G. W. Salisbury suggests a leveling off rather than a decline in research productivity among the largest state agricultural experiment stations. See G. W. Salisbury, *Research Productivity of the State Agricultural Experiment Station System: Measured by Scientific Publication Output* (Urbana, Ill.: University of Illinois Agricultural Experiment Station Bulletin 762, July 1980), pp. 27, 28.

13. Robert E. Evenson, "Comparative Evidence on Returns to Investment in National and International Research Institutions," in *Resource Allocation and Productivity in National and International Agricultural Research*, Thomas M. Arndt, Dana G. Dalrymple, and Vernon W. Ruttan, eds. (Minneapolis: University of Minnesota Press, 1977), pp. 237-64. S. M. Sehgal, "Private Sector International Agricultural Research: The Genetic Supply Industry," also in *Resource Allocation*, pp. 404-15.

14. Robert E. Evenson, Paul E. Waggoner, and Vernon W. Ruttan, "Economic Benefits from Research: An Example from Agriculture," *Science*, 205 (September 14, 1979), pp. 1101-07. See also A. Steven Englander and Robert E. Evenson, "Stability, Adaptability and Targeting in Crop Breeding Programs," a paper presented at the annual meeting of the American Agricultural Economic Society, Pullman, Wash., August 1979.

15. For a discussion of the significance of environmental variability for breeding for stability and adaptability, see Robert E. Evenson, John C. O'Toole, Robert W. Herdt, W. R. Coffman, and Harold Kauffman, "Risk and Uncertainty as Factors in Crop Improvement Research," in *Risk, Uncertainty and Agricultural Development*, James A. Roumasset, Jean-Marc Boussard, and Inderjit Singh, eds. (New York: Agricultural Development Council, and College Laguna, Philippines: Southeast Asian Regional Center for Graduate Study and Research in Agriculture, 1979), pp. 249-64.

16. *Report of the Advisory Committee for the U.S. Meat Animal Research Center* (Clay Center, Neb.: U.S. Meat Animal Research Center, October 1975). I am indebted to Dr. Robert R. Oltjen, director of the U.S. Meat Animal Research Center, for information on the center's current program and for a critical review of an earlier draft of this section. I should note that Dr. Oltjen does not agree with some of my comments regarding the limitations of the Clay Center location.

17. Hunter A. Dupree, *Science in the Federal Government: A History of Policies and Activities to 1940* (Cambridge, Mass.: Harvard University Press, 1957), p. 165.

18. Willis Peterson, "The Allocation of Research, Teaching and Extension Personnel in U.S. Colleges of Agriculture," *American Journal of Agricultural Economics*, 51 (February 1969), pp. 41-56.

19. The problems of economy and governance of small countries has generally been neglected. For a summary of a good deal of the literature, see the book by Robert A. Dahl and Edward R. Tufte, *Size and Democracy* (Stanford, Calif.: Stanford University Press, 1973).

20. In the mid-1970s, there were relatively few developing countries with as many as 50 agricultural scientists. The most recent estimate of the number of agricultural research scientists in developing countries is presented in Peter Oram, Juan Zapata, George Alibaruho, and Shyamal Roy, *Investment and Input Requirements for Accelerating Food Production in Low Income Countries by 1990* (Washington, D.C.: International Food Policy Research Institute, September 1979). For earlier estimates, see James Boyce and Robert E. Evenson, *National and International Agricultural Research and Extension Programs* (New York: Agricultural Development Council, 1975). Oram noted that two-thirds of the approximately 26,000 agricultural scientists working in developing countries are located in seven countries.

21. Consultative Group on International Agricultural Research, *Report of the Task Force on International Assistance for Strengthening National Agricultural Research* (Washington, D.C.: World Bank, 1978).

22. Adolf Weber, *Post Graduate Programs, Staff Development and Training Costs: Some Remarks and Observations*. (Nairobi, Kenya: Department of Agricultural Economics, University of Nairobi, 1976). See also International Agricultural Development Service, *Preparing Professional Staff for National Agricultural Research Programs* (New York: IADS, 1979).

23. For a useful description of the difficulties involved in linking the development of research and training institutions and of the effects of cycles in the support of such efforts on the institutionalization of research and training capacity in Argentina, Colombia, and Peru, see Jorge Ardilla, Eduardo Trigo, and Martin Piñeiro, *Human Resources in Agricultural Research: Three Cases in Latin America* (San José, Costa Rica: Instituto Interamericano de Cooperation para la Agriculture, March 1981).

Chapter 8

The Private Sector
in Agricultural Research

The relative contributions and appropriate roles of public-and private-sector agricultural research have generated a great deal of discussion and some occasional controversy in recent years.[1] Yet, our knowledge about the contribution of the private sector to the advancement of agricultural science and technology is quite limited. This chapter focuses, somewhat narrowly, on private-sector agricultural research in the United States. Most of the issues discussed are, however, relevant to agricultural research policy in other developed countries and in many developing countries.

The conventional view, shared by many public- and private-sector scientists and research administrators, is that there is substantial complementarity between public-sector and private-sector agricultural research. A relatively large share of the new knowledge and new technology generated by public-sector research reaches the ultimate users as inputs—machines, materials, and services—supplied by the private sector. A relatively small share is transferred or communicated directly to the users—whether they are farmers, consumers, or public agencies—without the private sector's participation.

In the United States, there has been growing pressure for a reexamination of the linkages among public-sector research and development activities.[2] Populist critics of the federal-state agricultural research system have suggested that private-sector objectives have been weighted too heavily in the selection of public-sector research portfolios. Official critics have argued that public-sector research in fields such as postharvest technology and mechanization duplicates or replaces research that appropriately belongs in the private sector and is wasteful of public resources. Industry representatives have occasionally been critical of state and federal research in the field of product development and have suggested that the research resources devoted to product development could be more effectively used for more basic research.

In this chapter, I focus on three areas in which the boundaries between the public sector and the private sector appear to be under pressure. These are (1) mechanization, (2) plant variety, and (3) insecticide research and development. First, however, I would like to attempt to specify my understanding of the principles that are appropriate in allocating research efforts between the public sector and the private sector. I would also like to examine briefly what is known about the allocation of resources to agricultural and food research in the private sector.

PUBLIC RESEARCH AND PRIVATE RESEARCH

Three criteria have been used to gauge the appropriate roles of the public and private sectors in agricultural research. The primary rationale for the public sector's investment in agricultural research has been that in many areas incentives for private-sector research have not been adequate to induce an optimum level of research investment—that the social rate of return exceeds the private rate of return because a large share of the gains from research are captured by other firms and by consumers rather than by the innovating firm. This is most obvious in the case of basic or supporting research in plant pathology and physiology, soil physics and chemistry, insect ecology, and many other areas. It also holds in many more-applied areas.

Even the largest farm could not capture more than a small share of the gains from improvements it might make in agronomic and farm management practices. Seed companies have not, in the past, been able to capture more than a small share of the gains from the development of new crop varieties.

A second criterion for public-sector investment in agricultural research is its complementarity with education. There is a strong synergistic interaction between research and education in the agricultural sciences and technology. This relationship is so strong that in many fields research productivity carries a strong penalty when conducted apart from graduate education. And graduate education could hardly be effective if both students and teachers were not engaged in research.

A third argument that has been made for public-sector research is that it has contributed to the maintenance or enhancement of a competitive structure in the agricultural production, input, and marketing sectors. There is, for example, considerable evidence that the flow of new technology from public-sector research and development has contributed to competitive behavior in the seed and fertilizer industries.

There is, however, no reason to believe that the optimum level of public-sector research investment implied by the three criteria would be the same. Where incentives for private research investment are particularly strong, the level of public-sector research implied by the training criterion could exceed the level implied by the social rate of return criterion. Where incentives for private research investment are weak, the level of public-sector research implied by the social rate of return criterion may substantially exceed the level implied by the training criterion. There would seem to be relatively few situations in which enhancement of market structure is capable of serving as the primary rationale for public-sector research investment. Any structural reform gains can best be regarded as a positive spillover from decisions made primarily on other grounds.

HOW MUCH DOES THE PRIVATE SECTOR SPEND?

Research and development expenditures by the private sector in support of the U.S. food and fiber system have been poorly documented. The best single set of data available are the 1965 estimates developed by the Agricultural Research Institute (ARI).[3] The most comprehensive description of the sources and flow of public and private research funds available for the United States is the data for 1976 assembled for the World Food and Nutrition Study. (See figure 8.1.) The data on public-sector research were reasonably firm. However, the private-sector research expenditure estimates (apparently based on data prepared by the Bureau of the Census for the National Science Foundation) appeared to omit several important areas of research.[4]

My own estimates based on the data from the 1965 and 1975 ARI studies and from trade associations' data suggest that private-sector research and development expenditures by firms engaged in agricultural input and agricultural processing and distribution industries must have been well in excess of $1 billion in 1976. A more recent estimate assembled by Illona Malstead suggested that private-sector research expenditures in support of the agricultural system by firms

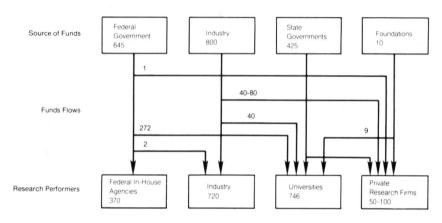

Figure 8.1. Research and Development Funds ($ millions) for the U.S. Food Research Systems, 1976. (From Commission on International Relations, National Research Council, *World Food and Nutrition Study: Supporting Papers*, vol. V, *Agricultural Research Organization* [Washington, D.C.: National Academy of Sciences, 1977], p. 22.)

in agricultural input and agricultural processing and distribution industries were in the neighborhood of $1.6 billion in 1979. (See table 8.1.)

The data on research and development presented in table 8.1 include expenditures on some items (market research and development, for example) that do not contribute directly to agricultural production or even consumer satisfaction. Yet, these are also important research expenditures that are not captured in the data in table 8.1. A study recently completed at the University of Wisconsin indicated that during the period between 1969 and 1977 less than 10 percent of the patents granted for processes and products in the food industry originated with the U.S. food industry.[5] Firms outside the food industry accounted for 37 percent; foreign firms, for 21 percent; individuals, for 31 percent; and the government, for 1 percent. Although similar data have not been assembled on farm machinery patents, it is well known that a relatively high percentage of inventions leading to patents in the farm machinery industry emerge outside of formal research and development laboratories and shops. My guess is that a complete accounting of research and development in support of the agricultural input and food processing and distribution industries for 1979 would show total expenditures in excess of $2 billion.

Table 8.1. Estimate of Industry Expenditures (in $ millions)
for Farming and Postharvest Efficiency

	1978	1979
Farm input Industries:	*751-846*	*814-909*
Plant breeding	55-150	60-155
Pesticides	290	339
Plant nutrients	3	3
Total plants	751-846	402-497
Animal breeding	49	55
Animal health (mostly veterinary drugs)	99	99
Animal feed and feed ingredients	30	133
Total animals	178	225
Farm equipment and machinery	*225*	*225*
Processing and distribution:		
Farm produce transport equipment	40	45
Food processing machinery	85	100
Food processing	350	400
Tobacco manufacturing	40-50	40-50
Natural fiber processing	10	20
Packaging materials	116	129
Total processing and distribution	641-651	734-744

Sources: Illona Malstead, "Agriculture: The Relationship of R&D to Federal Goals" (Washington, D.C.: 1980, mimeographed paper). Sources consulted in constructing the estimates included the Agriculturla Research Institute, the National Agricultural Chemical Association, the Animal Health Institute, the American Feed Manufacturers' Association, the Farm and Industrial Equipment Institute, the Southeastern Poultry Association, and the National Association fo Animal Breeders as well as individual company representatives.

In spite of the tentative data available, a few generalizations can be made:

- Private-sector research and development expenditures in support of the food system by firms in the agricultural input and food marketing and distribution industries have grown more rapidly than public-sector agricultural research since 1965. In 1965, private-sector research and development probably accounted for about 55 percent of the total public- and private-sector research in support of the food system. By 1979, the private sector's share was probably in the neighborhood of 65 percent. In both 1965 and 1979, research effort apparently was divided about equally between agricultural inputs and food marketing and distribution.
- The animal drug industry, with over 12 percent of its sales dollars allocated to research, and the pesticide industry, with about 10

percent of its sales dollars allocated to research, are the most research intensive of the agricultural input industries. The farm machinery industry, which allocates about 3 percent of its sales dollars to research, apparently is slightly above the average for all U.S. industry in its research and development intensity. The fertilizer industry, on the other hand, spends well below 1 percent of its sales dollars for research and development. The food and kindred products industry apparently allocates less than half of 1 percent of its dollars to research and development. Although percentage of sales is a commonly used measure, it tends to be somewhat biased because of the great variation in costs of raw materials as a share of the sales dollar among industries. A more adequate measure would be percentage of research and development expenditures per dollar of value added.

• Research and development activity in the agricultural input and food industries is focused primarily on product development. The food industry, for example, focuses its efforts on new product development but buys its process technology from suppliers. Similarly, the agricultural chemicals industry focuses its efforts on new products but not on the processes used to produce the products. The definitions of product innovation and process innovation are, however, quite arbitrary. A product innovation in the farm machinery industry becomes a process innovation when adopted by agricultural producers. The definition of research and development by the private sector is also somewhat arbitrary. Some of the activities classified as product development might be more appropriately classified as marketing.

• There are quite striking differences in the relative emphases given to the several fields of science and technology between the public and private sectors and, within the public sector, between the USDA and the state experiment stations. Close to two-thirds of private-sector research and development is concentrated in the physical sciences and engineering. Public-sector research is much more heavily concentrated in the biological sciences and technology. At the state agricultural experiment stations, approximately three-quarters of the research is in the biological sciences and technology. The share of the research dollar allocated to social science research related to agriculture is less than 3 percent in the private sector and less than 10 percent in the public sector.

PUBLIC FUNDING OF MECHANIZATION RESEARCH

The boundary between private- and public-sector research on mechanization has been a continuing area of concern and controversy in both developed and developing countries. Two issues have been prominent in these discussions. One is whether public-sector research duplicates or displaces private-sector research. A second is who gains and who loses as a result of the introduction of new technology.

The critics of public-sector mechanization research have adopted what might be termed a substitution view and the defenders, an organic view.[6] The substitution view emphasizes labor displacement; the organic view emphasizes contribution to production. These two views, though useful for polemic purposes, have little analytical content. Most of the operations that can be accomplished by mechanization in the past have been achieved at extremely low wage rates by hand labor or by some combination of hand labor, animal power, and simple machines. In Taiwan and Holland, for example, extremely high levels of cropping intensity were achieved before the introduction of mechanized motive power. The gains from mechanization emphasized by proponents of the organic view are rarely realized in practice in economies with low or stagnant wage rates.[7] However, in countries with higher wage rates, the labor force's constraints on production become more serious.

By and large, the development of mechanical equipment and motive power has been induced by long-term increases in the price of labor. (See chapter 2.) In industrial economies, mechanization in agriculture has been primarily a response to a declining agricultural labor force rather than a major cause of the decline. There is no doubt, however, that mechanization has contributed to the displacement of employment in specific locations and in specific commodities. The same equipment may play both roles. The cotton harvester was clearly a response to the loss of labor to urban-industrial employment. But, during the latter steps of its adoption, the cotton harvester may have displaced labor that was "left behind."[8]

In the United States, recent concern about the public funding of mechanization research has been focused by the controversy about the role of the University of California in the integrated development of mechanical and biological technology for the production and harvesting of tomatoes and a number of specialty crops.[9] The rationale for public-sector support for research and development of machinery in California has relied on two arguments. One is that many of the specialty crops grown there are unique to California and to other parts of the Southwest. Because of their limited acreage and the small market potential for machines used to produce these commodities, the argument has been made that there was little incentive for private-sector research and development. A second rationale has been made in terms of improving the ability of California farmers to compete with producers in other areas in the United States, as in the case of tomatoes, or with imports from other countries, as in the case of strawberries. Both arguments are, in principle, consistent with the traditional social rate of return criterion for public-sector support for agricultural research.

The history of the development of the tomato harvester extended over a period of about three decades. (See table 8.2.) Its development was speeded up by the end of the bracero program that had permitted Mexican citizens to enter the United States to harvest crops and to do other fieldwork. A combination of yield-increasing biological technology and labor-displacing harvest technology enabled California producers to capture a large share of the processed tomato market from the older producing areas in the Midwest and the East. Initially, this led to an increase in labor demand in tomato production. As the process continued, however, it led to the displacement of harvest laborers. The implications for state economic development were ambiguous. A study by Andrew Schmitz and David Seckler indicated that the gains to producers exceeded the losses to workers by a substantial margin.[10] But, since the losers typically were poor and the gainers were relatively well off, a major issue of equity is involved. And the inequity was exacerbated by the fact that, though the gains were sufficient to compensate for the losses, compensation was not made. (See chapter 13.)

In late 1979, U.S. Secretary of Agriculture Bob Bergland responded to the concern about public-sector support for mechanization by announcing that the USDA would no longer support research leading to the "replacement of an adequate and willing work force with machines."[11] Bergland clarified two general principles: The USDA will not put federal money into research when a careful review and analysis clearly indicate (1) that the direct and immediate benefits will go to a limited number of locales while neither serving the national interest nor benefiting the general public and (2) that the research poses a direct or an indirect threat to social stability, the natural resource base, the environment, the national security, or the economic well-being of a significant number of citizens. Bergland immediately qualified his remarks by indicating that he had no objection to research and development designed to ease the drudgery of work rather than to replace workers with machines—a distinction that is not feasible either technically or analytically. This analytical lapse does not negate the significance of Bergland's remarks, which set in motion a badly needed review of the appropriate roles of public- and private-sector research.

The importance of publicly supported research on agricultural

Table 8.2. Historical Events in the Development of the Tomato Harvester

Period	Historical Event
World War II	Labor shortage creates impetus for tomato harvester.
1941-42	Conveyer machine developed in Pennsylvania.
1942	A. M. Jongennel, a California tomato grower, suggests to G. C. Hanna that the university develop a tomato plant that could be harvested by machine.
1943	Professor Hanna at the University of California begins research for tomato plants with desirable properties. It was reported that a blacksmith in Holt, Calif. was also building a tomato picker for a canning firm in Stockton.
Late 1940s	Pear-shaped tomato plant that ripens at same point in time as is adaptable to machine harvest is released.
1949	Professor Coby Lorensen begins work on the tomato harvester at the University of California, Davis.
1951-52	Tomato growers in California experiment with conveyer systems.
1956	California Tomato Growers' Association grants funds to the University of California for work on the tomato harvester.
1958	Michigan State University team constructs a tomato harvester. University of Florida team develops conveyer belt machine; Food Machinery Corporation and H. D. Hume Company fund work on a tomato harvester at Purdue University.
1959	University of California successfully completes and licenses the Blackwelder Manufacturing Company to undertake its commercial manufacture. Blackwelder had been working closely with the university in the development of the tomato harvester.
1960	Blackwelder builds 15 harvesters. Five types of machines are tested, and 1,200 tons of pear-shaped tomatoes are harvested by machine.
1961	Mechanical tomato harvester first used ·commercially. There are 25 University

Table 8.2.—*Continued*

Period	Historical Event
	of California/Blackwelder machines in growers' hands; and 6 other firms test machines, including 2 large farm machinery corporations (Hume and Food Machinery Corporations).
	Professor Hanna releases the F-145 tomatoes at the University of California A strain selected from this variety is basic to the mechanization of tomatoes, in California.
1964	Public Law 78 (*bracero* program) is terminated.
1965	Tomato growers in California obtain special dispensation to import Mexican workers for the harvest. The first major action of the National Farm Workers' Association later to become the United Farm Workers, assumes the form of a grape strike in Delano.
1967	Federal minimum wage legislation extended to agricultural workers.
1970	Adoption of mechanical tomato harvester completed in California. Attempt by California Tomato Growers' Association to implement a government marketing order to control the supply of processing tomatoes fails.
1974	California Tomato Growers' Association is recognized by processors as grower bargaining association for negotiating forward pricing contracts.
1975	California law (Agricultural Labor Relations Act) grants agricultural employees the right to form unions and bargain collectively. Electronic sorter (which reduces the necessary labor on the harvester from about 15 to 5) used commercially in tomato harvest (on 30 machines).
1976	California law ensuring unemployment benefits for agricultural workers. United Farm Workers attempt to organize labor in the harvesting of tomatoes. Mass adoption of electronic sorter eliminates approximately 5,000 workers from the harvesting of tomatoes.

Source: Alain de Janury, Philip LeVeen, and David Runsten, *Mechanization of California Agriculture: The Case of Canning Tomatoes* (Berkeley, Calif.: University of California, Department of Agricultural and Resource Economics, September 1980), pp. 97-99.

mechanization has been blown completely out of proportion by both its critics and its defenders. The USDA Science and Education Administration has been able to identify only about $1 million in the USDA research budget that could lead to significant labor displacement. This is less than half of 1 percent of USDA-funded agricultural research. The harvest mechanization issue has been more important as a battleground for the forging and testing of political power than the economic significance of public-sector research on harvest mechanization would suggest. Neither the critics nor the defenders of public-sector mechanization research have been overly scrupulous in their adherence to careful analytical distinctions or in their respect for the accuracy of the empirical content of their arguments.

Those who have supported public-sector mechanization research have frequently attempted to interpret Bergland's remarks as an attack on mechanization rather than as a more-limited questioning of the rationale for public-sector funding of mechanization research. Those who have protested the mechanization research have seemed less concerned with the displacement of labor than with the failure of the University of California Experiment Station to consider farm workers as well as farm operators and processors as part of its clientele and the failure in California, and in the United States generally, to provide parity of treatment to farm laborers in social insurance and collective-bargaining legislation. (See chapter 13.)

The implication of the mechanization debate for public and private research on mechanization seems reasonably clear. The private sector has been an effective source of new mechanical technology. Lack of knowledge has seldom been a serious constraint, even in the developing countries, on advances in mechanical technology for agriculture. Some observers believe that the Blackwelder Company could have developed a fully effective tomato harvester by the early 1970s even without the University of California's participation.

Experience with the development of the mechanical cucumber harvester in Michigan is somewhat less conclusive. Agricultural engineers at Michigan State University collaborated with engineers at the H. J. Heinz Company and the FMC (Food Machinery and Chemical Corporation) in the development of a mechanical cucumber harvester. However, the commercially successful machine developed by the Wilde Manufacturing Company owed relatively little to MUS engineering research effort. In the development of both the tomato harvester and the cucumber harvester, the demand-side impetus for commercial development associated with the end of the bracero program appeared to be at least as important as the supply-side public-sector research effort.

The social rate of return rationale behind the federal government's substantial support for the research and development of mechanical equipment for agriculture is weak. The rationale for support by the state experiment stations must be based primarily on local rather than national benefits. It appears, therefore, that the rationale for most public-sector mechanization research at the land-grant universities must draw more heavily on the training criterion than on the social rate of return criterion.

DEVELOPMENT AND PROTECTION
OF PLANT VARIETIES

Plant variety development is another area in which the appropriate mix of public and private research has been the subject of continuous dialogue and occasional controversy. In the United States, the industry evolved along two relatively distinct lines. One branch supplied seed and other plant materials to home gardening; the other supplied seed for crop production. The private sector tended to be the predominant source of new varieties for the home gardens and horticultural crops. The public sector tended to be the dominant supplier of new varieties for field crops. This pattern began to change with the advent of hybrid corn. Control of inbred lines capable of serving as the parents for superior hybrids enabled the private sector to establish proprietary control over new hybrid corn varieties. In the mid-1970s, over 80 percent of the corn and sorghum varieties used in commercial production and approximately 70 percent of the sugar beet and cotton varieties were private varieties. Over 80 percent of the ryes, wheats, oats, soybeans, rices, barleys, peanuts, dry edible beans, and forage grasses were public varieties.[12]

In addition to their direct involvement in plant variety development through the public support of plant breeding, governments in most countries are also involved to at least some degree in seed multiplication and distribution. This involvement may include issuing seed certificates and variety recommendations (and/or controls) and protecting varieties. Some countries assume direct responsibility for seed multiplication and distribution through a government agency or public corporation.[13] (See figure 8.2.)

Seed-certification programs were developed in Europe and North America during the 19th century. Their objective was to improve and maintain the quality of seed offered for sale by settting standards and testing for varietal purity, seed viability, and freedom from weed contamination. Systems of variety recommendations by crop

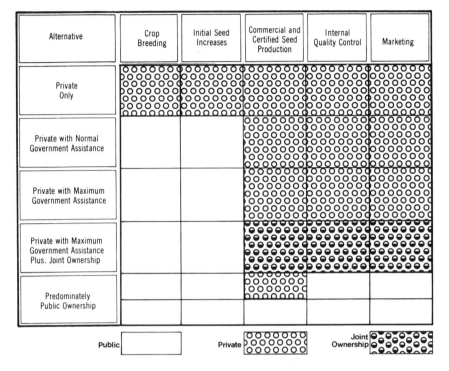

Figure 8.2. Alternative Methods for Developing Seed Enterprises. (Reprinted with permission from Johnson E. Douglas, ed., *Successful Seed Programs: A Planning and Management Guide* [Boulder, Colo.: Westview Press, for International Agricultural Development Service, 1980], p. 84. Copyright by International Agricultural Development Service, New York.)

research institutions and extension services were developed somewhat later in order to provide farmers with objective information, based on comparative yield trials, on varietal performance. A few countries, primarily in Europe, have begun to issue lists of permissible varieties. Varieties must demonstrate their "value for cultivation and use" in order to be included on the list. Varieties not on the approved lists cannot be offered for sale.

The rather complex involvement of the public sector in crop varietal development, seed certification, and varietal recommendations that prevails in the United States can be illustrated by using the state of Minnesota as a example. There are individual variations from state to state, but the general features are similar throughout the United States.

The Minnesota Agricultural Experiment Station, operated by the University of Minnesota, supports a substantial crop variety research and development program. When the performance of a new public variety of soybeans, for example, warrants seed multiplication and release to farmers, breeder seed is released to the Minnesota Crop Improvement Association for multiplication. The Crop Improvement Association, a nonprofit corporation whose owners are mostly farmers and small seed companies, has also been designated by the state legislature as the official seed-certifying agency in Minnesota.

The seed-certification program was designed to ensure the varietal purity and quality of seed developed by both public and private breeders. There are three stages in the certification program. The first stage is the production of foundation seed from breeder seed. In the case of public varieties, this first multiplication is carried out by the Crop Improvement Association. When adequate foundation seed is available, it may be purchased from the association and multiplied by approved seed growers. The third stage is the production of certified seed by approved certified seed growers. In order to ensure the quality of the seed grown by the seed growers, the Seed Improvement Association carries out field inspection of the seed crops and conducts laboratory tests for purity and viability on samples taken from the growers' processed seed before issuing certificates and labels.

A similar process is followed in certifying seed that is developed by a private breeder. However, for private seed, the Crop Improvement Association carries out only those inspection and testing activities involved in the certification process. Private breeders can produce and market noncertified seed as long as the labeling is accurate. The primary value of the certification program to private breeders is that it assures the buyer of seed quality. For hybrid corn and sorghum seed on which the private seed company attempts to develop its own brand name as a symbol of quality, the seed-certification process is not employed. However, most private seed companies continue to use the state certification programs for open-pollinated varieties.

The system described above has been remarkably effective in the generation and distribution of new seed varieties in the United States. It has also been an important factor in maintaining a competitive structure in the seed industry. However, it is highly dependent upon the level of public support for varietal development at the state agricultural experiment stations. It has been argued by the larger seed companies that allowing new plant varieties to be patented or given patent-equivalent protection would serve as an incentive for them to expand greatly their research and varietal development efforts. As evidence, they point to the substantial investments they have made in the development of hybrid corn and sorghum varieties as a result of the natural protection provided by the control of the inbred parent lines.

The establishment of patent or patentlike protection of "breeders' rights" in new varieties was well established in several European countries before World War II. Other countries passed similar legislation in the 1940s and 1950s. In 1961, the International Convention for the Protection of New Varieties of Plants was signed in Paris by several European countries. The International Convention established the International Union, which included 12 nations whose national laws grant similar rights to plant breeders. Nationals of each member country can obtain reciprocal rights in other member countries. In 1978, the International Convention was revised to make it possible for other countries to join the International Union. In addition to the present members, Canada, the United States, Ireland, Japan, Mexico, and New Zealand have signed the International Convention and have indicated their intention to join the International Union as soon as they have made appropriate revisions in their national legislation. In several countries, including the United States and Canada, this declaration of interest has stimulated a vigorous, and sometimes acerbic, debate about the advantages and dangers of legislation protecting plant varieties.

In the United States, the first legislation on plant variety protection was passed in 1930. The Plant Patent Act of 1930 extended patenting rights to breeders of a number of asexually reproduced plants. At that time, efforts to extend patenting privileges to breeders of sexually produced plants was rejected by farmers and scientists who feared that such legislation would inhibit the free exchange of genetic materials and lead to excessive concentration in the seed industry. Certified seed growers objected on the grounds that it would lead to concentration of seed production in the hands of larger firms capable of maintaining their own breeding programs.

In 1970, the U.S. Congress passed the Plant Variety Protection Act that had been developed by a committee of the American Seed Trade Association. The 1970 act covered seeds, transplants, and plants of about 350 species. Several "soup vegetable" species (tomatoes, carrots, cucumbers, okra, celery, and peppers) were omitted because of the objections made by canners and freezers. There was also substantial opposition to the act from scientsts and breeders at the state agricultural experiment stations and the U.S. Department of Agriculture who felt that adequate consideration had not been given to the scientific and legal considerations of such factors as the variability in crop performance and genetic drift under different environmental conditions and the exchange of information and germ plasm among public and private breeders.

In 1979 ahd 1980, hearings were held in both the U.S. House of Representatives and the Senate on a bill to amend and extend the 1970 Plant Variety Protection Act.[14] The amendments would bring the "soup vegetables" not included in the 1970 act under the provisions of the act, extend the period of protection from 17 to 18 years in conformity with the provisions of the International Convention, and tighten up several provisions of the act to facilitate its more effective administration.

Experience with the 1970 act resulted in a number of changes in perception regarding the effect of variety protection. It has been concluded by most participants in the debate that the act has encouraged expansion of plant-breeding efforts in the private sector. Fears that the act would lead to excessive litigation have not been realized. A good deal of the opposition to the variety protection by public-sector breeders has disappeared. And the canning and freezing industry did not register opposition to inclusion of the "soup vegetables" under the act. As of March 1979, the Plant Variety Protection Office of the USDA's Agricultural Marketing Service, which administers the act, had received 953 applications, of which 62 percent were for field crops, 23 percent for vegetables, and 10 percent for flowers. About 35 species of field crops were represented. About one-third of the applications were for public varieties and about two-thirds were for private varieties.

There remain a number of legitimate concerns about the implications of plant variety protection. The 1979 and 1980 House and Senate hearings served to focus these issues but did little to resolve them. Testimony presented by the USDA tended to be uninformative and at times inept. Testimony by opponents often relied more on populist rhetoric than on analysis. Testimony by the spokespeople from the seed industry regarding the favorable effects of the 1970 act in stimulating commercial plant-breeding activities rested more on simple assertions than on presentations of evidence, and the evidence that was presented did not always support the assertions being made.

Marketing restrictions

A major issue that ran through the hearings was whether the restrictive provisions that have emerged in the European seed-certification and marketing programs are a necessary and likely consequence of plant variety protection. European seed legislation has employed a "value for cultivation and use" criterion in order to shorten the list of varieties that may be offered for sale. Varieties are evaluated by official government tests. Failure to meet established standards of performance precludes the sale and use of the variety. Individual countries have developed national lists, and the Common Market or European Economic Community (EEC) has developed Common Catalogs of approved and recommended varieties. These lists are based on performance tests conducted by the national seed agencies. Two motivations appear to be involved. One is protecting farmers and market gardeners from planting seed of inferior varieties; the second is reducing the duplication resulting from using different names for the same varieties in different countries. Elimination of the duplication in seed names will make it possible to monitor more effectively trade among countries in breeder seed, foundation seed, and certified seed.

An important source of flexibility in the U.S. legislation is that it does not preclude the marketing of seed that does not go through the registration process. Under the U.S. legislation, a breeder applying for registration must indicate whether the new variety will be sold by variety name as a class of certified seed or by variety name only. When the certification option is chosen, it becomes unlawful for anyone except farmers to sell the variety unless the seed is certified. Certifying agencies can certify seed of a protected variety only for the owner of the certificate or authorized growers. In practice, there has been a somewhat surprising tendency to elect to follow the certification option and to forgo the advantages of greater flexibility in seed production and marketing from bypassing certification in order to take advantage of the assurance that the certification provides about seed quality and the legal advantage certification might provide if it became necessary for a breeder to take action to protect a registered variety from being marketed without authorization.

It seems quite clear that the restrictive features of European seed-marketing legislation are not a necessary complement to plant variety protection. It would be possible to operate the restrictive marketing provisions, as part of a seed-certification and marketing program, for example, in the absence of plant variety protection. An effective plant variety protection program could exist in the absence of restrictive seed-certification and marketing programs. Yet, there is a valid basis for concern that legal and administrative pressures could lead to exessive rigidity in the application of distinctiveness, uniformity, and stability criteria and impose an excessive burden on crop variety developers.

Variety conservation and development

A second major issue in the debate about plant variety protection is whether or not it poses a threat to the maintenance and conservation of genetic resources or to the exchange of information and breeding materials among public- and private-sector plant breeders. It is now generally accepted by those concerned with the conservation of genetic materials that it is the increased availability of higher-yielding crops, rather than varietal protection itself, that is the major threat to the continued existence of varietal diversity in the historic centers of origin of crop species. The appropriate response to this concern is more adequate support for crop exploration, for seed storage and preservation, and for associated taxonomic and cyto-genetic research. It is no longer reasonable to expect that the traditional landraces will be maintained in their original forms by subsistence farmers.

The question about the free flow of scientific information among public and private breeders has not been resolved. At present, a great deal of germ plasm is released by the USDA and the state experiment stations that does not have variety status. It is elite germ plasm or parental lines useful for breeding stock but not for immediate cultivation. It has no legal status under the Plant Variety Protection Act.

Experience at home and abroad indicates that private breeders and their companies want the public breeder to turn over superior germ plasm to them as soon as it is identified and that they rather than the breeder should produce the final product. This situation exists today in Denmark and Holland. In these countries, public breeders are not allowed to produce cultivars.[15]

A partial response to this concern is the U.S. legislation's assurance of the right to use protected varieties for research and breeding purposes. There are no restrictions on the use of a variety registered under the Plant Variety Protection Act in breeding programs of public or private breeders.

There can be no question about the importance of maintaining viable public-sector crop-breeding programs until it is possible to monitor and to evaluate the effects of plant variety protection on the performance of private-sector varietal improvement efforts. The experience with hybrid maize, on which proprietary inbred lines have provided even more secure protection than the provisions of plant variety legislation, is not entirely reassuring with respect to the capacity or the efficiency of private-sector breeding programs. Inbred lines developed by public-sector breeders continue to account for over 50 percent of hybrid maize seed production in the United States.[16] Private development of cotton varieties became important before the Plant Variety Protection Act was enacted. The rapid growth of both public and private wheat-breeding efforts since 1970 seems more related to advances in breeding methodology than to plant variety protection. Breeding effort and time is sometimes extended to economically unimportant traits such as glume color or ligule length in order to meet variety distinctiveness criteria. In order to achieve the genotypic purity needed to meet distinctiveness criteria, breeding efforts may have to be extended to several generations beyond the F^5 or F^8 generations that might be needed to realize the economically significant characteristics. Even in corn breeding, in which economic incentives have been the strongest, the private-sector companies continue to make only limited investments in the supporting sciences, such as genetics, plant pathology, plant physiology, and related areas.

The U.S. experience during the first 10 years under the Plant Variety Protection Act raises at least as many questions as it answers about whether or not the act is, in its present form, a fully effective instrument for inducing optimal private investment in varietal development. Plant breeding and related research in the supporting sciences continue to be an area in which the social rate of return to public-sector research remains high. (See chapter 10.) Thus, the social rate of return rationale for public support remains strong even in the countries that have attempted to strengthen private incentives through plant variety protection legislation. As institutional innovations provide more secure property rights and private-sector varietal development efforts continue to evolve, there will be a need to reevaluate continually the appropriate division of labor between public- and private-sector breeding programs.

This argument for the development and support of a substantial public-sector crop-breeding effort is even stronger in the developing countries than in the developed countries. The developing countries have more to gain than to lose by avoiding the complex institutionalization of the seed production and marketing structure that has emerged in some of the more-developed countries. It is to their advantage to avoid subjecting themselves to restrictions that would limit or delay access to the results of advances in either private- or public-sector crop breeding in the developed countries or in other developing countries.

INNOVATION AND REGULATION
IN INSECT CONTROL

Before the end of the 19th century, technical progress in American agriculture was directed primarily toward the achievement of increases in output per worker through advances in mechanical technology.[17] The closing of the frontier in the last decade of the 19th century created an impetus for more intensive efforts to develop yield-increasing technologies. By the mid-1920s, new biological and chemical technologies were beginning to exert a measurable impact on agricultural production. (See chapter 2.)

As the new chemical technologies were introduced, concern emerged about their effects on agricultural producers and consumers. The focus of much of this concern was on insecticides. The development of insecticide technology and the institutions designed to achieve more effective social control over the use of insecticides represent a particularly significant illustration of the induced technical and institutional innovation perspective outlined in chapter 2. The increased scarcity of land and water resources is continuing to create a demand for yield-increasing technologies in agriculture—for higher output per acre (or per hectare). The rising value that society is placing on human life and on environmental amenities has led to a demand for more effective social controls over the development and use of new technology. There is, therefore, a demand for both technical and institutional innovations that are capable of resolving the conflicts between the forces inducing yield-increasing techniques and the forces inducing more effective social controls.

The weights given to the two sources of demand for technical and institutional innovation depend on the priority that a particular society places on agricultural production and on health and amenity objectives. A poor society is likely to give greater weight to production objectives, while a wealthy society is likely to place greater weight on health and amenity objectives. The focus of this section is primarily on the attempts to resolve these conflicts in the United States.

First- and second-generation insecticides

The extensive use of chemical insecticides in agricultural production is a relatively recent phenomenon. In the United States, the discovery in the late 1860s that the arsenic compound Paris green could be used to control the Colorado potato beetle spawned a period of rapid growth in insecticide development and use. The first-generation of insecticides was dominated by several arsenic compounds (calcium arsenate, lead arsenate, white arsenic) and copper sulfate. Small amounts of organic insecticides such as pyrethrum, rotenone, and nicotine sulfate were also used. Use of commercial insecticides in agriculture was confined largely to high-value-per-acre crops such as fruits, vegetables, potatoes, and cotton. Evidence of the excessive use of insecticides on some fruits and vegetables was one of the important factors leading to the 1906 and 1938 food and drug legislation.[18] (See table 8.3.)

The second generation of insecticides began in 1939 with the discovery of the effectiveness of DDT (dichloro-diphenyl-trichloroethane) as an insecticide.[19] DDT was followed by the development of a series of new synthetic organic pesticides, including other chlorinated hydrocarbons, the organic phosphates, and the carbamates. The relatively low cost and effectiveness of these new materials led to their wide use for controlling insects in crop and forest production and controlling insect vectors of disease and nuisance insects. The basis for the effectiveness of the new insecticides was not only the high toxicity exhibited by many of the compounds. Some of the chlorinated hydrocarbons exhibited a persistence that increased their effectiveness and reduced the need for repeated application. Some of the more acutely toxic organic phosphates hydrolize rapidly and leave little or no residue after use. Others are systemic in growing plants and provide extended protection. Some of the carbamates that are effective in insect control have low acute toxicity to warm-blooded animals. The new synthetic organics thus included a broad selection of desirable insect-control characteristics for use in agricultural production and public health. By the mid-1950s, the use of the older first-generation insecticides had been sharply reduced.

During the 1950s and the 1960s, evidence accumulated to indicate that the benefits associated with the use of the new organic insecticides were obtained at a substantial cost. These costs were measured in terms of the development of insecticide resistance in target populations; the destruction of predatory, parasitic, and other beneficial insects; the direct and indirect effects on bird, fish, and other wildlife populations; and the acute and chronic effects on human health. During the middle and late 1950s, concerns about these effects were documented in the professional literature. In the early 1960s, public concern about these spillover effects was galvanized by Rachel Carson's description of the effects of the new insecticides in *Silent Spring*.[20] Although the message of the book has been characterized as overdrawn, there was no question that most of the concerns expressed in the book were valid.[21]

Table 8.3. Evolution of Pesticide Regulation, 1900-1975

- Residues of pesticides in or on food are subjected to regulation.
  - 1906: Food and Drugs Act establishes federal jurisdiction over food treated with pesticides and traded in interstate commerce.
  - 1938: The Federal Food, Drug, and Cosmetic Act (FFDCA) allows tolerance levels of chemicals on fresh fruits and vegetables but is unable to set tolerances due to complexity of FFDCA.
  - 1950-1952: The House Select Committee to Investigate the Use of Chemicals in Food and Cosmetics holds hearings. Recommended that FFDCA be amended to require that chemicals employed in or on food be subjected to safety testing such as required then for new drugs and meat products.
  - 1954: The Miller Amendment of FFDCA gives the Food and Drug Administration a new mechanism to set tolerances of pesticidal residues on food.
  - 1958: The Delaney Amendment to FFDCA declares food additives found to induce cancer in man or animal to be unsafe.
- The pesticide consumer receives protection against misbranded and adulterated pesticides.
  - 1910: Insecticide Act of 1910.
  - 1947: The Federal Insecticide, Fungicide and Rodentcide Act (FIFRA).
  - 1972: The Federal Environmental Pesticide Control Act (FEPCA) amends the 1947 FIFRA (additional amendments 1975). Most uses of DDT are eliminated by the EPA.
- Concern about the environmental impacts of pesticides becomes a significant policy issue.
  - 1962: Rachel Carson's *Silent Spring* sparks a still-lingering controversy.
  - 1963: The President's Science Advisory Committee issues *The Use of Pesticides*, which generally concurs with Carson's concerns.
  - 1969: The Secretary's Commission on Pesticides and their Relationship to Environmental Health (the "Mrak Commission") recommends limiting uses of DDT and DDD because of adverse environmental effects and restricting other persistent pesticides to "essential" uses.
  - 1970: The National Environmental Policy Act (NEPA) requires a statement of environmental impact for every major federal activity "significantly affecting the quality of the human environment."
  - 1975-1976: The National Academy of Sciences/National Research Council (Kennedy Committee) report on *Pest Control: An Assessment of Present and Alternative Technologies* recommends a major effort to develop alternatives to chemical pest control technologies.
- Long-standing occupational hazards associated with the use of pesticides are regulated at the federal level.
  - 1970: The Occupational Safety and Health Act (OSHA) establishes a public policy on the provision of safe and healthful working conditions.
  - 1972: FEPCA provides that pesticides are to be classified for "general use" and "restricted use." Restricted chemicals must be applied by or under supervision of a certified applicator.
- Response to concern that legislation and regulatory regimes have become unduly restrictive.
  - 1978: The Federal Pesticide Act of 1978 amended the 1972 FEPCA to permit conditional registration, relaxes requirements for registration of minor use pesticides, transfer enforcement responsibilities to states, and tighter label requirements.

Source: Adapted from National Academy of Science, *Pest Control: An Assessment of Present and Alternative Technologies*, "Contemporary Pest Control Practices and Projects: The Regent of the Executive Committee," vol. 1 (Washington, D.C.: National Academy of Science, 1975).

The search for a pesticide policy

The debates over pesticide policy and pest-control strategy in the 1960s and 1970s differed sharply from the pre-World War II debates.[22] The earlier debates had focused primarily on the direct effects of residues on human health and safety in insecticide manufacture and application. The later conflicts centered on the emergence of insecticide resistance among target insect populations and on a broad spectrum of environmental spillover effects. Moreover, the environmental movement developed the political capacity during the 1960s to challenge directly the dominance of the "iron triangle" —the U.S. House and Senate agricultural committees, the USDA and land-grant-university science establishment, and the organized agricultural interests—in the formulation of pesticide policy. As a result of this polarization, the debate was frequently cast in terms of a political battle between "agriculturalists" and "environmentalists" for the control of pesticide policy.

Professional debates within the entomological profession about pesticide policy were also intense. They were also more complex than the political debates. During the mid-1960s, substantial efforts were under way in the USDA and in some of the land-grant universities to develop pest-control techniques that were less dependent on the use of pesticides. Dr. Edward Knipling, who had participated in the testing of DDT for use in agricultural production and public health during the early 1940s, was a strong proponent of alternatives to chemical control. In the early 1950s, Knipling, then director of the USDA Entomological Research Division, and a group of his colleagues developed the sterile male concept that led to the elimination of the screwworm fly as a livestock pest in the Southeast and to its virtual elimination in the Southwest.[23]

By the late 1950s, there was general agreement among both agriculturally and ecologically oriented entomologists about the desirability of moving insect-control policies toward a reduced reliance on chemicals. The debates within the entomology community tended to center on the relative merits of (1) integrated pest-management (IPM) strategies, which would attempt to contain insect populations below an economic threshold, and (2) total population management (TPM), which proposed the suppression or eradication of total populations of insects over large areas through a combination of biological and chemical techniques.

The bureaucratic response to political pressures from the environmental movement, the agricultural interests, and the general public was to study the problem (table 8.3). If consensus regarding pesticide policy and pest-control research strategy could be extracted from the scientific community, the controversy over the institutional innovations needed to manage the development and use of pesticides might be muted.

The most significant of the several studies was the National Academy of Science/National Research Council's assessment of pest-control strategies (the Kennedy Committee report).[24] The study was organized under the Environmental Studies Board, rather than the Agricultural Board or the Agricultural Research Institute, in a deliberate effort to involve scientists with both environmental and agricultural orientations. A wide representation of agricultural, environmental, and behavioral scientists participated in the study. The committee examined control strategies for the entire spectrum of pest organisms, excluding only organisms directly pathogenic to humans. Perhaps the most significant perspective to emerge from the study was that problems inherent to chemical pest control, such as the development of resistance, were in and of themselves so serious that a major effort was needed to develop alternatives to existing chemical control technologies. The significance of this finding is that it was not necessary for the burden of justifying the stress on the importance of alternatives to chemical pest control and more effective regulatory regimes to rest primarily on environmental and health considerations. The long-term viability of agricultural pest control implied a need for alternative strategies. These conclusions were more firmly based in the case of insect-control than weed- or plant-pathogen-control strategies.

By the mid-1970s, there was broad agreement among both environmentally and agriculturally oriented entomologists that, to the extent feasible, less reliance should be placed on use of the synthetic organic pesticides. Development and use of nonpersistent and more selective (narrow-spectrum) insect-control agents was to be encouraged. Greater effort was to be devoted to the development of "biorational" chemical agents and microbial agents and biological control and "cultural" control procedures.[25] Substantial disagreement remained, however, over the precise meaning of these concepts and the potential for developing the concepts into effective insect management or control technologies.

It was recognized that each of these elements in the strategy had significant limitations. Many of the more specific and less persistent chemical agents were more toxic to humans and animals than the broad-spectrum chemicals. Only a few cases of demonstrated field effectiveness and commercial feasibility of the biorational chemical agents and microbial agents were avilable. Biological control strategies, including genetic modifications, as in the sterile male technique, were believed to be potentially applicable to no more than 30 to 50 percent of currently important insect pests. Lack of proprietary control and technical marketing problems suggested that there were limited incentives for the commercial development of biological controls.

It was believed, however, that many of the limitations on individual components in the strategy could be overcome by an "integrated" approach that would employ a combination of agents and management techniques. It was also clear to those who shared the emerging consensus that the strategies would require more adequate public support for research on insect population dynamics and behavior, more effective biorational and biological control technologies, and more effective institutional support for integrated pest management.

The legislative response to the concerns about the negative spillover effects on human health and the environment was a new set of institutional innovations designed to monitor and to regulate the development and use of insecticides. The history of these developments is illustrated in table 8.3. Among the most significant were the passage of the National Environmental Policy Act (1970) and the Federal Environmental Pesticide Control Act (1972) and the reorganization of major pesticide monitoring and control functions under the Envrionmental Protection Agency (1970). The mobilization of the political resources necessary to bring about these innovations was a direct product of the widespread concern about the impact on human health and the environment of the widespread use and misuse of the second generation of broad-spectrum insecticides.

The scientists, reformers, and legislators who worked together to bring about the new legislation and to establish the new administrative agencies at both the federal and state levels anticipated that their efforts would be followed by the introduction and use of a set of new third-generation pest-control methods, along the lines of the emerging scientific consensus discussed above, that would satisfy both the needs of agricultural producers and the concerns of the reform movement. These expectations have been partially realized. There has been a sharp decline in the use of chlorinated hydrocarbons and a modest decline in the intensity of insecticide application per crop acre.

Even under the best of circumstances, the new control strategy was, however, confronted with very serious economic constraints. The economic constraints on the development of the new third-generation controls stem from both natural and institutional sources. The new control agents tend to be specific to a single insect species or a single crop. This imposes a natural limit on the size of the market and on the potential return from the development of the more specific as compared to the broader-spectrum materials. The increase in the regulatory burden imposed by the new legislation and the administrative interpretation of legislative intent by the Environmental Protection Agency has imposed an institutional burden on the development of third-generation agents.[26] The definition of pesticides in the Federal Environmental Pesticide Control Act (1972) appeared to require EPA registration not only of chemical agents but also of living organisms. The EPA interpreted the legislation as requiring similar registration requirements for the traditional chemical insecticides, the biorational agents, and the microbial methods (regulation of the use of predatory and parasitic biological agents remained with the USDA). Even the insect-attacking bacterium, *Bacillus thuringiensis* (BT), which was enjoying some commercial success before the passage of the 1970 and 1972 legislation, was forced to go through an expensive retesting program that eventually led the small firm that had introduced it to withdraw from insect-control research and development. As of 1980, however, BT was being commercially marketed by two major companies (Abbott and Sandoz). The high cost of conducting tests and the confusion and delay in the processing of registration applications have also caused firms to discontinue the development of chemical agents that have desirable properties, such as low rates of application or a narrow spectrum of pesticidal activity. The impact of these natural and institutional constraints had led, at least in the short run, to the replacement of several of the more persistent but less toxic organochlorines with less persistent but more toxic organophosphates rather than with third-generation control agents or methods.

In addition to its more stringent regulatory regime for new pesticide material, the EPA announced in 1972 an intensive review of the toxicological properties and residue characteristics of the insecticides then in use. A procedure, termed rebuttable presumption against registration (RPAR), has been established to review pesticide registration on the basis of data that imply an acute or chronic hazard based on EPA-established criteria. When a material is placed on the RPAR list, it becomes subject to a hazard- and benefit-assessment process similar to that required for a new material. The final outcome of the RPAR process is a decision by EPA to continue unmodified registration, to require changes in application methods and use patterns (such as dosage rates), to restrict or to cancel use for some purposes, or to cancel registration for all uses. In recent years, the cost of such procedures has been estimated to run in the $500,000 range and above. Many firms have simply decided to discontinue materials with limited markets rather than to undergo the cost of reregistration. Because of the importance of some of these materials, the USDA and the state agricultural experiment stations have found it necessary to divert resources that could have been devoted to the development of new third-generation strategies to conducting the studies necessary to maintain registration of existing materials. They also have been devoting greater resources to the research and testing necessary to obtain tolerances and registrations for new minor-use insecticides and control methods (the IR-4 program).

Since the early 1970s, there has been a significant decline in the productivity of scientific efforts in pesticide research and development. This decline can be measured in terms of (1) diversion of scientific resources from synthesis and screening and field-testing and development to environmental testing and residue analysis and registration and administration; (2) increases in the costs, measured either in financial or scientific manpower terms, of developing new insecticides; (3) extension of the time between scientific discovery and registration; and (4) reduction in the number of new insecticide products and uses registered each year.[27] (See tables 8.4 and 8.5.) The information that would permit an assignment of the share of the decline that is due to increased regulatory stringency or to other factors such as the greater difficulty of developing new pest control agents that are sufficiently superior to existing materials to earn a place in the market is not yet available. There is little question, however, that changes in the regulatory regime has been a significant factor in the diversion of scientific and technical resources.

The objectives of the new environmental legislation of the 1970s and the objectives of the regulations promulgated by the EPA are valid. There is strong public demand for the human health and environmental quality objectives that the law and the agency are trying to achieve. There are also very large and growing demands on the productivity of agricultural resources. The objectives the institutional innovations seek to realize are, however, being achieved at a significant cost measured in terms of diversion of scientific effort and loss in agricultural productivity. This cost is being paid in terms of new environment- and health-compatible materials that have not been invented or brought to the market as well as the removal of existing materials from the market.

Table 8.4. Percentages of Research and Development Expenditures Allocated According to Function by American Pesticide Manufacturers[a]

Year	Synthesis and Screening (1)	Field-testing and Development[b] (2)	Toxicology and Metabolism[c] (3)	Formulation and Process Development (4)	Registration and Administration[d] (5)	Environmental Testing and Residue Analysis[e] (6)
1969	33	31	12	19	5	0
1970	31	32	13	18	6	0
1977	21	20	13	20	10	16
1978	22	20	13	21	9	15

Source: Council on Agricultural Science and Technology, Impact of Government Regulation on the Development of Chemical Pesticides for Agriculture and Forestry (Ames, Iowa: CAST, Report No. 87, November 1980).
a. From NACA industry surveys of 33 to 37 companies, 1967 to 1978.
b. Does not include residue analysis
c. Includes environmental and wildlife toxicology in 1976-1978.
d. Registration only until 1974; registration and administration thereafter.
e. Primarily environmental testing after 1970; environmental testing and residue analysis in 1976-1978.

Table 8.5. Time Required for Registration and Number of New Registrations of Agricultural Chemicals, 1969-1978

Year	Registration Time (NACA Date)		Registration Time (EPA Date)		NACA	EPA[c]	EPA[c]
	Submission to Approval (months) (1)	Discovery to Market (months) (2)	Total Time (months) (3)	Agency Review (months) (4)	New Agricultural Entities (number) (5)	New Agricultural Entities (number) (6)	New Agricultural Use Registrations (number) (7)
1969	18[a]	70[a]			10	9	3,437
1970	11	77			11	3	1,029
1977	29	110	19	7.2	3	2	730[b]
1978	32	69			2	3	450

Source: Council on Agricultural Science and Technology, Impact of Government Regulation on the Development of Chemical Pesticides for Agriculture and Forestry (Ames, Iowa: CAST, Report No. 87, November 1980).
a. Average for 1967 to 1971.
b. Fiscal year accounting.

Some of the problems created by the 1972 act are being corrected. A major emphasis of the Federal Pesticide Act of 1978 was to simplify and to shorten the registration process in order to relieve some of the excess regulatory burden resulting from the administrative and judicial interpretation of the 1972 act. The amendments permitted conditional registration and the waiving of efficacy requirements for new agents, relaxed requirements for registration of minor-use pesticides, provided a mechanism for the transfer of registration data among firms, transferred enforcement responsibility to the states, and tightened label requirements (table 8.3).

In December 1978, the EPA announced its intention to issue new registration guidelines specific to biorational and biological pest-control agents. Among the important features of the new guidelines are a sequential "tier-testing scheme" designed to assure that only the minimum data necessary to make sound regulatory decisions will be required. The procedure would eliminate the need for submission of extensive data for those biological pest control agents determined to be safe on the basis of the first tier of tests. The provisions of the 1978 act have not, however, been fully implemented. The new guidelines that were to be issued by the EPA in the fall of 1980 were still available only in draft form at the end of the year. The provisions designed to reduce the need for duplication of information needed for registration by compensating the originator of the data by another registration petitioner were still being contested in the courts. Although it is still too early to determine how sensitively these procedures will be administered, it seems clear that it is the intent of the EPA to provide a more favorable environment for the development of the third-generation insect-control strategy.[28]

The evaluation of an insect-pest control strategy capable of achieving consistency between agricultural production and environmental and health objectives will involve at least three major elements. One element must be the implementation of modified regulatory procedures that provide greater encouragement for the private sector to develop traditional chemical pesticides, biorational chemical agents, and biological agents that are compatible with health and environmental objectives. Explicit attention should be given to the special research, development, marketing, and use characteristics of the biorational and biological agents.

A second element will involve expanded support for research and development by public-sector institutions of biological and cultural control agents and procedures. This will involve additional public-sector support for research on the biology of insect predator and host populations, the identification of insect-control agents and the design of control technologies, the breeding of insect-resistant crop varieties, and the design of cultural practices to depress insect populations.

A third element will involve public-sector support for the design and operation of insect-population-management programs. Substantial progress has been made in institutionalizing capacity for integrated pest management of the insect pests affecting almonds in California and those affecting tree fruits in Michigan, for example. But for most crops and in most areas full operational potential for integrated pest management has yet to be realized.[29] The appropriate roles of publicly supported efforts by the state agricultural experiment stations and extension services and of privately organized laboratory, scouting, and consultation services are not yet clear. At present, the public support for the research and extension inputs into integrated pest management programs tends to be based on an unstable combination of user fees and project grants. Under some conditions, spillover and free-rider constraints imply that institutional innovations such as pest-control districts will be required for providing the incentives needed for effective integrated pest-management programs. And it seems quite clear that the effectiveness of both public and private efforts will depend on more adequate funding for publicly supported pest- and weather-monitoring and information systems and on public support for pest-management training.

PERSPECTIVE

The broad implications of the case studies presented in this chapter seem clear. Research directed toward advancing mechanical technology will remain a low priority in the allocation of public-sector research resources. Institutional innovations that provide greater incentives for private-sector variety development will result in a reallocation of public-sector research from plant breeding to supporting research in genetics, pathology, physiology, and related areas. The tensions that are inherent in attempts to achieve resolution of productivity, safety, and amenity objectives in pest control and management are likely to draw more public resources into research on pest-control technologies and institutions.

In areas examined in this chapter—mechanization, plant variety, and insecticide research and development—the appropriate balance between public-sector and private-sector research and development is being subjected to intensive scrutiny. Yet, it seems apparent that the systems approach to the design of integrated mechanical, biological, and chemical technologies will not become less important in the future. Integrated biological and chemical harvesting technologies have received some attention in the past. In the future, interdisciplinary approaches will increasingly be employed in the design of cropping systems to minimize soil loss, more efficient pest-management systems, and controlled environment growth systems, among other areas. It is unlikely that the scientific and technical skills involved in the design of such systems will be completely available within either public- or private-sector institutions. Institutional innovations capable of facilitating effective patterns of communication and collaboration between public- and private-sector research organizations must be developed and maintained. These innovations must be capable of ensuring the viability and integrity of research and development institutions in both sectors.

NOTES

1. This chapter represents a revision and an expansion of a paper I presented to the annual meeting of the Agricultural Research Institute, St. Louis, October 15, 1980. I am indebted to Arnold Aspelin, Don Duvick, Molly M. Frantz, David Godden, Maureen K. Hinkle, J. C. Headley, Harley J. Otto, John H. Perkins, Robert W. Romig, Andrew Schmitz, and W. E. Splinter for comments they made on an earlier draft of this chapter.

2. For a useful review, see chapter 6, "Industry Groups," in Don F. Hadwiger, *The Politics of Agricultural Research* (Lincoln, Neb.: University of Nebraska Press, in press).

3. H. L. Wilcke and H. B. Sprague, "Agricultural Research and Development by the Private Sector of the United States," *Agricultural Science Review*, 5 (1967), pp. 1-8; H. L. Wilcke and J. L. Williamson, *A Survey of U.S. Agricultural Research by Private Industry* (Washington, D.C.: Agricultural Research Institute, 1977).

4. The NSF-U.S. Bureau of the Census data are reported in U.S. National Science Foundation, *Research and development in Industry, 1977* (Washington, D.C.: U.S. Government Printing Office, November 1979 and earlier issues). The National Science Foundation's estimate of research expenditures for the food and kindred products, agricultural chemicals, and farm machinery and equipment industries in 1965 was $316 million. The ARI's 1965 estimates exceeded the NSF's by 45.6 percent. If this same ratio prevailed in 1975, the estimated expenditures, using the ARI's definition, would be $922 million. The 1975 ARI survey reported expenditures by the firms surveyed of $530 million. However, the 1975 survey by the ARI was much less complete than the 1965 survey and does not lend itself to the detailed analysis that was possible with the 1965 survey. Since the ARI disposed of the 1965 survey files, they are no longer available for detailed analysis and comparison with more complete survey materials that might become available in the future. Our knowledge of the rates of return to private-sector agricultural research remains undeveloped. For a discussion of some of the theoretical issues involved in measuring returns see Willis L. Peterson, "A Note on the Social Returns to Private Research and Development," *American Journal of Agricultural Economics*, 58 (May 1976), pp. 324-26.

5. Willard F. Mueller, John Culbertson, and Brian Peckhorm (with Julie Croswell and

Philip Kaufman), *Market Structure and Technological Performance in the Food Manufacturing Industries* (Madison, Wis.: University of Wisconsin, College of Agricultural and Life Sciences, January 1980).

6. This distinction is based on Hans P. Binswanger, *The Economics of Tractors in South Asia: An Analytical Review* (New York: Agricultural Development Council; and Hyberabad, India: International Crops Research Institute for the Semi-arid Tropics, 1980). I have used the term "organic view" instead of the term "net-contributor view" employed by Binswanger.

7. Binswanger concluded, from his very detailed review in *The Economics of Tractors in South Asia*, that "the tractor surveys fail to provide evidence that tractors are responsible for substantial increases in intensity, yields, timeliness, and gross returns on farms in India, Pakistan and Nepal. Such benefits may exist but are so small that they cannot be detected and statistically supported even with massive survey research efforts. This is in sharp contrast to new varieties or irrigation where anybody would be surprised if he failed to find statistically significant yield effects, even in fairly moderate survey efforts" (p. 73).

8. One of the few rigorous tests of the displacement hypothesis concludes that the adoption of the cotton harvester in the American South can largely be explained by the replacement, rather than the displacement, of labor. See Willis Peterson and Yoav Kislev, "The Cotton Harvester in Retrospect: Labor Displacement or Replacement?" (St. Paul: University of Minnesota, Department of Agricultural and Applied Economics, mimeographed paper, January 1981).

9. See Wayne D. Rasmussen, "Advances in American Agriculture: The Mechanical Tomato Harvester as a Case Study," *Technology and Culture*, 9 (October 1968), pp. 531-43, for a description of the systems approach to the integrated development of biological and mechanical technology.

10. Andrew Schmitz and David Seckler, "Mechanized Agriculture and Social Welfare: The Case of the Tomato Harvester," *American Journal of Agricultural Economics*, 52 (November 1970), pp. 469-77. See also W. H. Friedland and H. E. Baxter, "Destalking the Wily Tomato: A Case Study in Social Consequences in California Agricultural Research" (Davis, Calif.: University of California, Department of Applied Behavioral Research, 1975); Alain de Janvry, Philip Le Veen, and David Runsten "Mechanization in California Agriculture: The Case of Canning Tomatoes" (Berkeley, Calif.: University of California, Department of Agricultural Economics, mimeographed paper, September 1980).

11. Bob Bergland, "The Federal Role in Agricultural Research," a paper presented at USDA Science and Education Administration Conference, Reston, Va., January 31, 1980 (USDA Office of the Secretary, 262-80). See also Eliot Marshall, "Bergland Opposed to Farm Machinery Policy," *Science*, 708 (May 9, 1980), pp. 578-80.

12. D. G. Hanway, "Agricultural Experiment Stations and the Variety Protection Act," *Crops and Soils*, 30 (5), pp. 5 and 6; and 30 (6), pp. 5-7.

13. For useful guides to the technology and organization of the seed industry, see Johnson E. Douglas, *Successful Seed Programs: A Planning and Management Guide* (Boulder, Colo.: Westview Press, 1980), and Norman W. Simmonds, *Principles of Crop Improvement* (London: Longman, 1979), pp. 205-41. For an excellent review of the development and policy issues in the area of breeders' rights, see S. O. Fejer, "The Problem of Plant Breeders Rights," *Agricultural Science Review*, 4 (3) (1966), pp. 1-7. For a populist but not always an accurate account, see P. R. Mooney, *Seeds of the Earth: A Private or Public Resource?* (London: International Coalition for Development Action, 1979). I have also benefited from David Godden's "Plant Variety Rights: Some Economic Issues," a paper presented to the annual conference of the Australian Agricultural Economics Society, Adelaide, February 1980.

14. U.S. House of Representatives, *Plant Variety Protection Act Amendments*, hearings

before the Subcommittee on Department Investigations, Oversight, and Research of the Committee on Agriculture, Serial No. 96-CCC (Washington, D.C.: U.S. Government Printing Office, 1980). Much of the testimony at the Senate hearings duplicated the testimony presented at the House-hearings.

15. R. K. Downey, "Plant Breeders' Rights—A View Point of a Plant Breeder in the Public Sector," in *Plant Breeders' Rights*, Eighth Annual Meeting of the Canada Grains Council, Winnipeg, April 5 and 6, 1977. See also R. G. Anderson, "Plant Breeders Rights— Promise or Problem," a paper presented to Seminario Panamericano de Semilas, Buenos Aires, Argentina, November 20, 1980.

16. M. S. Zuber, "Corn Germ Plasm Base in the U.S.—Is It Narrowing, Widening or Static?" *Proceedings of the 30th Annual Corn and Sorghum Research Conference* (December 9-11, 1975), pp. 277-87; D. N. Duvick, "Major United States Crops in 1976," *Annals of the New York Academy of Sciences*, 287 (February 25, 1977), pp. 86-96.

17. In preparing this section, I have drawn on a literature review by L. Upton Hatch, *U.S. Pesticide Policy: A Review* (St. Paul: University of Minnesota, Department of Agricultural and Applied Economics, mimeographed paper, March 1981.

18. For a useful review of the history of pesticide innovation and regulation prior to World War II, see James Wharton, *Before Silent Spring: Pesticides and Public Health in Pre-DDT America* (Princeton, N.J.: Princeton University Press, 1974). In the 1930s, public attention was focused on the problem of food adulterants and additives and pesticide residues by Arthur Kallott and F. J. Schlink, *100,000,000 Guinea Pigs* (New York: Vanguard Press, 1932).

19. John H. Perkins, "Reshaping Technology in Wartime: The Effect of Military Goals on Entomological Research and Insect-Control Practices," *Technology and Culture*, 19 (April 1978), pp. 169-86.

20. Rachel Carson, *Silent Spring* (New York: Fawcett, 1962). See also Frank Graham, Jr., *Since Silent Spring* (Boston: Houghton Mifflin, 1970).

21. For a balanced scientific perspective, see the review by LaMont E. Cole, "Rachel Carson's Indictment of the Undue Use of Pesticides," *Scientific American*, 207 (December 1962), pp. 13-17. For an early attempt to assess the benefits and costs of pesticide use in economic terms, see J. C. Headley and J. N. Lewis, *The Pesticide Problem: An Economic Approach to Public Policy* (Baltimore: Johns Hopkins University Press, 1967).

22. In this section I draw heavily on John H. Perkins, "Strategies for Pest Control: The NAS Report," *Agricultural Research Institute Proceedings and Minutes*, 28 (October 1979), pp. 105-21; and John H. Perkins, "The Quest for Innovation in Agricultural Entomology," in *Pest Control: Cultural and Environmental Aspects*, David Pimental and John H. Perkins, eds. (Boulder, Colo: Westview Press, 1980), pp. 23-80.

23. Beginning in the mid-1950s, the USDA entomology research program, under Dr. Knipling's leadership, began to try to place greater emphasis on alternatives to the chemical control of insects. It was not until the mid-1960s, however, that the Congress provided the financial support needed to strengthen research on the ecological effects of pesticides and on alternative methods of pest control. See E. F. Knipling, *The Basic Principles of Insect Suppression and Management* (Washington, D.C.: U.S. Department of Agriculture, Agricultural Handbook No. 512), pp. 9-11.

24. National Research Council, Environmental Studies Board, *Pest Control: An Assessment of Present and Alternative Technologies*, vols. 1-4 (Washington, D.C.: National Academy of Sciences, 1975). See also the earlier "Mrak Commission" report: U.S. Department of Health, Education, and Welfare, *Report of the Secretary's Commission on Pesticides and Their Relation to Environmental Health*, parts 1 and 2 (Washington, D.C.: U.S. Department of Health, Education, and Welfare, December 1969).

25. The term "biorational" has been used to refer to the use of chemical analogues for

insect pheromones (sex attractants) and hormones (growth regulators). For a more complete discussion, see Carl Djerassi, Christina Shih-Coleman, and John Diekman, "Insect Control of the Future: Operational and Policy Aspects," *Science*, 186 (November 15, 1974), pp. 596-607. In the EPA regulatory literature, the term "biological pesticides" is used to include both "true biological agents" and "naturally occurring biochemicals such as plant growth regulators, insect pheromones and hormones."

26. For a journalistic discussion of this topic, see William Tucker, "Of Mites and Men," *Harper's*, 257 (August 1978), pp. 43-58. See also Dajarassi, Shi-Coleman, and Diekman, "Insect Control of the Future." See also the retrospective review of *Silent Spring* by Paul R. Ehrlich, *Bulletin of the Atomic Scientists*, 35 (1979), pp. 34-36.

27. For a more complete discussion, see Council on Agricultural Science and Technology, *Impact of Government Regulation on the Development of Chemical Pesticides for Agriculture and Forestry* (Ames, Iowa, Report No. 87, November 1980).

28. See the announcement by the Office of Pesticide Programs, "EPA Regulation of Biological Pesticides" (Washington, D.C.: Environmental Protection Agency, December 4, 1978). A draft of the new guidelines was released for discussion on September 29, 1980.

29. For an optimistic review of the status and potential of integrated pest management, see Dale R. Bottrell, *Integrated Pest Management* (Washington, D.C.: Council on Environmental Quality, December 1979). See also Office of Technology Assessment, *Pest Managerial Strategies*, vols. 1 and 2 (Washington, D.C.: Office of Technology Assessment, 1979). For a careful scientific assessment, see Carl B. Huffaker, ed., *New Technology of Pest Control* (New York: Wiley, 1980). For a discussion of some of the institutional issues, see A. Dan Tarlock, "Legal Aspects of Integrated Pest Management," in *Pest Control: Cultural and Environmental Aspects*, David Pimental and John H. Perkins, eds. (Boulder, Colo.: Westview Press, 1980), pp. 217-36. See also G. E. Allen and J. E. Bath, "The Conceptual and Institutional Aspects of Integrated Pest Management," *Bioscience*, 30 (October 1980), pp. 644-58.

Chapter 9

Institutional and
Project Funding of Research

In this chapter, I analyze the implications of two alternative research funding schemes.[1] The first, the institutional research support system, provides funds to support the research program of a particular research institution. The selection of the research program is developed by the administration and staff of the experiment station or research institute. The second, the project research grant system, provides support through project grants to individual scientists or research teams. The allocation of research effort typically is determined by the granting agency.

Examples of institutional support for research are numerous. Institutional research was the traditional instrument employed to support federal and state mission-oriented research in the fields of defense, agriculture, natural resource exploration, industrial standards, and related areas in the United States prior to World War II. The program of federal support for agricultural experiment stations on a formula basis is a prototype of the institutional research system.[2]

In the private sector, the Ford Foundation has provided institutional support for the research program of Resources for the Future since the early 1950s. Institutional grants were also made by the Ford Foundation to the Social Science Research Council to enable the council to operate the Foreign Area Fellowship Program, a program of project grants. The U.S. Agency for International Development (USAID) provides institutional grants in support of the research programs of the international agricultural research centers that are part of the Consultative Group on International Agricultural Research.

The competitive grant program of the National Science Foundation, in which grant requests received from individual scientists and research teams are evaluated by peer panels, is a prototype of the project research system. Most of the major foundations and a number of the federal agencies operate project research grant programs in addition to institutional research programs. The project research grant mechanism emerged as a major instrument for linking academic research with mission-oriented federal agencies in the United States during the late 1940s and early 1950s.[3]

Each of the two research management schemes has strengths and weaknesses. The system constraints and the reward system in the two programs are very different from each other. The next section of this chapter presents an analysis of the behavior of individual scientists in an environment characterized by the availability of centralized institutional grant systems. In the following section, the behavior of research administrators under the two research support systems is analyzed. Finally, some inferences are drawn regarding the implications of the two methods of supporting research for a research system's efficiency. The analysis provides a theoretical framework for the empirical evaluation of the gains and losses from the two systems of research management. The inferences from the theoretical analysis appear consistent with experience. Nevertheless, they should be regarded as tentative until the results of more careful empirical research become available.

THE RESEARCH SCIENTIST

First, the implications of the two systems of research support for the behavior of the individual scientist should be considered. What objectives does the individual research scientist attempt to maximize? And how do the project and institutional research grant systems impinge on the behavior of the individual scientist?

The research scientist has been depicted as both a hero and a villain.[4] My perspective is more modest.[5] The typical research scientist, as a result of inclination and conditioning, is prepared to accept a rather high degree of deferred gratification within the professional reward system. In the immediate postdoctoral years, the scientist is usually willing to defer immediate financial reward for an appointment that assures continued professional development, preferably documented by evidence of research productivity in the form of published papers. If professional productivity is accompanied by reasonable advancement in rank and earnings, the initial research orientation is reinforced.

If research productivity lags or is not accompanied by advancement in salary or rank, there is often a shift in emphasis toward research that has a more immediate or short-term payoff. This shift toward more-applied research may also be associated with the scientist's movement to another institution whose program is more oriented toward applied research. During this process, the mid-career scientist may also develop certain entrepreneurial skills. These skills may run in the direction of a capacity to generate research support from funding agencies and/or to mobilize the interest and energies of colleagues to focus their efforts around problems of scientific or technical diversification that require a team effort. Development of entrepreneurial skills often comes at the expense of disciplinary capacity and direct involvement in research.

This description is, of course, highly simplified. And our ability to develop formal analytical models that are rich enough to interpret this simplified description of the scientist's behavior is inadequate. However, the description does help to identify several key elements of the individual scientist's objective function that appear to be directly related to behavior under the institutional and project grant systems. The simplified behavioral model of the scientist's behavior has the following characteristics.

- Each researcher maximizes some utility function by simultaneously allocating time among teaching, entrepreneurial activities such as seeking research support, actively work on research, and a set of nonwork-related leisure activities.
- Research output is determined by the level of research support available to the researcher and the amount of time spent actively working on research. The researcher faces diminishing returns both in the production of research funds and in the conduct of research.
- The transaction costs incurred by the individual researcher in obtaining project research support are greater than those incurred in obtaining institutional research support. However, the level of available institutional research support will often be a funding constraint. That is, institutional research funds may be rationed and become unavailable before the researcher achieves the equilibrium allocation of time that maximizes utility.
- Income is a positive function of research output and teaching and/ or extension (or public service) output depending on the way in which a particular scientist's appointment is defined.

The goal of the analysis is to determine the characteristics of the demand for research funds. Although several factors influence the individual scientist's demand for research funds, only the cost of obtaining grant and formula funds is considered in the analysis. Other factors include, of course, the maximum amount of institutional funds available, the preferences (utility function) of the researcher, the production function of research output, and the rewards from research output. Varying the cost of obtaining research funds allows examination of the impact of alternative research management schemes.

The results of the model depend critically on the observation that the transactions costs of the project research system are significantly higher than those of the institutional research system. In other words, the marginal productivity of time spent seeking institutional research funds is significantly higher than that spent seeking project research funds. Most university scientists can quote examples of the colleague who has spent much more time preparing grant requests for the support of summer research than was spent actually carrying out the grant-supported research. It also seems consistent with experience that the greater the degree of centralization of research management, the greater the costs of grant seeking and of the review and allocation process. Several implications can be drawn from the view of the research scientist's motivation outlined in this section.

First, at any institution there are some scientists whose research productivity is not constrained by the limit on institutional research funds. This group includes individuals with low research motivation who are very productive in nonresearch activities relative to research. It also includes those with a very high marginal product of effort in research relative to the research funds available to them. The former group specializes in teaching, extension, and administrative activities and the latter group includes the pencil-and-paper theorists.

Second, there probably are a number of individuals at any institution who would demand more institutional research funds if such funds were not rationed. But they demand no competitive funds because the marginal cost is greater than the increment to income (or utility) the added funds would bring. The time, effort, and uncertainty associated with the preparation and submission of grant proposals make such proposals not worth the effort.

Third, some researchers try to put more personal effort into their research activity in order to increase their research output. Such effort is subject to diminishing returns as the labor intensity of their research effort increases in response to the limitation on funds and the high cost of competitive funds. This increase in labor intensity on the part of the researcher may also involve changes in the research output mix. It may result in more theoretical work, more use of secondary data, smaller and fewer experiments, and smaller and fewer instances of primary data generation.

Fourth, another effect of the discontinuous demand for research funds is an increase in the labor devoted by the scientist to other aspects of his or her appointment: teaching, extension, administration. Consulting activities may also expand. The extent of this diversion of effort to nonresearch activities depends on the strength of the individual's demand for income and the marginal wage rate for the nonresearch activities. Thus, the limit on institutional research funds and the high cost of competitive funds not only increase the labor intensity of the research enterprise but they also have the effect of increasing the labor intensity of *all* other aspects of the individual's job.

Fifth, those scientists who are characterized by strong professional motivation or demand for income seek more project research funds. The share of the scientist's time devoted to grantsmanship rises and the share devoted to research declines. For those researchers who have strong professional motivation and a high level of energy, research output may be high, relative to researchers whose activity is constrained by institutional funding limitations, even though substantial time is devoted to grant-seeking activity.

Finally, given a distribution of grantsmanship skills among scientists, the greater the difference between the marginal costs of institutional research and project research funds, the greater the likelihood that specialization of function will arise among scientists. Those who are more skilled as grantsmen specialize in grantsmanship and reduce their actual research effort. As they acquire more project research funds, they add scientists who are more skilled in research to their research team.

THE RESEARCH ADMINISTRATOR

An assessment of the implications of a project research grant system relative to an institutional research grant system must consider the effects of the two systems on the behavior of the research administrator as well as on the behavior of the individual scientist. This section considers how the project and institutional research grant systems impinge on the behavior of the research administrators. It is particularly concerned with the impact on the behavior of the director of an agricultural experiment station, a research center, or a laboratory located within a university environment.

There is a dearth of systematic knowledge about the decision-making processes used by research administrators. Most of the knowledge that there is is based on casual observation and introspection. Nevertheless, it does seem feasible to specify some of the elements that enter into the decisions of research administrators and scientists.

The typical research manager tends to have a view of the world that places a heavy weight on the value of new knowledge and new technology and places a low weight on both the direct and indirect costs of research. The administrator visualizes an almost endless frontier waiting to be discovered—and with limited financial, physical and professional resources. The administrator's standing, both within his or her own institution and among outside collegiate and clientele constituencies, is directly related to the ability to assemble or to develop a research staff that is recognized for the quality of its work or its value to clientele constituencies. Within public-sector institutions, where the salary structure is bureaucratically determined and has little flexibility at the top, prestige considerations carry greater weight. In the private sector, where the output of the research laboratory is evaluated more directly in terms of the enhancement of the firm's profits, productivity is given more weight.

The net effect of these considerations leads a research director to measure success in terms of the capacity to acquire additional resources and the ability to utilize these resources productively. The measurement of the quality or the value of research output at the level of the individual scientist or research team is highly subjective, and the management of research enterprises is highly collegial. These factors often lead administrators to place greater emphasis on the quality of the major input, professional personnel, relative to the value they place on research output. Emphasis on more effective monitoring of research output is greatest in those cases in which there is strong clientele pressure. Clientele pressure on research management is reasonably strong in state and federal agricultural research programs in the United States because of the close feedback loop between farmers, legislators, and research institutions. It is even stronger in most private-sector research institutions.

The above description of the elements that enter into the objective function of the research administrator or manager is not inconsistent with the utility function of the bureau manager that has been suggested in the literature on bureaucratic behavior.[6] In that literature, it is assumed that the bureau manager's utility is a function of the bureau's output and the bureau's discretionary budget. In the case of the agricultural experiment station or the agricultural research institute, one can interpret bureau size in terms of research staff and the output of applied research that is valued by the research institution's clientele. Discretionary budget can be interpreted in terms of funds to support more fundamental (basic or supporting) research and for related professional activities (seminars, symposiums) that serve to enhance the capacity of the research staff or the prestige of the research unit. It represents funds that are not required to meet salaries and other overhead items.

If incremental growth in research funding is primarily in the form of project rather than institutional research support, as suggested in the introductory section, one effect will be to reduce the discretionary resources available to state experiment station directors. A higher proportion of institutional support funds will have to be devoted to salary and overhead items. Capacity to mobilize resources for problems of significance at the state or regional level will be reduced.

This description of the utility function of the research manager involves an even greater simplification of a complex reality than the description of the utility function of the individual research scientist. It does, however, appear to capture important elements of the research manager's motivation.

I can now proceed to examine the effects of the institutional research grant and the project research grant systems on the behavior of individual scientists and research administrators and the efficiency or the productivity of the research system.

SOME RESEARCH MANAGEMENT
AND POLICY IMPLICATIONS

The analysis presented above suggests that the institutional research management system entails lower transaction costs than the project research management system. On the other hand, the project research management system facilitates greater control of the research output mix from a central (national) level. Advocating one system in preference to the other involves a trade-off between central control and system efficiency.

System efficiency

It would be surprising, however, if optimization of the objective functions of individual research scientists and administrators would, under most circumstances, lead to system efficiency. System efficiency is a function of the institutional environment, including the structure of incentives and constraints, in which scientists and administrators carry out their professional responsibilities. When one examines the implications for the behavior of scientists and research administrations under the two systems, some rather clear-cut empirical generalizations concerning system efficiency emerge.

First, the external project research grant system diverts efforts by individual scientists from research to grant-seeking and other related entrepreneurial activities and imposes higher research administration costs on both the research-funding and performing institutions. (See figure 9.1.) As a case in point, in 1978 the U.S. Department of Agriculture, which then administered a competitive grant program of $15 million, received over 1,100 research proposals involving funding requests for over $200 million. Similar ratios have been noted for other grant programs. In addition to the amount of time devoted to the preparation of unfunded grant proposals, there is also the very substantial amount of time devoted to peer review and administration. A careful comparative study in the field of materials research concluded that institutional grants involved much less total administrative cost per dollar spent (by the federal government and the university) than individually funded project grants.[7]

Excessive allocation of scientific effort to grant-seeking activity is clearly induced by a major structural feature of the competitive grant system. To the individual researcher, the supply of project research funds appears relatively elastic (with respect to effort devoted to grant seeking). Each individual project is small relative to the resources available to the granting agency. In the aggregate, however, the supply of research funds is relatively inelastic in the short run. An increase in the number of project submissions results in an increase in the share of research resources devoted to grant seeking relative to research and an increase in the bureaucratic resources devoted to grant management. It also may result in a smaller average size of individual grants and fragmentation of research effort.

Second, in a system in which institutional support is limited primarily to personnel support for core scientific staff (such as tenured professors) and capital equipment (such as laboratory space and computing equipment), incremental research costs must be covered by project grants. Over time, a research institute committed to solving a particular scientific and technological problem (adapting soybeans to shorter growing season environments, for example) may find its staff responding more to the priorities of external funding

Granting institution seeks increased granting authority.
↓
Granting institution provides information with respect to grant opportunities.
↓
Scientist and staff (or department) prepare a proposal.
↓
Proposal is transmitted to granting institution.
↓
Granting institution receives proposal.
↘
Granting institution chooses peer review and mails grant to study panel.
↓
Study panel makes recommendation.
↙
Granting institution decides on $$ award or rejects the proposal.
↓
Granting institution decides on priority area.
↓
Scientist (et al.) gets grant for X years.
↘
Equipment must be acquired.
↓
Lab must be made available or constructed.
↓
Paperwork must be done.
↓
Teaching slack is created and must be dealt with in some way.
↙
Scientist does the research project until grant expires.
↓
Scientist returns to teaching duties.
↓
Laboratory, equipment, and other resources particular to the project are closed out.

Figure 9.1. Current Grants Process. (Adapted from a chart prepared by W. Keith Huston, Minnesota Agricultural Experiment Station, 1979.)

institutions rather than concentrating its effort on the crop improvement mission. It is not difficult to imagine a situation in which university or research institute administrators begin to value their research staff less for the significance of the scientific and technological knowledge they produce than for the overhead generated by research grants or contracts.

This problem appears to be most acute in situations in which institutional research support has been closely linked to, or hidden by, reduced teaching loads. In the 1950s and 1960s, many universities used the expansion of undergraduate education to support the expansion, almost surreptitiously, of their institutional research support.[8] This has created difficult problems in a period of declining or shifting undergraduate enrollment. Institutional support for agricultural research has not been as closely coupled to undergraduate enrollment as in many other areas. There are, however, substantial pressures in some states from university administrators and state legislative committees to conform to university-wide standards with respect to student-teacher ratios. In the future, effective allocation of institutional research resources will require the development of budgeting mechanisms that more effectively uncouple the institutional support for teaching and research activities.

Research mix

One major advantage that is sometimes claimed for a project grant system is its flexibility in redirecting research effort into areas of new scientific interest or technical concern. Research systems that rely primarily on institutional support have been characterized as limited in their capacity to reallocate scientific and financial resources from traditional areas of concern or staff capacity to new areas. The state agricultural experiment stations in the United States have been criticized for devoting excessive attention to problems of increasing food and fiber production and for being unresponsive to environmental and human concerns in a period of changing social values.[9] In the United Kingdom, the national agricultural research institutes were criticized in the Rothschild report for devoting too much attention to basic research and for being unresponsive to the needs of the farmer and the food industry.[10]

A second argument that is sometimes made for a competitive project grant system is that it creates a marketlike environment for both talent and ideas. This argument frequently is coupled with arguments in favor of a peer review system to assure competent scientific judgment and effective quality control in the allocation of research resources.[11]

Both arguments have a great deal of intuitive appeal. This appeal should not be allowed to obscure the difficulties of using a project grant system for organizing a research effort with a technology development objective.[12] Neither should it be allowed to obscure the possibility that a centralized project research grant can itself be a powerful force for conformity. I have great difficulty, for example, in visualizing how a peer panel might have reacted in 1900 to a proposal from a young physicist named Einstein, who had just published his first professional paper, for support of a project that had as an objective the development of a unified theory embracing mechanics, optics, and electrodynamics—and without resort to conventional laboratory techniques!

THE USDA'S COMPETITIVE RESEARCH GRANT PROGRAM

The sources of funds available to U.S. state agricultural experiment stations have consisted primarily of federal funds appropriated to the states on a formula basis and funds appropriated for agricultural research by state legislative bodies. (See table 9.1.) Although federal formula funds are granted to the states on a matching basis, in recent years most states have supported agricultural research at a level that substantially exceeds the federal matching requirements. In addition to federal and state funds, many state agricultural experiment stations also obtain substantial contract and grant support from private industry, private foundations, the U.S. Department of Agriculture, and other federal and state sources. The contract and grant support from other federal agencies and other sources has tended to increase slightly more rapidly than the traditional sources of funds.

The last decade has been characterized by a growing lack of confidence in research decision-making processes. In the United States, the agricultural research establishment has been viewed as being unresponsive to environmental, distributional, and humanitarian concerns. New clientele groups have attempted to move concerns such as nutrition, rural development, environmental impact, soil conservation, and problems of hired workers higher on the research agenda.

Partially in response to this criticism, the U.S. Congress has established a federally funded, competitive, research grant program open to all scientists and administered by the Competitive Research Grants Office of the USDA's Science and Education Administration.[13] The fiscal year 1978 appropriation act made a total of $15 million available for competitive research grants. The executive budget for fiscal year 1979 proposed that the competitive grants program be increased by an additional $15 million. (See table 9.2.)

Table 9.1. Amount and Relative Importance of Research Performed by Funding Sources for State Agricultural Experiment Stations, the 1890 Colleges and Tuskegee Institute, and Forestry Schools, and USDA Agencies, 1967 and 1977-1979.

Sources	1967 Amount ($ millions)	1967 Percentage	1977 Amount ($ millions)	1977 Percentage	1978 Amount ($ millions)	1978 Percentage	1979 Amount ($ millions)	1979 Percentage
State Research:								
U.S. Science and Education Administration (USDA)[a]	53.8	12.5	118.8	11.5	134.5	11.7	139.0	11.8
Other federal cooperative grants and Agreements (USDA)	10.3	2.4	12.6	1.2	16.5	1.4	19.2	1.6
Other federal agencies	24.1	5.6	55.6	5.4	57.9	5.0	57.9	4.9
Total state research from federal sources	88.2	20.5	187.0	18.1	208.9	18.2	216.1	18.3
State appropriations	118.6	27.6	341.2	33.1	374.9	32.7	392.6	33.2
State sales	13.5	3.1	39.1	3.8	40.1	3.5	44.9	3.8
Total state research from state sources	132.1	30.7	380.3	36.9	415.0	36.2	437.5	37.0
Other sources	13.1	3.0	54.6	5.3	57.1	5.0	61.4	5.2
Total state research	233.4	54.3	621.9	60.3	681.0	59.4	715.0	60.5
Federal Research:								
Science and Education Administration Economic Statistics and	144.7	33.6	288.6	28.0	315.3[b]	27.5	318.1[b]	26.9
Cooperative Service	14.6	3.4	28.9	2.8	33.3	2.9	36.8	3.1
Forest Service	37.2	8.7	92.4	8.9	110.8	9.6	106.7	9.0
Other sources	—	—	—	—	6.6[c]	0.6	6.0[c]	0.5
Total federal research	196.5	45.7	409.9	39.7	466.4	40.6	457.6	39.5
Total state and federal research	429.9	100.0	1,031.8	100.0	1,147.4	100.0	1,182.6	100.0

Source: USDA/CRIS printout. FY 1978 and FY 1979 are preliminary. I am indebted to Roland Robinson of the USDA Science and Education Administration and to B. R. Eddlemann of the National Agricultural Research Planning and Analysis Project, Mississippi State University, for assistance in assembling this data.
a. Includes cooperative grants and agreements administered through the USDA Science and Education Administration's Cooperative Research Program (USD/ SEA/CR). In 1978, $7.2 million and, in 1979, $7.9 million were received by the state agricultural experiment stations under the Competitive Grants Program. The funds received by states differ from the funds appropriated by the amount of direct and indirect federal administrative charges.
b. In 1978 and 1979, $0.6 million was received by the USDA Science and Education Administration directly administered Agricultural Research Program (USDA/ SEA/AR) under the Competitive Grants Program.
c. Includes funds to land grant colleges not channeled through state experiment stations, to private universities and nonprofit research institutions, and to private for-profit institutions under the Competitive Grants Program.

Table 9.2. Funds Appropriated for Agricultural Research under Cooperative Research Programs of the USDA Science and Education Administration[a]

Program	1977 Amount ($ millions)	1977 Percentage	1978 Amount ($ millions)	1978 Percentage	1979 Amount ($ millions)	1979 Percentage	1980 Amount ($ millions)	1980 Percentage
Hatch Act[b]	98.0	76.0	109.1	69.0	109.1	62.6	118.6	62.8
Regular	73.1	56.7	81.1	51.3	81.1	46.5	87.9	46.5
Regional	21.7	16.8	24.5	15.5	24.5	14.0	26.9	14.2
McIntire-Stennis[c]	8.2	6.4	9.5	6.0	9.5	5.4	10.0	5.3
1980 colleges and Tuskegee Institute	13.4	10.4	14.2	9.0	16.4	9.4	17.8	9.4
Special research grants	6.3	4.9	7.2	4.6	16.2	9.3	17.7	9.4
Competitive research grants	—	—	15.0	9.5	15.0	8.6	16.0	8.4
Animal health and disease	—	—	—	—	5.0	2.9	6.0	3.2
Rural development	1.5	1.2	1.5	0.9	1.5	0.9	1.5	0.8
Federal administration	1.7	1.3	1.7	1.1	1.7	9.7	1.5	0.8
Total	129.0	100.0	158.2	100.0	174.4	100.0	189.0	100.0

Source: USDA, SEA, and CSRA memoranda to state agricultural experiment station directors.

a. Prior to 1978 called the Cooperative State Research Service (CSRS).

b. Total includes regular and regional funds distributed to states plus federal administration and penalty mail.

c. Cooperative forestry research.

This was offset by a reduction of approximately $11 million in Hatch Act formula funding plus reductions of approximately $1 million in McIntire-Stennis Cooperative Forestry Research, $2 million in special research grants, and $1.5 million in rural development research. The 1979 appropriations act that was finally passed by the Congress restored the cuts in the Hatch Act funds that had been recommended by the administration and continued the competitive research grant program at the $15-million level. The executive budget for fiscal year 1980 continued Hatch funding at the 1979 level and proposed an increase in the competitive grants program from $15 to $30 million.

Strong support for a program of competitive project research grants to be administered by the USDA had been made in two reports sponsored by the National Research Council.[14] The National Research Council's recommendations reflected, in part, a judgment that the productivity of agriculture and agriculture-related research could be enhanced by making USDA research support available to scientists in departments and institutions that had not been eligible for support under the formula funding arrangements.

Administrative officers and scientists at the state agricultural experiment stations had also been generally supportive of the move to expand funds for competitive grants. Both the National Research Council and the leadership of the state experiment stations had, however, expected that an expansion of funds for competitive research grants would take place in an environment of expanded support for agricultural research. Apparently, they did not anticipate the trade-off between the competitive grant and formula funding that emerged in the fiscal year 1978 executive budget proposal.[15]

The argument for expanding support for agricultural research typically has drawn on two sources of support. Agricultural scientists and science administrators have pointed to the technical constraints that must be overcome to meet future food and fiber requirements. They also have argued that technical change leading to lower production costs represents one way of balancing the conflicting claims of farmers for higher incomes and consumers for restraint in food price increases. Economists have buttressed these arguments with an expanding body of empirical research that has documented the high rates of return to past agricultural research. (See chapter 10.) In recent years, they also have worked closely with research administrators to provide *ex ante* rate of return projections. Both the constraint and the rate of return approaches suggest that there is substantial underinvestment in agricultural research both in the United States and in most other countries where such studies have been conducted.

AN APPRAISAL OF THE TWO SYSTEMS
OF RESEARCH SUPPORT

I have noted that the effects of a system that appears optimal to the individual scientist or to the individual research manager in a world characterized by limited institutional support and substantial project research support alternatives are inducing both excessive allocation of professional resources to grant seeking and contributing to the disintegration of the capacity to undertake major mission-oriented applied research programs. These two sources of inefficiency can be reduced by a national agricultural research policy that emphasizes institutional research support for mission-oriented applied research and for basic research directly related to the needs of applied research programs.

There is substantial evidence to support the claim of efficiency for the institutional support system. High rates of return have been attributed to the state and federal agricultural research systems in the United States, to a number of older research institutions in formerly colonial countries (such as the Rubber Research Institute of Malaysia), and to the older units of the CGIAR-sponsored international agricultural research system. (See chapter 10.) It would be extremely difficult to imagine that the long-term research effort required to develop the high-yielding clones that have revolutionized productivity in the Malaysian rubber industry could have been accomplished on the basis of a series of project grants from a colonial research secretariat in London. (See chapter 4.)

The inferences drawn from agricultural research are consistent with the experience of a number of highly productive industrial research programs. Edwin Mansfield and his associates have documented rates of return to industrial research in the same range as the rates of return to public-sector agricultural research.[16] The more productive private research programs typically have been those that have combined long-term sustained support by a firm with a sufficiently broad product line to be able to utilize a substantial share of the product of a major in-house research program.

Finally, one can point to the productivity of a number of long-term institutional research support activities by the private foundations. The support of the Carnegie Institution for the fundamental studies on inheritance in maize conducted by George H. Schull is a classic example. The Rockefeller Foundation's support for the research program of the Office of Special Studies in the Mexican Ministry of Agriculture over several decades (from the mid-1940s to the early 1960s) established the basis for the research program of international research centers that are part of the Consultative Group on International Agricultural Research system. The Ford Foundation's institutional support for the research program of Resources for the Future has been a major factor in establishing resource economics as a major field of economic research in the United States.

Long-term institutional research support also can become a source of inefficiency. Institutional research programs are subject to the danger of becoming too conventional or losing a sense of urgency with respect to their mission. It was noted earlier that project grant support has been defended on the basis that it forces a research system to be more responsive to new issues on the research agenda. I find no fault with this argument as long as project research support remains relatively small and as long as it encourages the exploration of new opportunities within the broad research mission of an institution.

There are also other devices for offsetting geriatric tendencies in a mature research institute. The development of cost-sharing arrangements between public research institutes and clientele groups, or user representation on boards of directors or advisory committees, and the use of periodic peer review panels are among the possibilities. An important factor in the case of the state agricultural experiment stations has been their location within a university environment. The interaction between graduate training and research and the opportunities to draw on professional capacity in related fields have contributed to research productivity.

Another argument that must be dealt with is whether or not a competitive project research support system is an effective way of taking advantage of the research capacities that exist outside institutionally funded research programs. A major argument in favor of the new USDA project grant program is that it would be able to draw on professional resources in departments that do not receive experiment station funding and in institutions that were not part of the land-grant system. An analogy is often made with the project grant programs of the National Science Foundation, which often supports the research of individual scientists in institutions that have very little institutional research capacity.

This argument is only partially compelling. The United States has been reasonably successful in evolving a dual system of colleges and universities in which those institutions that are capable of organizing effective graduate training and research activities are sharply differentiated from those that do not possess such a capacity. The major research univerisities probably have a greater capacity to manage a program of research grants based primarily on the quality of individual projects than a central granting agency, such as the USDA or the National Science Foundation.

For the colleges and universities that do not have substantial graduate programs, faculty research must be justified primarily on the basis of its contribution to the viability of the teaching programs. A limited commitment of faculty effort to scholarship and research contributes to the vitality of undergraduate teaching programs. My experience on university and agency research review panels has led me to believe that institutional support for a program of small grants, even in institutions that are primarily committed to an undergraduate education mission, would be more efficient than a grant program that is centralized in a Washington agency.

The current argument about the merits of institutional and project research support is more appropriately cast in terms of the relative mix of the two systems of support than in the absolute merits of either system. Nevertheless, I must insist that the issue of efficiency in the allocation and use of research resources is important. The proudctivity of agricultural research, based on historical rates of return, has been high (See chapter 10.) This places a heavy burden on those who would argue for a substantial shift of resources in the U.S. system from institutional to project support to demonstrate that such a shift would either enhance the productivity at the existing level of research support or draw substantial new resources into the agricultural research system.

RESEARCH STRATEGIES FOR
AN IMPERFECT WORLD

What are the policy options available at the level of the individual agricultural experiment station or research institute when confronted with a world in which institutional support is severely limited and incremental project research grant funds are increasingly available from external sources? This is a question that is of even greater relevance to research directors in most developing countries than it is to administrators in U.S. agricultural experiment stations. There are three alternative strategies.

One alternative is a research entrepreneurship strategy. This strategy, as it is followed by some experiment station directors, utilizes most institutional support funds from federal formula funds and state matching funds, primarily to cover scientific staff and staff-related costs. Staff members are then encouraged to "prospect" for research program support among public and private agencies that make research grants or that contract research. This is the standard pattern for research-oriented academic departments or schools that do not have access to sources of substantial institutional support.

This research entrepreneur model has some important advantages for the individual experiment station or research institute. It permits the recruitment of a larger research staff than a strategy in which institutional support is reserved for research program support. It probably results in a selection process in which staff members with research entrepreneurial ability are attracted to research stations that emphasize a research entrepreneurship strategy. It provides research administrators with an independent judgment of the quality of the staff's research effort. Quality is inferred not only from publication in peer-reviewed journals but also from the amount and source of project research funds attracted.

There are also costs to both the individual station and the research system. It was noted earlier that the development of a research entrepreneurship capacity, particularly when it is developed relatively early in a scientist's career, may be competitive with the development of a capacity to advance scientific knowledge. There are also serious institutional repercussions for a research entrepreneurship strategy when the supply of grant or contract funds in areas that are important to the institution's central thrust declines. This may be a particularly serious problem for a state agricultural experiment station at which research effort is expected to pay off in terms of state economic and social development objectives.

A second alternative is an in-house strategy. In an in-house strategy, research is limited to those research programs and activities that can be supported by federal formula funds, state matching funds, and special appropriations, endowments, and other forms of relatively unrestricted long-term institutional support. An advantage of the in-house approach is that it enables the research director to assemble a staff of scientists who are motivated primarily toward the development and exercise of scientific capacity rather than entrepreneurial capacity. It permits a focus on relatively long-term technology development and fundamental research. And it provides a greater opportunity for the scope and direction of the research program to be set by the experiment station or research institute, rather than by the granting agencies.

There are also costs to an in-house strategy. In the presence of a strong director, the research decision-making process may become too authoritarian. The security of research funding may result in a research program that is too routine, in one that does little more than attempt to fill in the gaps in the literature or to meet the information needs of clientele groups. In the presence of a weak director, the research resource allocation system may become too political— too responsive to the pressures of strong department chairpeople or research scientists who generate strong local clientele support. An in-house strategy may also impose extreme limits on the size of the research unit.

A third alternative that may be the most efficient strategy in an "nth best" world would include elements of both the research entrepreneurship strategy and the in-house strategy. A successful mixed strategy would include three elements.

The first involves the recognition that the aggregate supply of research resources is likely to be more responsive to the efforts of research directors, or of directors acting as a group, than to the efforts of individual researchers. This may imply that the entrepreneurial (political) activities of experiment station directors and deans of agriculture may be more productive in their efforts to expand the availability of research resources than in their roles as allocators of in-house research resources.

The second element involves the retention of sufficient control over in-house research resources to provide seed money for young researchers. This initial institutional support can help develop a record of research productivity that will make more accurate judgments of the research and entrepreneurial capacities.[17] It can also be used to back the high-risk or speculative research of serious researchers of proven capacity that may later serve to attract external support. It is generally recognized in the research community that "one cannot expect a proposal to be funded until a considerable amount of work has been done on the project."[18]

The third element involves the allocation of the balance of in-house funds to salary and related costs for the scientific staff on the expectation that most mid-career and senior staff members have a reasonable capacity to attract external funds.

NOTES

1. This chapter is a revision and an expansion of a paper by Maury E. Bredahl, W. Keith Bryant, and Vernon W. Ruttan, "Behavior and Productivity Implications of Institutional and Project Funding of Research," *American Journal of Agricultural Economics*, 61 (August 1980), pp. 371-83. I wish to express my appreciation to my coauthors for giving me

their permission to use the paper in this volume. I have benefited from comments made on an earlier draft of this paper by Emerson Babb, Robert Evenson, Walter Fishel, Edward Foster, Don Hadwiger, R. J. Hildreth, Keith Huston, Yoav Kislev, Michael Lea, Willis Peterson, Jean Robinson, Roland Robinson, Theodore W. Schultz, and Burt Sundquist.

2. Funds allocated to the states under the Hatch Act, except for funds reserved for cooperative regional research efforts, are allocated to the state agricultural experiment stations by a formula based on the number of farms and the size of the rural population in each state. The several sources of federal funds for state agricultural research are identified in the latest annual report of the USDA Cooperative State Research Service. Factors affecting the support given for state agricultural experiment stations have been analyzed by Willis L. Peterson, "The Allocation of Research, Teaching and Extension Personnel in U.S. Colleges of Agriculture," *American Journal of Agricultural Economics*, 51 (February 1969), pp. 41-55; and by Wallace E. Huffman and John A. Miranowski, "An Economic Analysis of Expenditures on Agricultural Experiment Station Research," *American Journal of Agricultural Economics*, 63 (February 1981), pp. 104-18.

3. Bernard R. Stein, "Public Accountability and the Project Grant Mechanism," *Research Policy*, 2 (1973), pp. 2-16.

4. For a discussion of the agricultural research scientist as a hero, see E. C. Stakman, Richard Bradfield, and Paul C. Mangelsdorf, *Campaigns against Hunger* (Cambridge, Mass.: Harvard University Press, 1967). For a discussion of the scientist as a villain, see Wendell Berry, *Culture and Agriculture: The Unsettling of America* (New York: Avon, 1978).

5. The perspectives on the motivation and behavior of agricultural scientists and science administrators expressed in this and the next section are based primarily on my own observations and introspection. As this study was being completed, several reports on a major research project on the sociology of agricultural scientists and science administrators became available. See Lawrence Busch, William B. Lacy, and Caroline Sachs, "Research Policy and Process in the Agricultural Sciences: Some Results from a National Study" (Lexington, Ky.: Department of Sociology, Kentucky Agricultural Experiment Station, University of Kentucky [RS 66], July 1980); Lawrence Busch, "Structure and Negotiation in the Agricultural Sciences," *Rural Sociology*, 45 (Spring 1980), pp. 26-48.

6. Vernon W. Ruttan, "Bureaucratic Productivity: The Case of Agricultural Research," *Public Choice*, 35 (1980), pp. 529-47.

7. James G. Ling and Mary Ann Hand, "Federal Funding in Materials Research," *Science*, 201, (September 12, 1980), pp. 1203-7. According to Ling and Hand, "The performance of the 20 materials research laboratories (MRL's) at universities funded with institutional grants by the National Science Foundation, Department of Energy, and National Aeronautics and Space Administration was evaluated in comparison with 15 other universities (non-MRL's) receiving individually funded projects for materials research. Performance was measured by peer review and citation frequency analysis of publications, subjective evaluation of research achievements and researcher reputation by a panel of experts, review of equipment purchases and utilization, and analysis of administrative costs. The study concludes that there are no significant differences between the MRL's and non-MRL's with respect to innovation, interdisciplinarity, utilization of specialized equipment, concentration of funding, rate of turnover, duration of research areas, and level of effort per research paper. The MRL's have a greater number of major achievements and attract researchers with higher reputations. The MRL's tend to emphasize experimental work, and in about 70 percent of the materials research areas sponsored by the National Science Foundation there is no overlap between the two groups. Institutional grants involve much less total (federal plus university) administrative cost per grant dollar than project grants" (p. 1204).

8. Nathan Keyfitz, "The Impending Crisis in American Graduate Schools," *The Public Interest*, 52 (Summer 1978), pp. 85-97.

9. Don Paarlberg, "A New Agenda for Agriculture," *Policy Studies Journal*, 6 (Summer 1978), pp. 504-6.

10. Tilo L. V. Ulbricht, "Contract Agricultural Research and Its Effect on Management," in *Resource Allocation and Productivity in National and International Agricultural Research*, Thomas M. Arndt, Dana G. Dalrymple, and Vernon W. Ruttan, eds. (Minneapolis: University of Minnesota Press, 1977), pp. 381-93.

11. See R. Bowers, "The Peer Review System on Trial," *American Scientist*, 63 (6) (November-December 1975), pp. 624-26; Stephen Cole, Leonard C. Rubin, and Jonathan R. Cole, "Peer Review and the Support of Science," *Scientific American*, 237 (October 1977), pp. 34-41; and T. Gustafson, "The Controversy over Peer Review," *Science*, 190 (December 12, 1975), pp. 1060-66.

12. A useful illustration of the ineffectiveness of grant programs in advancing technology was provided by the efforts to develop the knowledge and technology of aquaculture in the 1960s and 1970s. "Instead of a strong, centralized, multidisciplined NASA-type assault on the problems, we settled for a scattered burst of unrelated, small-scale, individualistic little research projects which produced hundreds of graduate degrees, tons of paperwork and very few pounds of fish. At the same time, a handful of visionary and articulate entrepreneurs persuaded a succession of equally religious investors to initiate commercial aquaculture ventures which, with a few exceptions, were mostly premature disasters" (Carl N. Hodges and Wayne L. Collens, "Food Production with Saline Water," in *Food and Climate Review, 1979*, S. K. Levin, ed. [Boulder, Colo.: Aspen Institute for Humanistic Studies, January 1979], p. 65).

13. U.S. Department of Agriculture, Science and Education Administration, "Competitive Grants for Basic Research: Plant Biology and Human Nutrition," *Federal Register*, 43 (March 7, 1978), pp. 9432-40; and Gary A. Strobel, "A New Grants Program in Agriculture," *Science*, 199 (March 3, 1978), p. 935.

14. National Research Council, *Report of the Committee on Research Advisory to the USDA* (Springfield, Va.: National Technical Information Service, 1973); and National Research Council, *World Food and Nutrition Study: The Potential Contributions of Research* (Washington, D.C.: National Academy of Sciences, 1977).

15. Dr. M. Rupert Cutler, assistant secretary for conservation, research, and education of the USDA, appeared to be surprised by the results of his own budgeting efforts. In response to agriculture committee's question on this point, he responded "that [the] apparent relationship was unintended. By that I mean the relationship between beginning a competitive grant research program open to all agricultural scientists and the level of the Hatch Act request." Dr. Cutler went on to explain that many of the programs of the USDA (entitlement and regulatory programs) are legislatively mandated. "Given a budget ceiling, the remaining funds available for agricultural research programs are fixed. Thus the only available method to initiate the competitive grants system was the reduction of other research areas." (Robert M. Cutler, "Statement," in *Appraisal of Title 14 (Research), Agricultural Act of 1977*, hearings before the Subcommittee on Department Investigations, Oversight, and Research of the Committee on Agriculture, Serial No. 95-RR, House of Representatives, 95th Congress, 2nd session, February 23, 1978 [Washington, D.C.: U.S. Government Printing Office, 1978], pp. 91-92.)

16. Edwin Mansfield, John Rapoport, Anthony Romeo, Edmund Villani, Samuel Wagner, and Frank Husic, *The Production and Applications of New Industrial Technology* (New York: Norton, 1977), pp. 144-66.

17. G. W. Salisbury has emphasized the importance of the priority given to support young scientists by the Wisconsin Alumni Research Foundation as an important factor in the exceptionally high research productivity at the Wisconsin Agricultural Experiment

Station. See G. W. Salisbury, *Research Proudctivity of the State Agricultural Experiment Station System: Measured by Scientific Publication Output* (Urbana, III.: University of Illinois Agricultural Experiment Station Bulletin 762, July 1980), pp. 28-29.

    18. Richard A. Muller, "Innovation and Scientific Funding," *Science*, 209 (August 22, 1980), p. 880.

Chapter 10

The Economic Benefits from Agricultural Research

In earlier chapters there has been repeated reference to the contribution of research to productivity growth in agriculture.[1] It has been pointed out that the significance of technical change is that it permits the substitution of knowledge for resources or of inexpensive and abundant resources for scarce and expensive resources or that it releases the constraints on growth imposed by inelastic resource supplies.

There can be little argument with such generalizations. But how much does agricultural research contribute to either productivity or output growth? Can the impact of research on productivity and output growth be measured? And can its impact be valued?

Since the mid-1950s, considerable effort has been devoted to attempts to answer these questions. The answers have been very satisfying to agricultural research managers. The answers indicate that under a wide range of circumstances the economic returns to investment in agricultural research have been very high in comparison to almost any other investment available to society. The first part of this chapter is devoted to a review of the body of evidence on returns to agricultural research. The second part of the chapter is more speculative. It attempts to respond to the question: why have the returns to agricultural research remained so high? In particular, why has the United States failed to take fuller advantage of the growth dividends from agricultural research?

PRODUCTIVITY GROWTH

The beginning of modernization in agriculture is signaled by the emergence of sustained growth in productivity. During the initial stages of development, productivity growth is usually accounted for by improvement in a single partial productivity ratio, such as output per unit of labor or output per unit of land. In the United States and other countries of recent settlement, such as Canada, Australia, New Zealand, and Argentina, increase in labor productivity in output per worker has carried the main burden of growth in total productivity. In countries that entered the development process with relatively high population-land ratios, such as Japan, Denmark, and Germany, increases in land productivity—in output per hectare—have initially been largely responsible for growth in total productivity.

As modernization progresses, there is a tendency for growth in total productivity—output per unit of total input—to be sustained by a more balanced combination of improvement in partial productivity ratios—in output per worker, per hectare, and per unit of capital. Thus, among the countries that have the longest experience of agricultural growth, there tends to be a convergence in the patterns of productivity growth. (See chapter 2.)

What is implied by the concept of productivity growth?[2] In the above paragraphs, the term has been used in several ways. *One way it was not used was as a simple indicator of growth in production!* When the term "productivity" is used accurately, it always refers to a ratio—to the ratio of output to the input of a single input or to the ratio of output to an aggregation of several inputs. When the term "productivity growth" is used, it refers to changes in such ratios over time.

Crop production per acre is a traditional productivity measure used in agriculture. It is a partial productivity ratio. Output per unit of labor—labor productivity—is another commonly used partial productivity ratio. Energy yield per unit of energy applied is another partial productivity ratio that has received a good deal of attention. The major limitation of all partial productivity ratios as indicators of technical change is that improvement in one partial productivity ratio may simply reflect the effect of a decline in another partial productivity ratio. Gains in crop yield per unit area may reflect the combined effects of increases in fertilizer use per unit area and improvements in crop response to fertilizer. In this case, the partial productivity ratio, increase in output per unit area, will overstate the improvement due to technical change.

A more inclusive measure of productivity is needed to measure the gains in efficiency resulting from technical change. The closest approach to an indicator that can be used to measure differences or changes in efficiency is output per unit of total input. In this measure, the several inputs used in agricultural production—land, labor, buildings, and equipment and current inputs such as fertilizers, pesticides, and fuels—are aggregated and changes in output are related to changes in total input. When total productivity measures are constructed carefully, they can be interpreted as reasonably accurate measures of the contribution of technical change to the growth of agricultural production. A more cautious interpretation is that total productivity measures the residual growth in output that remains unexplained after the effects of the inputs included in the total input measure are accounted for.

The changes in two partial productivity measures, land productivity and labor productivity, and in total productivity are illustrated for U.S. agriculture for the period 1950 to 1978 in figure 10.1. During the 1950s and early 1960s, all three productivity measures grew rapidly. During the late 1960s, the rate of growth of land productivity and total productivity slowed down. During the 1970s, these two productivity indexes appear to have renewed their upward trend. Note also that the labor productivity index grew more rapidly than the total productivity index throughout the entire period. Part of the growth in labor productivity is due to higher capital investment per worker. The total productivity index grew at a slower rate because the services of the capital equipment, along with labor and other inputs, are included in the input index.

In tables 10.1 and 10.2, changes in total productivity and in the two partial productivity growth rates are presented for the United States and Japan for the period since 1870. The tables illustrate the point made earlier in this section. Prior to the mid-1950s, productivity growth in Japanese agriculture was dominated by growth in land productivity. Prior to the 1940s, productivity growth in U.S. agriculture was dominated by growth in labor productivity. Note also that both countries experienced periods of relatively slow productivity growth. During the first quarter of the 20th century, the rate of growth in labor productivity declined in the United States. Total inputs grew more rapidly than output. Total productivity declined.

The question of why productivity growth has been so rapid in a few countries and why it has remained stagnant in so many countries in the last several decades is a central issue for development policy. In the next section, the evidence on the contribution of research to productivity growth is examined. Until recently, the available evidence was drawn primarily from the U.S. experience. Within the last decade, a good deal of evidence on the contribution of research to productivity growth has been available from other countries.

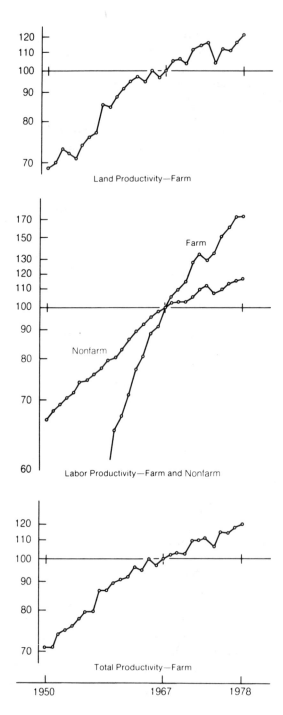

Figure 10.1. Partial and Total Productivity Measures for U.S. Agriculture, 1950-1978 (1967 = 100). (Used with permission from Robert E. Evenson, Paul Waggoner, and Vernon W. Ruttan, "The Economic Benefit from Research: An Example from Agriculture," *Science*, 205 [September 14, 1979], p. 1104. Copyright 1979 by the American Association for the Advancement of Science.)

Table 10.1. Average Annual Rates of Change (Percentage per Year) in Total Output, Inputs, and Productivity in U.S. Agriculture, 1870-1979

Item	1870-1900	1900-1925	1925-1950	1950-1965	1965-1979
Farm output	2.9	0.9	1.6	1.7	2.1
Total inputs	1.9	1.1	0.2	−0.4	0.3
Total productivity	1.0	−0.2	1.3	2.2	1.8
Labor inputs[a]	1.6	0.5	−1.7	−4.8	−3.8
Labor productivity	1.3	0.4	3.3	6.6	6.0
Land inputs[b]	3.1	0.8	0.1	−0.9	0.9
Land productivity	−0.2	0.0	1.4	2.6	1.2

Sources: Data from USDA, Changes in Farm Production and Efficiency (Washington, D.C.: 1979); and D. D. Durost and G. T. Barton, Changing Sources of Farm Output (Washington, D.C.: USDA Production Research Report No. 36). February 1960. Data are three-year averages centered on the year shown for 1925, 1950, and 1965.
a. Number of workers, 1870-1910; worker-hour basis, 1910-1971.
b. Cropland use for crops, including crop failures and cultivated summer fallow.

Table 10.2. Average Annual Change in Total Output, Inputs, and Productivity in Japanese Agriculture, 1880-1975

Item	1880-1920	1920-1935	1935-1955	1955-1965	1965-1975
Farm output	1.8	0.9	0.6	3.5	1.5
Total inputs	0.5	0.5	1.2	1.5	0.7
Total productivity	1.3	0.4	−0.6	2.0	0.8
Labor inputs	−0.3	−0.2	0.6	−2.7	−4.1
Labor productivity	2.1	1.1	0.0	6.2	5.6
Land inputs	0.6	0.1	−0.1	0.1	−0.7
Land productivity	1.2	0.8	0.7	3.4	2.2

Sources: Data from Saburo Yamada and Yujiro Hayami, "Agricultural Growth in Japan, 1880-1970," in Agricultural Growth in Japan, Taiwan, Korea and the Philippines, Yujiro Hayami, Vernon W. Ruttan, and Herman Southworth, eds. (Honolulu: University Press of Hawaii, 1979), pp. 33-58; Saburo Yamada, "The Secular Trends in Input-Output Relations of Agricultural Production in Japan, 1878-1978," a paper presented at the Conference on Agricultural Development in China, Japan, and Korea, Academica Sinica, Taipei, December 17-20, 1980.

THE CONTRIBUTION OF RESEARCH TO PRODUCTIVITY

The results of a large number of studies of the contribution of research to productivity growth have been assembled in table 10.3. Almost all of the studies indicate high rates of return to investment in agricultural research—well above the 10 to 15 percent (above inflation) that private firms consider adequate to attract investment. It is hard to imagine many investments in either private- or public-sector activities that would produce more favorable rates of return.

Table 10.3. Summary Studies of Agricultural Research Productivity

Study	Country	Commodity	Time Period	Annual Internal Rate of Return (%)
Index Number:				
Griliches, 1958	USA	Hybrid corn	1940-1955	35-40
Griliches, 1958	USA	Hybrid sorghum	1940-1957	20
Peterson, 1967	USA	Poultry	1915-1960	21-25
Evenson, 1969	South Africa	Surgarcane	1945-1962	40
Barletta, 1970	Mexico	Wheat	1943-1963	90
Barletta, 1970	Mexico	Maize	1943-1963	35
Ayer, 1970	Brazil	Cotton	1924-1967	77+
Schmitz and Seckler, 1970	USA	Tomato harvester, with no compensation to displaced workers	1958-1969	37-46
		Tomato harvester, with compensation of displaced workers for 50% of earnings loss		16-28
Ayer and Schuh, 1972	Brazil	Cotton	1924-1967	77-110
Hines, 1972	Peru	Maize	1954-1967	35-40[a] 50-55[b]
Hayami and Akino, 1977	Japan	Rice	1915-1950	25-27
Hayami and Akino, 1977	Japan	Rice	1930-1961	73-75
Hertford, Ardila, Rocha, and Trujillo, 1977	Colombia	Rice	1957-1972	60-82
		Soybeans	1960-1971	79-96
		Wheat	1953-1973	11-12
		Cotton	1953-1972	none
Pee, 1977	Malaysia	Rubber	1932-1973	24
Peterson and Fitzharris, 1977	USA	Aggregate	1937-1942	50
			1947-1952	51
			1957-1962	49
			1957-1972	34
Wennergren and Whitaker, 1977	Bolivia	Sheep	1966-1975	44
		Wheat	1966-1975	−48
Pray, 1978	Punjab (British India)	Agricultural research and extension	1906-1956	34-44
	Punjab (Pakistan)	Agricultural research and extension	1948-1963	23-37
Scobie and Posada, 1978	Bolivia	Rice	1957-1964	79-96
Pray, 1980	Bangladesh	Wheat and rice	1961-1977	30-35
Regression Analysis:				
Tang, 1963	Japan	Aggregate	1880-1938	35
Griliches, 1964	USA	Aggregate	1949-1959	35-40
Latimer, 1964	USA	Aggregate	1949-1959	not significant

Table 10.3 —*Continued*

Study	Country	Commodity	Time Period	Annual Internal Rate of Return (%)
Peterson, 1967	USA	Poultry	1915-1960	21
Evenson, 1968	USA	Aggregate	1949-1959	47
Evenson, 1969	South Africa	Sugarcane	1945-1958	40
Barletta, 1970	Mexico	Crops	1943-1963	45-93
Duncan, 1972	Australia	Pasture Improvement	1948-1969	58-68
Evenson and Jha, 1973	India	Aggregate	1953-1971	40
Cline, 1975 (revised by Knutson and Tweeten, 1979)	USA	Aggregate	1939-1948	41-50[c]
		Research and extension	1949-1958	39-47[c]
			1959-1968	32-39[c]
			1969-1972	28-35[c]
Bredahl and Peterson, 1976	USA	Cash grains	1969	36[d]
		Poultry	1969	37[d]
		Dairy	1969	43[d]
		Livestock	1969	47[d]
Kahlon, Bal, Saxena, and Jha, 1977	India	Aggregate	1960-1961	63
Evenson and Flores, 1978	Asia— national	Rice	1950-1965	32-39
			1966-1975	73-78
	Asia— International	Rice	1966-1975	74-102
Flores, Evenson, and Hayami, 1978	Tropics	Rice	1966-1975	46-71
	Philippines	Rice	1966-1975	75
Nagy and Furtan, 1978	Canada	Rapeseed	1960-1975	95-110
Davis, 1979	USA	Aggregate	1949-1959	66-100
			1964-1974	37
Evenson, 1979	USA	Aggregate	1868-1926	65
	USA	Technology oriented	1927-1950	95
	USA	Science oriented	1927-1950	110
	USA	Science oriented	1948-1971	45
	Southern USA	Technology oriented	1948-1971	130
	Northern USA	Technology oriented	1948-1971	93
	Western USA	Technology oriented	1948-1971	95
	USA	Farm management research and agricultural extension	1948-1971	110

Source: Robert E. Evenson, Paul E. Waggoner, and Vernon W. Ruttan, Economic Benefits from Research: An Example from Agriculture," *Science*, 205 (September 14, 1979), pp. 1101-7. Copyright 1979 by the American Association for the Advancement of Science.
a. Returns to maize research only.
b. Returns to maize research plus cultivation "package."
c. Lower estimate for 13-, and higher for 16-year time lag between beginning and end of output impact.
d. Lagged marginal product of 1969 research on output discounted for an estimated mean lag of 5 years for cash grains, 6 years for poultry and dairy, and 7 years for livestock.

Sources for Table 10.3: The results of many of the studies reported in this table have previously been summarized in the following works.

Thomas M. Arndt, Dana G. Dalrymple, and Vernon W. Ruttan, eds., *Resource Allocation and Productivity in National and International Agricultural Research* (Minneapolis: University of Minnesota Press, 1977), p. 6, 7.

James K. Boyce and Robert E. Evenson, *Agricultural Research and Extension Systems* (New York: Agricultural Development Council, 1975), p. 104.

Robert Evenson, Paul E. Waggoner, and Vernon W. Ruttan, "Economic Benefits from Research: An Example from Agriculture," *Science*, 205 (September 14, 1979), pp. 1101-7.

Robert J. R. Sim and Richard Gardner, *A Review of Research and Extension Evaluation in Agriculture* (Moscow, Idaho: University of Idaho, Department of Agricultural Economics Research Series 214, May 1978), pp. 41, 42.

The sources for individual studies are

H. Ayer, "The Costs, Returns and Effects of Agricultural Research in São Paulo, Brazil" (Ph.D. dissertation, Purdue University, 1970).

H. W. Ayer and G. E. Schuh, "Social Rates of Return and Other Aspects of Agricultural Research: The Case of Cotton Research in São Paulo, Brazil," *American Journal of Agricultural Economics*, 54 (November 1972), pp. 557-69.

N. Ardito Barletta, "Costs and Social Benefits of Agricultural Research in Mexico" (Ph.D. dissertation, University of Chicago, 1970).

M. Bredahl and W. Peterson, "The Productivity and Allocation of Research: U.S. Agricultural Experiment Stations," *American Journal of Agricultural Economics*, 58 (November 1976), pp. 684-92.

Philip L. Cline, "Sources of Productivity Change in United States Agriculture" (Ph.D. dissertation, Oklahoma State University, 1975).

Jeffrey S. Davis, "Stability of the Research Production Coefficient for U.S. Agriculture," (Ph.D. dissertation, University of Minnesota, 1979).

R. C. Duncan, "Evaluating Returns to Research in Pasture Improvement," *Australian Journal of Agricultural Economics*, 16 (December 1972), pp. 153-68.

R. Evenson, "The Contribution of Agricultural Research and Extension to Agricultural Production" (Ph.D. dissertation, University of Chicago, 1968).

R. Evenson, "International Transmission of Technology in Sugarcane Production" (New Haven, Conn: Yale University, Mimeographed paper, 1969).

R. E. Evenson and P. Flores, *Economic Consequences of New Rice Technology in Asia*, Los Banos, Laguna, Philippines: International Rice Research Institute, 1978.

R. E. Evenson and D. Jha, "The Contribution of Agricultural Research Systems to Agricultural Production in India," *Indian Journal of Agricultural Economics*, 28 (1973), pp. 212-30.

P. Flores, R. E. Evenson, Y. Hayami, "Social Returns to Rice Research in the Philippines: Domestic Benefits and Foreign Spillover," *Economic Development and Cultural Change*, 26 (April 1978), pp. 591-607.

Z. Griliches, "Research Costs and Social Returns: Hybrid Corn and Related Innovations," *Journal of Political Economy*, 66 (1958), pp. 419-31.

Z. Griliches, "Research Expenditures, Education and the Aggregate Agricultural Production Function," *American Economic Review*, 54 (December 1964), pp. 961-74.

Y. Hayami and M. Akino, "Organization and Productivity of Agricultural Research Systems in Japan," in *Resource Allocation and Productivity in National and International Agricultural Research*, Thomas M. Arndt, Dana G. Dalrymple, and Vernon W. Ruttan, eds. (Minneapolis: University of Minnesota Press, 1977), pp. 29-59.

R. Hertford, J. Ardila, A. Rocha, and G. Trujillo, "Productivity of Agricultural Research in Colombia," in *Resource Allocation and Productivity in National and International Agricultural Research*, Thomas M. Arndt, Dana G. Dalrymple, and Vernon W. Ruttan, eds. (Minneapolis: University of Minnesota Press, 1977), pp. 86-123.

J. Hines, "The Utilization of Research for Development: Two Case Studies in Rural Modernization and Agriculture in Peru" (Ph.D. dissertation, Princeton University, 1972).

A. S. Kahlon, H. K. Bal, P. N. Saxena, and D. Jha, "Returns to Investment in Research in India," in *Resource Allocation and Productivity in National and International Agricultural Research*, University of Minnesota Press, 1977), pp. 124-47.

M. Knutson and Luther G. Tweeten, "Toward an Optimal Rate of Growth in Agricultural Production Research and Extension," *American Journal of Agricultural Economics*, 61 (February 1979), pp. 70-76.

R. Latimer, "Some Economic Aspects of Agricultural Research and Extension in the U.S." (Ph.D. dissertation, Purdue University, 1964).

J. G. Nagy and W. H. Furtan, "Economic Costs and Returns from Crop Development Research: The Case of Rapeseed Breeding in Canada," *Canadian Journal of Agricultural Economics* 26, (February 1978), pp. 1-14.

T. Y. Pee, "Social Returns from Rubber Research on Peninsular Malaysia" (Ph.D. dissertation, Michigan State University, 1977)

W. L. Peterson, "Return to Poultry Research in the United States," *Journal of Farm Economics*, 49 (August 1967), pp. 656-69.

W. L. Peterson and J. C. Fitzharris, "The Organization and Productivity of the Federal State Research System in the United States," in *Resource Allocation and Productivity in National and International Agricultural Research*, Thomas M. Arndt, Dana G. Dalrymple, and Vernon W. Ruttan, eds. (Minneapolis: University of Minnesota Press, 1977), pp. 60-85.

C. E. Pray, "The Economics of Agricultural Research in British Punjab and Pakistani Punjab, 1905-1975" (Ph.D. dissertation, University of Pennsylvania, 1978).

C. E. Pray, "The Economics of Agricultural Research in Bangladesh," *Bangladesh Journal of Agricultural Economics*, 2 (December 1979), pp. 1-36.

A. Schmitz and D. Seckler, "Mechanized Agriculture and Social Welfare: The Case of the Tomato Harvester," *American Journal of Agricultural Economics*, 52 (November 1970), pp. 569-77.

G. M. Scobie and R. Posada T., "The Impact of Technical Change on Income Distribution: The Case of Rice in Colombia," *American Journal of Agricultural Economics*, 60 (February 1978), pp. 85-92.

A. Tang, "Research and Education in Japanese Agricultural Development," *Economic Studies Quarterly*, 13 (February-May 1963), pp. 27-41 and 91-99.

E. B. Wennergren and M. D. Whitaker, "Social Return to U.S. Technical Assistance in Bolivian Agriculture: The Case of Sheep and Wheat," *American Journal of Agricultural Economics*, 59 (August 1977), pp. 565-69.

In addition to the studies listed in the table, there have been several other important research impact studies in which results are reported in a cost-benefit rather than an internal rate of return format.

L. L. Bauer and C. R. Hancock, "The Productivity of Agricultural Research and Extension Expenditures in the Southeast," *Southern Journal of Agricultural Economics*, 7 December 1975), pp. 177-22.

J. S. Marsden, G. E. Martin, D. J. Parham, T. J. Risdill, and B. G. Johnston, *Returns on Australian Agricultural Research: The Joint Industries Assistance Commission – CSIRO Benefit-Cost Study of the CSIRO Division of Entomology* (Canberra: Commonwealth Scientific and Industrial Research Organization, 1980).

H. Graham Purchase, "The Etiology and Control of Marek's Disease of Chickens and the Economic Impact of a Successful Research Program," in *Virology in Agriculture: Beltsville Symposium in Agricultural Research-I*, John A. Romberger, ed. (Montclair, N.J.: Allanheid, USMUN, 1977), pp. 63-81.

The contributions of research to increased agricultural productivity have been studied primarily by two methods. The estimates listed under the "index number" heading in table 10.3 were computed directly from the costs and benefits of research on, for example, hybrid corn. Benefits were estimated by using accounting methods to measure the increase in production attributed to hybrid corn. The contribution of research was usually measured as the residual after all other factors that contributed to increased production were accounted for. The calculated returns represent the average rate of return per dollar invested over the period studied, with the benefits of past research assumed to continue indefinitely. Benefits are defined as the benefits retained in the form of higher incomes to producers or passed on to consumers in the form of lower food prices.

The estimates listed under the "regression analysis" heading are computed by a different method, which permits estimation of the incremental return from increased investment rather than the average return from all investment. Further, this method can assign parts of the return to different sources, such as scientific research and extension advice. Because regression methods are used, the significance of the estimated returns from research can be tested statistically. The dependent variable is the change in total productivity, and benefit is defined as the value of the change in productivity. The independent variables include research variables, which reflect the cost of research and the lag between investment and benefit. The objective of the regresssion procedure is to estimate that component of the change in productivity that can be attributed to research.

The estimates from regression analysis can be exemplified by the study prepared by A. Tang for Japan for 1880 to 1938. The money spent on research in Japan during this period were estimated to have increased the total productivity of Japanese agriculture and to have yielded a significant annual return of 35 percent.

The effects of the time and type of research have been analyzed in greater detail in the 1979 studies by Robert Evenson for the United States. These results, along with the regression equations used in the study, are presented in table 10.4.[3] Changes in the productivity of American agriculture from 1868 to 1971 were related to the research performed by the state agricultural experiment stations and the U.S. Department of Agriculture. The effects of agricultural extension and the education of farmers were also taken into account.

During the period between 1868 and 1926, an estimated 65 percent annual rate of return was realized on this investment. From 1927 to 1950, the research was divided into two types. The first was called technology oriented and was defined as research in which new technology was the primary objective. This included plant breeding, agronomy, animal production, engineering, and farm management. The second type was called science oriented. Its primary objective was answering scientific questions related to the production of new technology. Science-oriented research included research in phytopathology, soil science, botany, zoology, genetics, and plant and animal physiology at the state experiment stations or the U.S. Department of Agriculture. The science-oriented research analyzed here is conducted in institutions in which it is closely associated with technology-oriented research. It is possible that the results might not apply, or would apply with a longer time lag, to science-oriented research isolated by organizational or disciplinary boundaries.

From 1927 to 1950, technology-oriented research yielded a rate of return of 95 percent. During the same 23 years, science-oriented research yielded a 110 percent rate of return, even more then technological research. The years 1927 to 1950 were a period of substantial biological invention, exemplified by hybrid corn, and improvements in the nutrition of plants and animals and in veterinary medicine. It was also a period of rapid mechanization. It is important to notice in the equations in table 10.4 that science-oriented research does not have a significant independent effect. The high payoff to science-oriented research is achieved only when it is directed toward increasing the productivity of technology-oriented research.

Research conducted in one state changes productivity in other states. This is referred to as "spillover." It has been estimated that, for 1927 to 1950, 55 percent of the change in productivity attributed to technology-oriented research from a typical state was realized within that state. The remaining 45 percent was realized in other states with similar soils and climate. The spillover from science-oriented research was considerably greater. The observations of 1948 to

Table 10.4. Estimated Impacts of Research and Extension
Investments in U.S. Agriculture

Period and Subject	Annual Rate of Return (%)	Percentage of Productivity Change Realized in the State Undertaking the Research
1868-1926:		
All agricultural research	65	not estimated
1927-1950:		
Technology-oriented agricultural research	95	55
Science-oriented agricultural research	110	33
1948-1971:		
Technology-oriented agricultural research		
South	130	67
North	93	43
West	95	67
Science-oriented agricultural research	45	32
Farm management and agricultural extension	110	100

Source: Robert E. Evenson, Paul E. Waggoner, and Vernon W. Ruttan, "Economic Benefits from Research: An Example from Agriculture," *Science*, 205 (September 14, 1979), pp. 1101-7. Copyright 1979 by the American Association for the Advancement of Science. *Note*: The regression equations, standard errors of parameters (in parenthesis), coefficients of determination (adjusted for degree of freedom), and numbers of observations (N) are as follows:
1868-1926

(1) $P = 45.29 + .521\ INV + .813\ RES + 3.04\ LANDQ$
      $(.162)\quad (.171)\quad (23.38)$

   $R^2 = .634; N = 40$ years

1927-1950

(2) $LN(P) = 1.40\ LN(INV) + .106\ LN(TRES) + .0000053\ LN(TRES)*(SRES)$
      $(.24)\qquad (.037)\qquad (.0000033)$

   $R^2 = .503; N = 24$ years x 4 regions

1948-1971

(3) $LN(P) = .0331\ LN(TRES-S) + .0119\ LN(TRES-N) + .0187\ LN(TRES-W)$
      $(.0085)\qquad (.0085)\qquad (.0089)$

$+ .2061\ LN(TRES)*SRES + .3540\ LN(ED) - .0394\ LN(EXT)$
$(.0710)\qquad (.0426)\qquad (.0097)$

$-.0116\ LN(EXT)*ED + .1821\ LN(TRES)*EXT$
$(.0021)\qquad (.0230)$

   $R^2 = .569; N = 23$ years x 48 states

(4) $LN(P) = .0299\ LN(TRES-S) + .0040\ LN(TRES-N) + .0113\ LN(TRES-W)$
      $(.0090)\qquad (.0090)\qquad (.0090)$

      $+.5639\ LN(TRES)*SRES + .5855\ LN(ED) - .02539\ LN(EXT)$
      $(.0104)\qquad (.0369)\qquad (.0102)$

$$-.0196 \ LN(EXT*ED) + .1369 \ LN(TRES)*EXT + .00148 \ LN(TRES)*SUB$$
$$(.0021) \qquad\qquad (.0044) \qquad\qquad (.00017)$$
$$R^2 = .595; N = 23 \text{ years} \times 48 \text{ states}.$$

Each equation also included region and time period dummy variables.

The 1948-1971 equations also included a business-cycle variable and a cross-sectional scaling variable.

Variables: P: total productivity index; INV: index of inventions; RES: stock of all agricultural research with time weights; LAND: land quality; TRES: stock of technology-oriented research with time and pervasiveness weights (S' W' N, for South, West, North); SRES: stock of science-oriented research; ED: schooling of farm operators; EXT: extension and farm management research stocks: LN is natural logarithm; *indicates variables multiplied.

1971 for individual states allowed still more detailed analysis. Technological research continued to yield returns of over 90 percent. The payoff to research was especially high in the South, where research had lagged in earlier periods. Science-oriented research remained profitable as it interacted with technological research, but it was less profitable than during 1927 to 1950.

Evidence concerning the effects of the education of farmers and the availability of extension advice on productivity can also be obtained from the equations used to estimate the results presented in table 10.4. The schooling of farm operators had a direct positive effect. The effect of extension education and farm management advice is more complex. It was particularly beneficial in those states with both considerable technological research and farmers with little schooling. The effect of these interactions, combined with the direct effects of extension, was positive.

The effect on productivity of decentralization of scientists to substations was also captured by the regression equations in table 10.4. There has been considerable debate on how a shift in the distribution of scientists between the central state stations and substations would affect the productivity of technological research. (See chapter 7.) In the regression equation, the fraction in the substation is multiplied by technological research. The interaction was positive and significant, indicating that decentralization has had a positive effect on the productivity of state research systems.

THE CHARACTER OF PUBLIC AGRICULTURAL RESEARCH IN THE UNITED STATES

Three characteristics are prominent in the American agricultural research establishment. It is articulated, decentralized, and undervalued. *Articulation* implies systematic interrelation of parts to form an integrated whole. *Decentralization* implies dispersion of authority and function at the regional or state level in contrast to centralization of authority and function. And *undervaluation* describes an invest-ment that yields a high return at a low cost; it indicates too little investment.

From their first settlements in America, the Europeans articulated science with farming. The first governor of Connecticut was a mem-ber of the Royal Society and reported his experiments with maize in the Royal Society's *Philosophical Transactions* in 1678. A Virginian reported the effect of soil upon tobacco quality in the transactions in 1688. During the first half of the 19th century, agricultural soci-eties were formed, and they helped scientists report to Americans the results of Liebig's theories on soil fertility. In 1835 a farmer became commissioner of patents and introduced scientific agricul-ture into the federal government via his office.

After seeing a Saxon Landwirtschaftlich Versuchsstation during his student years, a Yale professor led Connecticut into establishing the first American experiment station. A dozen years later, the U.S. Congress confirmed the existing articulation and encouraged decen-tralization by enacting the Hatch Act for "experiment respecting the principles and applications of agricultural sciences," which resulted in stations being established in every state.

The director of the first station wrote a timeless specification for an agricultural scientist: "Unites the requisites of the philosopher and the man of business" and possesses a practical knowledge of agriculture "that he may be able to elucidate and elevate it by sci-ence." This articulation of theory with practice is exemplified by the scientist who invented double-cross hybrid corn, produced hybrid seed without detasseling, stated the principle of heterosis, belonged to the National Academy of Sciences, and answered garden questions in the *Rural New Yorker*.

In experiment stations, the articulation can be seen in the multi-disciplinary training of the faculty of departments such as soils, agronomy, and entomology. The articulation can also be seen in the association between experiment stations and the extension services created by the Smith-Lever Act. In the agricultural research establish-ment, there are connections and communications among theoretical research, practical research, and farm production.

The Hatch Act assured the existence of a decentralized agricultural research system. It spread state agricultural experiment stations across the nation. Further, most scientists of the U.S. Department of Agriculture are dispersed, often in cooperation with the state sys-tems. A distribution of researchers exposes scientists to the problems of farmers, gives farms and extension workers easy access to specialists and their libraries, spins off talent and ideas, and gives a region the technological capacity essential to development.

Decentralization strengthened the articulation between science and farming. Although few experiment station directors have emulated the 19th century director of the Wisconsin agricultural experiment station who emphasized his affinity for farmers by wearing overalls when he sought money from the state legislature, most continue to pay careful attention to balancing effectively the mix in their research portfolios between science-oriented research and technology-oriented research that is expected to pay off in terms of state economic growth. The oversight exercised by legislative bodies has been weakened in much of the scientific enterprise. Yet, the articulation among legislatures, rural constituencies, and agricultural scientists has remained strong. Research by the U.S. Department of Agriculture continues to respond to priorities expressed by the Congressional agricultural and appropriation committees.

State and regional interests are reflected in the use of the federal funds appropriated for use by the state stations. Approximately two-thirds of the federal funds available to the states is allocated by formula, rather than in the form of project grants, to the state experiment stations. There is probably even greater articulation between constituency interests and research support in the case of the other funds of the state stations. Approximately two-thirds of the funds available to state experiment stations are based on state appropriations that reflect the evaluation of each station by state legislators. (See table 9.1.)

The outcome of the articulation can be seen in the congruence between the value of a crop or an animal and the money spent on research by the state stations and the U.S. Department of Agriculture. (See figure 10.2.) As one would expect, more money is spent on research on livestock relative to their value than on field crops; and most is spent on horticultural crops, which are produced on small acreages, are subject to severe pest and disease problems, and have high requirements for quality. Within each of the three groups, however, there is congruence between research and value of the product. The degree of congruence grew closer in the period between 1965 and 1975 even though research expenditures per dollar value of crops declined.

WHY DOES AGRICULTURAL RESEARCH REMAIN UNDERVALUED?

The high rate of return estimates for investment in agricultural research obtained in most of the studies summarized in table 10.3

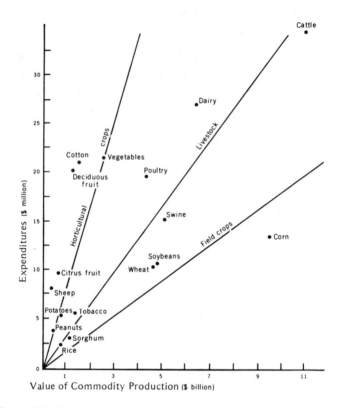

Figure 10.2. Congruence between Research Expenditures and Value of Commodity Production in 1975 (in constant 1967 dollars).

force to the surface several questions. Why does society underinvest rather than overinvest in public-sector agricultural research? Several hypotheses can be suggested. Before doing so, however, it is worth addressing the question of the accuracy of the rate of return estimates.

Accuracy of rate of return estimates

How accurate are the rate of return estimates? When first confronted with the results presented in table 10.3, many skeptics simply refuse to believe the results.

The presentation of the results of the early hybrid corn and sorghum studies in the form of "external" rather than "internal" rate of return estimates did result in considerable confusion concerning the interpretation and in skepticism regarding the validity of the estimates.[4] There have also been several major methodological criticisms of the earlier rate of return studies. One is that they have often failed to take into consideration the complementary technical inputs and the related marketing and extension education costs incurred in order to realize the productivity gains resulting from the adoption of the new technology. A second criticism is that the rate of return estimates are highly sensitive to assumptions about the form of the supply curve shift that is associated with adoption of the new technology. Some critics have also argued that an exaggerated impression of the rate of return has resulted from the selection of a few cases of spectacular research success, such as hybrid corn.[5]

It is clear that the assumptions employed in several of the early rate of return studies, particularly those conducted within the index number tradition, did lead to exaggerated rate of return estimates. The assumptions about the role of complementary inputs and the slope of the supply curve were particularly critical in Griliches's hybrid corn studies. The criticisms do not apply with the same force to the more recent index number studies or to the studies that have been conducted within the production function tradition. Most of the more recent index number studies have assumed a divergent shift in the supply curve. Any bias introduced by this assumption has the effect of underestimating the true rate of return. The production function studies have explicitly taken into account the complementary effects of related inputs, and the newer index number studies have typically given more careful attention to budgeting the costs of complementary inputs. A number of studies are now available within both traditions that estimate rate of return to national research systems rather than to individual commodities. There is also a tendency, since the important study by Schmitz and Seckler of tomato harvesting in California, to consider the distributional implications of agricultural research.

A review of the body of literature summarized in tables 10.3 and 10.4 impresses one with the increasing degree of sophistication that the authors of the more recent studies have displayed in responding to the limitations of the earlier studies. The effect of more careful model specification, more complete measurement of costs, and greater caution in estimating benefits has led, in my judgment, to results that tend to underestimate rather than overestimate the returns to agricultural research.[6]

In my judgment, there are two major reasons for the continued high rate of return to public-sector agricultural research in the

United States. One is the efficient allocation of resources to research; the other is the underinvestment due to a lack of congruence between the costs and the benefits from research.

Efficient allocation of resources to research

As noted earlier in the discussion of decentralization and articulation, the United States is characterized by both federal and state agricultural research systems. In recent years, over two-thirds of the funding for the state systems has been appropriated by the individual state legislatures. Most of the federal support for state agricultural research is allocated to each state by a formula based on the number of farms and the size of the rural population in that state. Furthermore, the research activities of the national agricultural research system administered by the U.S. Department of Agriculture are conducted, because of the location-specific nature of agricultural research, at widely dispersed locations and often in cooperation with the state research system.

The United States has, therefore, a public-sector agricultural research system that, in addition to being an industry leader that operates under a decentralized management system (the USDA/SEA/AR), includes 50 state-level "firms" whose output is viewed by its clientele as an input into state economic development. It seems reasonable to expect that an agricultural research system organized along lines roughly consistent with the competitive model helps explain why U.S. public-sector agricultural research is induced to select a high payoff research portfolio. There is a short feedback loop among the experiment station, farm and agribusiness clientele, and the appropriation process in state legislatures. This appears to induce a pattern of portfolio choice behavior on the part of state experiment stations that is consistent with the behavior expected from firms in a competitive market. These pressures induce the director of the Minnesota agricultural experiment station, for example, to allocate the resources available to the station in a manner that will enable Minnesota farmers to remain competitive with Iowa corn producers, Illinois soybean producers, and Wisconsin dairy farmers.

The "competitive" organization of the U.S. agricultural research industry may help explain the research resource allocation process leading to the selection of an "efficient" research portfolio. However, it does not go very far in helping one to understand the underinvestment in agricultural research implicit in the high rates of return suggested in tables 10.3 and 10.4. If the USDA-state agricultural research system is efficient and if it continues to achieve high rates of return, why does the political process, motivated by organized producers and their legislative allies, continue to undervalue, and to underinvest in, public-sector agricultural research?

Regional spillover effects

A second possible explanation of the underinvestment in agricultural research lies in the spillover effects and the resulting lack of congruence between the costs and the benefits of agricultural research. There are two important dimensions to the problem.

One dimension stems from the fact that the research that is paid for by one state can be expected to have an impact on productivity growth in other states (table 10.4). Much of agricultural research, particularly at the applied end of the research and development spectrum, is highly specific to particular agroclimatic regions. The results presented in table 10.4 indicate that there are substantial spillover effects among states and nations in the areas of basic research, supporting research, and even applied research. In the case of wheat, for example, of the total area planted to crosses released by public agencies (9,615,896 acres), 38.6 percent (3,709,100 acres) represented varieties developed in other states. The proportions were particularly high among the major varieties for: Blueboy (North Carolina), 90.6 percent; Caprock (Texas), 72.4 percent; TAM W-101 (Texas), 71.6 percent; and Twin (Idaho), 65.8 percent. Blueboy was raised in 21 states, far more than any other variety. Virtually all states "borrowed" varieties developed by public agencies in other states.[7]

It might be argued that it should be the precise function of the USDA's Agricultural Research Service to focus its efforts in those areas of basic and applied research in which geographic spillover limits the incentive for individual states to invest in research. However, it is my impression, and that of others, that the USDA's agricultural research system is even more oriented toward applied research than the state system is.[8] This may reflect the interests of the commodity-based clientele groups that tend to support the research program of the USDA in contrast to the more general area-based support for the state agricultural experiment stations.

The formula funding arrangement requires the states to match the federal support for state research. This represents, in effect, partial compensation to the individual states for the benefits other states, and the nation, receive from state agricultural research. The plausibility of the spillover hypothesis is supported by the differences among small and large states in their support for agricultural research.

The small states, which tend to capture the smallest share of the benefits from research in their own states, tend to allocate only enough funds to agricultural research to meet federal matching requirements. The large states appropriate substantially greater amounts for agricultural research than are required to match the federal formula funding. For example, in 1975 the California agricultural experiment station received $2.4 million in federal formula funds and $33.9 million in state appropriations. The Illinois station received $2.3 million in federal formula funds and $6.5 million in state appropriations.[9]

Another test of the spillover hypothesis would be to compare rates of return among commodity research areas and between technology- and science-oriented research. The results of crop research tend to be more location specific than the results of animal research. And the results of technology-oriented research tend to be more location specific than the results of science-oriented research. It would seem reasonable to expect lower rates of return (and less underinvestment) for the more location-specific areas. The two studies listed in table 10.3 that lend themselves to this test are the results reported by Bredahl and Peterson (1976) and by Evenson, Waggoner, and Ruttan (1979). Bredahl and Peterson's results indicate, as expected, lower marginal returns to cash grain research than to dairy and other livestock research. The results of the comparison between science-oriented and technology-oriented research are less conclusive. During the period between 1927 and 1950, the rate of return to science-oriented research was higher than that to technology-oriented research. Between 1948 and 1976, the rate of return to technology-oriented research was higher than that to science-oriented research.

It seems quite clear, however, that the present matching arrangement does not induce an optimum level of state appropriations for agricultural research. The significant spillover of benefits from state agricultural research to other states implies that the optimum level of investment by an individual state is below the level that would be optimum if it were evaluated at a regional or national level. Other formula arrangements might be considered that would more effectively compensate the individual states for the benefits that spill over into other states and that would induce a more efficient level of state appropriations for agricultural research. One possibility would be to revise the formula to require federal matching of state appropriations rather than the present method that requires state matching of federal appropriations.

Spillover from producers to consumers

A second dimension of the spillover effects of agricultural research is the transfer of the gains from research from producers to consumers. The manner in which the gains from technical change in agriculture are partitioned between the producers and consumers of a particular commodity depends on the slopes of the demand and supply curves for the product and the rates at which the two curves are shifting to the right over time.[10] In a market characterized by highly elastic demand or by rapid growth in demand, producers are able to retain a relatively large share of the gains from technical change. In a market characterized by inelastic demand and by slow growth in demand, most of the gains from technical change are passed on to consumers in the form of lower product prices. (See chapter 11.) In high-income countries with low rates of growth in demand, such as the United States, some Western European countries, and Japan, the gains from productivity growth in agriculture in the past have been rapidly transferred from producers to consumers.

Under competitive market conditions, the early adopters of the new technology in the agricultural sector have tended to gain while the late adopters are forced by the product market "treadmill," which grinds out lower prices as adoption proceeds, to adopt the new technology in order to avoid even greater losses than if they retained the old technology. One effect of the treadmill phenomenon, to the extent that it is recognized by farmers, is to limit the economic motivation for support of agricultural research to a relatively small population of early adopters of new technology. The early adopters also tend to be the most influential and politically articulate farmers. Support for agricultural research has not been able to achieve as broad a base among the farm population as support for commodity price programs. Apparently, the benefits from commodity price support programs, which have the effect of slowing the transfer of the gains from technical change from producers to consumers, are perceived to have a more immediate impact and to be more broadly shared within the agricultural community than the benefits from agricultural research.

How effective is the transfer of productivity gains from the farm to the nonfarm sector in generating consumer support for agricultural research? Studies of factors affecting the level of state funds for agricultural research indicate that differences in state nonfarm income have been even more important than differences in state farm income in accounting for variations in state support for agricultural research.[11] On a more disaggregated basis, however, differences in farm income were more important than differences in nonfarm income in accounting for variations in state support for departments engaged in production research (such as agronomy and animal science) while differences in nonfarm income were more important in accounting for variations in support for departments with a public affairs or consumer orientation (such as agricultural economics and horticulture). This suggests that both consumers and producers tend to support those agricultural research activities with which they have the most direct contact. The relatively sophisticated arguments based on relative shifts in demand and supply functions and on changes in producers' and consumers' surpluses have apparently been difficult to translate into a language that generates political support from organized producers or consumers. The large gains to consumers as a group are, when divided among individual consumers, too small to induce sustained consumer support for production research.[12] Support tends to emerge during periods of sharply rising prices and to dissipate rapidly during periods of relative price stability.

SOME IMPLICATIONS

The U.S. public-sector agricultural research system appears to be relatively efficient in the allocation of research resources. The funds allocated to agricultural research result in high rates of return and explain a significant share of productivity growth. The rates of return are high compared to almost any form of public-sector investment.[13]

The U.S. agricultural research system is far less effective in resource acquisition than in the use of research resources. The analysis presented in this chapter suggests that both its efficiency in the allocation of research resources and its capacity to mobilize resources are strongly related to its decentralized organization. A highly centralized system might be expected to be less efficient in allocating resources to research but more effective in resource acquisition. If the system were more highly centralized, a combination of a loss in efficiency in research resource allocation and an increase in funding for research might be expected to drive the returns to agricultural research to more conventional levels.

A clear implication of this chapter is that the design of institutional innovations to facilitate the allocation of resources to the provision of bureaucratic services in areas in which there is substantial underinvestment by the public sector should remain an important area of economic inquiry. The redesign of the formula for federal support of state research is one possibility that should receive careful attention. As the international assistance agencies expand their support for national agricultural research systems, it would seem desirable to give considerable attention to the development of an incentive structure that would induce an optimum level of funding from both national and international sources.

NOTES

1. This chapter depends very heavily on two papers: Robert E. Evenson, Paul E. Waggoner, and Vernon W. Ruttan, "Economic Benefits from Research: An Example from Agriculture," *Science*, 205 (September 14, 1979), pp. 1101-7; and Vernon W. Ruttan, "Bureaucratic Productivity: The Case of Agricultural Research," *Public Choice*, 35 (1980), pp. 529-47.

2. For a technical discussion of the problems involved in agricultural productivity measurement, see American Agricultural Economics Association Task Force on Measuring Agricultural Productivity, *Measurement of U.S. Agricultural Productivity: A Review of Current Statistics and Proposals for Change* (Washington, D.C.: U.S. Department of Agriculture, ESCS Technical Bulletin No. 1614, February 1980). Total and partial productivity measures for the agricultural sector in the United States are published annually by the USDA. For the latest report, see Economics, Statistics, and Cooperative Service, *Changes in Farm Production and Efficiency: 1978* (Washington, D.C.: U.S. Department of Agriculture, Statistical Bulletin No. 628, January, 1980). For a discussion of the methodological problems of constructing productivity growth estimates in developing countries, see Yujiro Hayami, Vernon W. Ruttan, and Herman M. Southworth, eds., *Agricultural Growth in Japan, Taiwan, Korea and the Philippines* (Honolulu: University Press of Hawaii, 1979).

3. For a more detailed discussion of the results presented in table 10.4, see Robert E. Evenson, "A Century of Agricultural Research and Productivity Change in U.S. Agriculture: A Historical Decomposition Analysis," a paper presented to the Symposium on Agricultural Research and Extension Evaluation, Moscow, Idaho, May 21-23, 1978.

4. When the "external" rate of return method is used, the flows of costs and benefits are accumulated (or discounted) to a point in time using a rate of interest (k) that is intended to reflect the opportunity costs of capital. The research costs are expressed as an accumulated capital sum. The benefits (value of inputs saved) are also accumulated to the same point of time but are then expressed as a perpetual flow. The external rate of return is obtained by dividing the annual flow of benefits by the accumulated costs (past research expenditures) and expressing the results as a percentage. The external rate of return (r) estimates are directly translated into a benefit-cost (B/C) ratio by B/C = r/100k. If the external rate of return is 75, the B/C ratio is 7.5. Thus, the B/C raio and the external rate of return are just two ways of expressing the same concept. Both the B/C ratio and the external rate of return are highly sensitive to the rate of interest that is chosen to reflect the opportunity cost of capital. The "internal" rate of return avoids this problem. It is the rate of interest that makes the accumulated present value of its flow of costs equal to the discounted flow of returns at a given point in time. The external rate of return to hybrid corn research, estimated by Griliches who used a 5 percent oppotunity cost for capital, of 743 percent per year converts to an internal rate of return (both calculated to 1955) of 37 percent. Even though there is a rather large difference between the external and the internal rates of return, both are based on the same data and assumptions and are but alternative ways of expressing the same flow of costs and benefits. For a further discussion of the

internal and external rates of return calculations, see Willis Peterson, "The Returns to Investment in Agricultural Research in the United States," in *Resource Allocation in Agricultural Research*, Walter L. Fishel, ed. (Minneapolis: University of Minnesota Press, 1971), pp. 139-62.

5. This criticism and several of the others reported in this paragraph are discussed by W. S. Wise, "The Role of Cost Benefit Analysis in Planning Agricultural R & D Programmes," *Research Policy*, 4 (July 1975), pp. 246-61. They were aired rather thoroughly at a 1975 Airlie House Conference. Webster and Ulbricht were particularly vigorous in their discussion of the limitations of the earlier studies. See the report of the discussion in the summary paper by Thomas M. Arndt and Vernon W. Ruttan, "Valuing the Productivity of Agricultural Research: Problems and Issues," in *Resource Allocation and Productivity in National and International Agricultural Research*, Thomas M. Arndt, Dana G. Dalrymple, and Vernon W. Ruttan, eds. (Minneapolis: University of Minnesota Press, 1977), pp. 1-25. R. K. Lindner and F. G. Jarrett, "Supply Shifts and the Size of Research Benefits," *American Journal of Agricultural Economics*, 60 (February 1978), pp. 48-58, have pointed out that the assumption of a convergent or parallel shift in the supply curve will result in the overestimation of the rate of return if, in fact, the shift is divergent. Conversely, assumptions of a divergent shift will result in the underestimation of the rate of return if the shift is actually parallel or convergent.

6. Jeffrey S. Davis, "A Comparison of Alternative Procedures for Calculating the Rate of Return to Agricultural Research Using the Production Function Approach" (St. Paul: University of Minnesota, Department of Agricultural and Applied Economics Staff Paper P79-19, May 1979). Davis computed marginal internal rates of return to U.S. agricultural research for 1964 by using the estimating equations employed by several earlier investigators. He concluded that the procedures used to arrive at the estimates (in table 10.3) by Evenson (1968), Bredahl and Peterson (1976), and Lu and Cline (1977) resulted in the modest underestimation of the rate of return.

7. Dana G. Dalrymple, *The Development and Spread of Semi-dwarf Wheat and Rice Varieties in the United States* (Washington, D.C.: U.S. Department of Agriculture, Agricultural Economic Report No. 455, June 1980).

8. National Research Council, *Report of the Committee on Research Advisory to the U.S. Department of Agriculture* (Springfield, Va.: National Technical Information Service, 1973).

9. Annual data on the sources of funds available to state agricultural experiment stations have been published annually by the Cooperative State Research Service, *Funds for Research at State Agricultural Experiment Stations and Other Cooperating Institutions* (Washington, D.C.: U.S. Department of Agriculture, 1975).

10. For a discussion of the spillover mechanism, see Willard W. Cochrane, *Farm Prices, Myth and Reality* (Minneapolis: University of Minnesota Press, 1958). For estimates of the gains to consumers, see M. Akino and Y. Hayami, "Efficiency and Equity in Public Research: Rice Breeding in Japan's Economic Development," *American Journal of Agricultural Economics*, 57 (February 1975); pp. 1-10. In the United States, the short-run price elasticity of demand for domestically consumed food at the farm level is approximately -0.10 and for exports, approximately -1.0. Since approximately 20 percent of the U.S. farm output is exported, the overall elasticity of demand is in the neighborhood of -0.25, according to Fred C. White and Joseph Havlicek, Jr., "Impacts of Alternative Levels of Investment in Agricultural Research and Extension" (Blacksburg, Va.: Virginia Polytechnic Institute and State University, Department of Agricultural Economics, mimeographed paper, 1978.

11. See W. L. Peterson, "The Allocation of Research, Teaching, and Extension Personnel in U.S. Colleges of Agriculture," *American Journal of Agricultural Economics*, 51

(February 1969), pp. 41-55; and Wallace E. Huffman and John A. Miranowski, "An Economic Analysis of Expenditures on Agricultural Experiment Station Research," *American Journal of Agricultural Economics*, 63 (February 1981), pp. 104-18.

12. According to M. Olson, Jr., *The Logic of Collective Action: Public Goods and the Theory of Groups* (Cambridge, Mass.: Harvard University Press, 1965), the constraints on group mobilization are particularly relevant to the issue of mobilizing consumers in support of agricultural research. "First, the larger the group, the smaller the fraction of the total group benefit that any person acting in the group interest receives. . . . Second, . . . the larger the group, . . . the less likelihood that any small subset of members . . . will gain enough from getting the collective good to bear the burden of providing even a small amount of it. . . . Third, the larger the number of members of the group, the greater the organization costs and thus the higher the hurdle that must be jumped before any of the collective good at all can be obtained" (p. 48).

13. Robert Haveman reported, for example, that of 147 water resource projects constructed in 10 southern states between 1946 and 1962, only 9 had ex ante rates of return above 20 percent (in *Water Resource Investment and the Public Interest* [Nashville, Tenn.: Vanderbilt University Press, 1965]). In his more recent work, *The Economic Performance of Public Investments: An Ex Post Evaluation of Water Resource Investment* (Baltimore: Johns Hopkins University Press, 1972), Haveman suggested that there is a substantial upward bias in the *ex ante* estimates. There has been a tendency to assume that the payoff to agricultural extension activities would be lower than the payoff to research. However, several recent studies have suggested rates of return to extension in the same range as to agricultural research. See Wallace E. Huffman, "The Productive Value of Human Time in U.S. Agriculture," *American Journal of Agricultural Economics*, 58 (November 1976), pp. 672-83; and "Returns to Extension: An Assessment" (Ames, Iowa: Iowa State University, Department of Agricultural Economics, mimeographed paper, 1978). See also A. Halim, "The Economic Contribution of Schooling and Extension to Rice Production in Laguna, Philippines," *Journal of Agricultural Economics and Development*, 7 (January 1977), pp. 33-46. Similar rates of return have also been reported for the statistical services of the U.S. Department of Agriculture by Yujiro Hayami and Willis Peterson, "Social Returns to Public Information Services: Statistical Reporting of U.S. Farm Commodities," *American Economic Review*, 62 (March 1972), pp. 119-30.

Chapter 11

Research
Resource Allocation

Should research be planned? The answer to this question often depends on the interpretation that the respondent attaches to planning.[1] The response is frequently confounded by the respondent's perception of the response to a second question: Who will have the authority for research planning? Researchers have often suggested, and with good reason, that the rates of return presented in table 10.3 place a major burden of proof on those who urge the use of more-formal planning methods to demonstrate that the resources devoted to planning will yield higher returns than the resources devoted to research.

Nevertheless, central management and planning staffs have been strengthened in most major national agricultural research systems during the 1960s and 1970s. (See chapter 4.) These more-intensive planning efforts have often been mandated by the legislative, financial, or planning bodies that are responsible for allocating resources to research. In the United States, for example, the Food and Agricultural Act of 1977 mandated the establishment of the Joint Council on Food and Agricultural Sciences in order to foster improved planning of federal and state agricultural research and the National Agricultural Research and Extension Users Advisory Board in order to attempt to reflect the priorities of both the users of research and those affected by research.

The planning staffs responsible for research resource allocation have not found it easy to respond to the expectations that their efforts would contribute both to greater efficiency in the use of research resources and to greater relevance in research resource allocation.

They have been pressed to respond to a succession of styles in analysis and planning: project and priority weighting or scoring of research objectives in the mid-1960s, program planning and budgeting in the late 1960s, systems analysis and simulation in the early 1970s, and the rhetoric of technology assessment in the mid-1970s. By the late 1970s, the program planning and budgeting methodology, which had temporarily fallen into disrepute as a result of the gap between promise and performance, had been resurrected under the rubric of zero-based budgeting.[2]

As planning activity has intensified, planning objectives have become more diverse. Concern with distributional impacts and environmental spillover has been added to the traditional concerns of quality of research performance and contributions to the productivity of the agricultural sector. It was no longer adequate to justify agricultural research in terms of making "two blades of grass grow where one grew before" or of increasing the productivity of farm workers by making it possible for a farm worker to cultivate twice as many acres in a day.

As research-planning staffs have struggled with the demands placed on them, it has become increasingly obvious that effective research planning requires close collaboration among natural and social scientists and among agronomists, engineers, and planners. This is because any research resource allocation system, regardless of how intuitive or how formal in its methodology, cannot avoid making judgments about two major questions.

What are the possibilities of advancing knowledge or technology if resources are allocated to a particular commodity, problem or discipline? If, for example, resources are allocated to the transfer, development, or enhancement of nitrogen-fixing capacity to grasses, what is the probability of success? The answers to such questions can only be answered with any degree of authority by scientists who are on the leading edge of the research discipline or problem being considered. The intuitive judgments of research administrators and planners rarely are adequate to answer such questions.

What will be the value to society of the new knowledge or the new technology if the research effort is successful? If efforts to develop nitrogen-fixing capacity in maize are successful, for example, will it become an efficient source of plant nutrition when evaluated in relation to the economic and environmental costs or other forms of nitrogen fertilizer? The answers to these questions require the use of formal economic analysis. The intuitive insights of research scientists and administrators are no more reliable in answering questions of value than the intuitive insights of research planners are in evaluating scientific or technical potential.

Many of the arguments about research resource allocations flounder on the failure of the participants to clearly recognize the distinction between these two questions and the differences in expertise and judgment that must be brought to bear on responding to them.

The purpose of this chapter is not to provide a technical exposition of research-planning methodology. Rather, it is to provide research administrators with a guide to, and an evaluation of, what can be expected from the planning methodologies available to them.

In the next section of the chapter, I discuss the parity or congruence model of research resource allocation that is implicit in much discussion of research resource allocation. I then give explicit attention to some of the considerations that determine the economic value of agricultural research and the distribution of gains from the productivity growth resulting from research. This is followed by a section that reviews some of the methodologies that might be used to select the individual research programs or projects that make up the research portfolio of a national agricultural research system or of an autonomous research institute.

This chapter is focused on the relatively narrow issue of the allocation of resources to production research. In chapter 13, I attempt to deal with some of the broader issues of social policy and agricultural research.

THE PARITY MODEL OF RESEARCH RESOURCE ALLOCATION

In chapter 10, substantial differences were noted in research expenditures relative to the values of individual commodities in the United States. In the cases of wheat and soybeans, the research expenditures in 1975 amount to about $2 per $1,000 of product. In the case of cotton, the research expenditures amount to about $15 per $1,000 of product. Were these ratios efficient? Were the expenditures on cotton too high? Were the expenditures on wheat and soybeans too low? When such figures are cited, there is often a presumption that the commodity characterized by a low research/output ratio is not getting its fair share of the research dollar. At a more sophisticated level, there is usually an implication that the return from an additional dollar would be highest if invested in research on the commodity with the lower research/output ratio. In either case, the critic usually has in mind, at least implicitly, what might be termed a "parity" model of research resource allocation.

The parity perspective is a useful first step in any analysis of research resource allocation. There are, however, two assumptions that are usually implicit in its application. The first assumption is that the opportunities for productive scientific effort or productivity-enhancing technical change are equivalent in each commodity and resource category. The second is that the value of a scientific or technical innovation is proportional to the value of the commodity or the value of the contribution of a particular resource to production.[3]

No one believes that either assumption is valid. Yet, in the absence of specific knowledge of research opportunities and payoffs, application of the parity model may not be entirely inappropriate. But even the parity rule may not be as simple to apply as one might suspect.

In agriculture the research resource allocation process involves a four-way allocation of resources: (1) among commodities, such as wheat, cotton, and beef; (2) among resource categories, such as soil and water, agricultural chemicals, labor, and management; (3) among stages or levels, such as industrial inputs, farm production, postharvest technology and markets, and community services; and (4) among disciplines, such as genetics, economics, and human nutrition. Even the concept of economic importance can easily become muddy. Should research resources be allocated between wheat production and bakery products in proportion to the market value of wheat and the market value of bakery products? Or should they be based on ratio of the value of wheat to the value added to the wheat by the baking industry? Should research dollars be allocated to livestock and livestock products in proportion to their market value or only in proportion to the value added by livestock to the feed they consume?

Whether one accepts the parity model as a primary criterion for the allocation of research resources or as a point of departure for the further fine tuning of research resource allocation, it would seem exceedingly important to be able to account for research expenditures in a manner that would permit accurate measures along the lines of the four dimensions outlined above. This would then permit the following parity or congruence calculations:

- A comparison of the ratio of research expenditure by commodity to the value added in farm production for each commodity.
- A comparison of the ratio of research expenditure by factor (or

resource) input to the cost or economic value of the factor (or resource) in production.

- A comparison of the ratio of research expenditure to the value added at each stage in the food production chain from purchased inputs to the consumer.
- A comparison of the ratio of research expenditure in each field of science to the value added for each commodity, factor, and stage.

Similar accounts should be accumulated in terms of scientist-years. This would permit comparisons of scientific effort as well as expenditures by factors, commodities, and stages. Capital investment in the form of facilities and major items of equipment should be reported separately from annual personnel and other operating costs.

The compilation of a set of research parity accounts does not imply a judgment that research resources should be allocated by a parity rule. It does suggest that an explicit rationale should be developed for any departures from a parity rationale. Reasonable bases for such departures are not difficult to develop. A favorable judgment concerning the production potential of soybeans in 1940 or sunflowers in 1970 would have been a sound basis for a relatively high ratio of research to value added. A judgment that private-sector investment in farm machinery research and development is adequate could be a sound basis for a low ratio of research expenditures to the value of machinery services in farming. The rationale for a judgment that the ratio of research expenditures to the value of natural resource inputs should be higher than to the value of labor and managerial inputs may not be as obvious.

Regardless of the desirability of using the parity model as a first step in the analysis of research resource allocation, its use is feasible for only relatively gross comparisons even in countries with relatively well developed data systems. The U.S. Department of Agriculture, for example, publishes data on the value of commodity marketings. The data it publishes on the value of crop production are incomplete. The value of forage crops that are fed on the farm on which they are produced is not reported. Neither are data on the value added in the production of livestock and livestock products separately reported.

The research expenditure classifications that have been developed also have severe limitations. They seem to have been put together without a clear conception of their potential analytical uses. A three-way classification (1) among activities, (2) among commodities or resources, and (3) among fields of science was developed as part of the 1966 *National Program* study.[4] (See table 11.1.) A modification of this classification serves as a basis for the USDA's Current Research

Information System (CRIS). The original CRIS classifications came reasonably close to meeting the criteria outlined above in its commodity and field of science (discipline) classifications. The natural resource subcategory, however, includes both resource inputs (soils) and commodities (forest products). Information on interdisciplinary research efforts, such as integrated pest management, is difficult to identify and to retrieve. And the activity classification seems to have no clear-cut rationale for its categories.

Criticisms similar to those I have made with respect to the U.S. system also apply to a number of other systems. The analytical uses of information systems typically have become apparent after the systems have been put in place rather than at the design stage.

The next sections of this chapter are devoted to a more systematic review of some of the criteria that might be used as a basis for departures from the parity model of research resource allocation.

ALLOCATION OF RESOURCES TO RESEARCH

An implicit assumption of the parity model of research resource allocation is that the benefits from research are proportional to the size of the research budget and to the economic significance of the commodity sector or resource (factor) input to which the research effort is directed.[5] Such an assumption is clearly naive. Yet, one of the attractions of the parity model is that it does leave implicit rather than make explicit the specification of the size and distribution of the benefits from research.

Research administrators and policymakers have traditionally attempted to avoid overly precise specification of research objectives. They have preferred to be able to mobilize support from farmer clientele and clientele representatives by emphasizing the contributions of research to the reduction of production costs, to increases in yields, or to the adaptation of crops to different environments. Administrators have preferred, at the same time, to emphasize to other constituencies the gains to consumers in the form of lower food costs or the contributions of agricultural exports to the solution of balance-of-payments difficulties.

In nations characterized by strong organization among farmers, benefits to producers tend to be emphasized and benefits to consumers, muted. In countries where farmers are poorly organized, consumer benefits tend to be emphasized. Thus, in many developing countries with large but politically inert rural populations, the primary emphasis in establishing a claim to research resources is often

Table 11.1. Classification Codesheet for Report of Agricultural Research for Fiscal Year Ending June 30, 1965

A. ACTIVITY	B. COMMODITY OF RESOURCES	C. FIELD OF SCIENCE
Conservation, Development and Use of Soil, Water, Forest, and Related Resources	*Natural Resources*	*Biological*
1. Resources description and inventory	1. Soil and land	1. Biochemistry and Biophysics
2. Resource conservation	2. Water	2. Biology— Environmental, Systematic, and Applied (Botany, Ecology, Zoology, etc.)
3. Resource development and management	3. Watersheds and river basins	
4. Evaluation of alternative uses and methods of use	4. Air and climate	
	5. Recreational resources	3. Biology—Molecular
	6. Timber and forest products	4. Entomology
Protection of Man, Plants, and Animals from Losses, Damage, or Discomfort Caused by	7. Range	5. Genetics
	8. Wildlife and fish	6. Immunology
5. Insects	*Crops and Crop Products*	7. Microbiology
6. Diseases, parasites, and nematodes	9. Citrus and subtropical fruit and tree nuts	8. Nematology
7. Weeds	10. Deciduous and small fruits	9. Nutrition and Metabolism
8. Fire and other hazards	11. Potatoes	10. Parasitology
	12. Vegetables	11. Pathology
Efficient Production and Quality Improvement	13. Ornamentals and turf	12. Pharmacology
	14. Corn	13. Physiology
9. Biology of plants and animals	15. Grain sorghum	14. Virology
10. Improving biological efficiency of plants & animals	16. Rice	
11. Increasing consumer acceptability of farm and forest products	17. Wheat	
	18. Other small grains	*Physics*
12. Mechanization and improvement of physical efficiency	19. Pasture	
13. Management of labor, capital, and other inputs to maximize income	20. Forage crops	15. Chemistry — Analytical
	21. Cotton	16. Chemistry — Inorganic
	22. Cottonseed	17. Chemistry — Organic
Product Development and Processing	23. Soybeans	18. Chemistry — Physical
	24. Peanuts	19. Engineering
14. Chemical and physical properties of food products	25. Other oilseed crops	20. Geology and Geography
15. Developing new and improved food products & processes	26. Tobacco	21. Hydrology
16. Chemical and physical properties of non-food products	27. Sugar crops	22. Mathematics and Statistics
17. Developing new and improved non-food products and processes.	28. Miscellaneous & new crops	
	Animals and Animal Products	
	29. Poultry	
	30. Beef cattle	
	31. Dairy cattle	
	32. Swine	
	33. Sheep and wool	
	34. Other animals	
	35. Bees and honey	
	Manmade Resources Used on Farms or by People	
	36. General purpose farm supplies and facilities, including equipment, structures, fertilizers, and pesticides	
	37. Clothing and textiles	

Table 11.1—Continued

A. ACTIVITY	B. COMMODITY OF RESOURCES	C. FIELD OF SCIENCE
Efficient Marketing, Including Pricing and Quality	38. Food	23. Meteorology
18. Identification, measurement & maintenance of quality	39. Housing, household equipment & non-textile furnishings	24. Physics
19. Improving economic & physical efficiency in marketing, including analysis of market structure and functions	*Human Resources, Organizations, and Institutions*	*Social and Behavioral*
20. Analysis of supply, demand and price, including interregional competition	40. People as individual workers, consumers, and members of society	25. Anthropology
21. Developing domestic markets, including consumer preference and behavior	41. The family and its members	26. Economics
22. Foreign trade, market development, and competition	42. The farm as a business enterprise	27. Education and Communications
Improvement of Human Nutrition and Consumer Satisfaction	43. Communities, areas, and regions, including counties and States and their institutions and organizations	28. History
23. Nutritional values, consumption patterns, and eating quality of foods	44. Agricultural economy of United States & sectors thereof, including interrelationships with the total economy	29. Law
24. Quality of family living, including management and use of time, money, and other resources	45. Agricultural economy of foreign countries and sectors thereof, including interrelationships with the total economy.	30. Political Science
Development of Human Resources and of Economies of Communities, Areas, and Nations	46. Farmer cooperatives	31. Psychology
25. Description, inventory, and trends	47. Other marketing, processing, and farm supply firms	32. Sociology
26. Economic development and adjustment	48. Marketing systems and sectors thereof	
27. Improvement of social well being, including social services and facilities and adjustment to social and economic changes	49. Research which cannot be allocated to one or more of the above commodities of resources	
28. Evaluation of public programs, policies & services		
29. Research which cannot be allocated to one or more of the above activities		

Source: U.S. Department of Agriculture and Association of State Universities and Land Grant Colleges, *A National Program of Research for Agriculture* (Washington, D.C.: U.S. Government Printing Office, 1966), p. 26.

placed on meeting national food needs. In many developed countries with relatively small but politically articulate agricultural constituencies, discussion of research benefits tends to emphasize the gains to agricultural producers. The contribution of agricultural research to foreign exchange earnings does not go unnoticed when a minister of agriculture discusses budget issues with a minister of finance or a director of the budget.

In spite of these differences in rhetoric, however, research administrators generally are very uneasy about attempts to implement "demand-oriented" research programs, that is, to direct research efforts toward specific social or economic objectives. They are much more comfortable with a "supply orientation," that is, with research efforts that attempt to take advantage of perceived opportunities for scientific or technical advance. This orientation toward the supply, or opportunity, side of the equation rather than toward the demand, or value, side, in my judgment, has often led to a lack of effectiveness on the part of research administrators in dealing with budget offices, legislative committees, and special-interest groups. There are an infinite number of interesting scientific problems, but not all of them are important. The effective research administrator must be able to resolve the interests of scientists in exploring the endless frontier of interesting problems with the legislative demands that funds be allocated to those areas that are most important.

In the next four sections I give particular attention to the factors that determine the distribution of the gains, and the losses, from research and that influence the incentives to support research.

Gains to producers and consumers in a closed economy

In a closed economy in which exports and imports are limited, it is fairly easy to sort out the gains to producers and consumers from agricultural research that enhances productivity growth. Initial gains from lower unit production costs are realized by farmers who first adopt the new technology. As the new technology is diffused and production increases, a larger and larger portion of the unit-cost reduction will be shared with consumers.

When the rate of growth in productivity (measured in terms of output per unit of total input or the rate of decline in unit costs) is less than the rate of growth in demand (from population and income growth), prices rise, but less rapidly than if there were no technical change. Although the consumers gain from less-rapid price increases, most of the gains are realized by producers. When the rate of growth in productivity is slightly more rapid than the growth of demand, prices decline and the gains from lower costs are shared by producers and consumers.

When the rate of growth in productivity is substantially more rapid than the rate of growth in demand, all of the gains from productivity growth may be transferred from producers to consumers in the form of lower prices. Thus, the share of the productivity growth retained by producers declines and the share transferred to consumers increases as the rate of productivity growth rises relative to the rate of growth in demand. The rate of transfer of productivity gains to consumers, in the form of lower prices, also is greater when consumer demand is unresponsive (that is, inelastic) with respect to price.

In a closed economy, when the primary objective of research investment is to assist consumers, research effort should be directed to commodities for which demand is relatively unresponsive to changes in income and prices. These are poor people's foods—basic carbohydrates, such as cassava, maize, rice, and wheat, and vegetable proteins, such as beans. If the primary objective is to improve the income of farmers, research should be directed toward commodities for which demand is responsive to income growth and price changes. These are the products consumed by the middle classes and the rich —animal proteins and commodities that add diversity and interest to the diet, such as vegetables and fruits, or that are associated with other forms of conspicuous consumption, such as ornamental horticultural crops and animals used for sport and recreation.

Gains to producers and consumers in an open economy

In an open economy, a larger share of the gains from research leading to unit cost reduction or productivity growth is captured by producers than in a closed economy. The extent to which the gains are retained by producers depends, in addition to domestic demand elasticities, on the amount of a particular commodity that a country exports relative to the size of the world market. When exports are small, a rise in productivity can permit expansion of exports without having a noticeable effect on the world price. In this case, essentially all of the gains are retained for the producers. When the country accounts for a significant share of the world exports of a commodity, expansion of exports can be expected to cause prices to fall, just as in a closed economy. In this case, the gains are shared with foreign as well as domestic consumers. Producers of competing products in other countries that do not experience comparable productivity growth lose.

The effect of productivity growth on producers and consumers of a commodity that is partially produced at home and partially imported is somewhat different than in the case of a commodity that is exported. When productivity growth enables the country to expand domestic production and partially eliminate imports, most of the gains are realized by producers. When the productivity gains are sufficient to enable the country to make the transition from importer to exporter, domestic consumers may realize substantial gains. Although the country is an importer, prices are typically somewhat above world market prices as a result of shipping and related costs. When the country becomes an exporter, prices typically decline to below the world market price in order to absorb shipping costs to the new export markets. Unless exports expand rapidly, producers may not realize sufficient gains to offset the price reduction resulting from the transition. All of the gains, and perhaps some of the losses to producers, are realized by domestic consumers.

If the objective of research is to increase income to producers, research should be committed first to those commodities that are currently exported and second to those commodities on which research would enable the country to achieve a substantial export market. If the objective is to transfer the gains to consumers, research should be focused on commodities in which the country imports a small share of its consumption. A small increase in production, relative to consumption, in this situation, will result in a transition to a net export position, or at least to a potential export position, and push domestic prices below world market prices.

Saving land, saving labor, and saving energy

The allocation of resources to research also involves choices about the importance of releasing the constraints imposed by resource suppliies—of saving land and labor, for example. These choices are implicit in decisions about expanding or contracting research on problems such as soil and water conservation, soil fertility, photosynthetic efficiency, mechanization, energy use, and labor efficiency and management. What criteria are available to determine the relative balance between agricultural research directed primarily toward decreasing labor requirements per hectare or research directed toward increasing the amount of product that can be produced per hectare?

Historically, the answer to this question has been quite clear. In countries such as the United States, Canada, and Brazil that had relatively abundant land resources and a strong demand for labor in industry, the primary thrust was toward improvements in mechanical technology that would enhance labor productivity. Only after expansion of the area available for cultivation became limited was attention turned to the development of technologies to expand output per hectare. The mechanical revolution in American agriculture began in the middle of the 19th century. The biological and chemical revolution did not begin until after the first quarter of the 20th century.

In countries such as Japan and Denmark that had abundant labor and relatively limited or poor land resources, the primary thrust in agricultural technology was toward increased output per hectare. In Japan, the emphasis was placed on increases in crop yields. In Denmark, the emphasis was placed on increased crop yields per hectare and on technologies that facilitate intensification in the production of livestock and livestock products. In both countries, research designed to enhance labor productivity was delayed until the agricultural labor force began to decline in response to the rising demand for labor in the nonagricultural sectors of the economy.

In addition to the changing relative prices of land and labor, declining real prices of energy gave a further impetus to the invention of technologies that permitted the substitution of mineral fuels for organic sources of energy: tractors that utilized petroleum-based fuels were substituted for animal power that utilized farm-produced feed. The declining price of energy, embodied in chemical fertilizers, encouraged the development of crop varieties capable of responding to higher levels of nutrition.

Since the early 1970s, the world has entered into a period of great uncertainty with respect to changes in the relative prices of labor, land, and energy. The end of the era of cheap energy, like the end of the era of cheap land, is inducing a reallocation of research efforts. But the new sources of productivity growth have not yet been clearly identified. Until the new trends become more evident than they have been during the 1970s, the appropriate allocation of research effort among land-, labor-, and energy-saving alternatives will remain uncertain. In this environment of great uncertainty, an efficient research portfolio will include a wide range of options. It should avoid becoming locked into a commitment, or a "fix," on any single option. It is, for example, too early to be able to make firm judgments about the proportion of nitrogen supply that can be expected from advances in biological nitrogen fixation, low-pressure nitrogen systems, and/or improvements in the efficiency of conventional high-pressure nitrogen technology. As long as this uncertainty remains, resources should be allocated to the exploration of the possibilities for each of the several potential options.

Research for large farms and small farms

During the 1960s and 1970s, agricultural research institutions in both developed and developing countries have been widely criticized for focusing their research efforts on the problems of large farms and for neglecting research that would be beneficial to small farmers. Research designed to improve labor productivity has been criticized on the grounds that it leads to displacement of workers by machines. An attempt has been made in the state of California to legislate restrictions on mechanization research at the University of California.

Some of this criticism is valid. Some of it is ideologically motivated. Much of it is confused. The long-term thrust of research and development on mechanical technology in American agriculture is a response to the rising real price of labor. It has been primarily a response to a labor shortage rather than a source of labor displacement. But this does not mean that the concern with premature mechanization may not be valid in specific situations. Subsidies to mechanization have occurred in several forms. In the United States, tax laws that provide for investment tax credits and accelerated depreciation schedules have driven a wedge between private profitability and economic efficiency. In Brazil and India, access to foreign exchange on excessively favorable terms has at times biased the choice of technology in favor of mechanization and often in favor of large-scale rather than smaller, intermediate-scale equipment.[6] When the choice of technology is subsidized in this manner, it has the effect of inducing related research, in fields such as agronomy and farm management, designed to improve the efficiency and to speed the diffusion of a technology that is itself not appropriate.

But this is only part of the issue of technology for small farms. Is it possible to design technologies that are specifically suited to the needs or the factor endowments of small farms? I find it very difficult to think of examples of technology that would have greater benefits—that is, result in greater unit cost savings—for small farms than for large farms. But it is not too difficult to think of technologies that are roughly neutral in their impact. Indeed, much of the yield-increasing biological and chemical technology is roughly neutral with respect to size. The effect of a new pesticide on rice yield may be no different when the pesticide is applied by aerial sprayers than when it is applied by backpack sprayers. The choice of application technology in this case would depend on the price of labor relative to the price of capital equipment.

There is, however, one way in which research benefits may be biased in favor of small farms in some cases. That is in the choice of commodity emphasis. In many countries in Latin America, beans are produced on small farms and beef is produced on large farms. A decision to improve the productivity of bean production does, therefore, have the effect of biasing the direct impact of gains from productivity growth in favor of small farms. Similarly, a decision to conduct research on beef does bias the direct impact in favor of large producers.

A choice in favor of beans relative to beef also directs the gains that get transferred to low-income consumers, many of whom are small farmers or hired laborers. How much of the gains from productivity growth will be retained by the farmers who initially adopt the new technology will depend on the relationship between growth of productivity and growth of demand, which was outlined in the sections on closed and open economies.

ALLOCATION OF RESOURCES IN RESEARCH

In the introduction to this chapter, it was noted that there are two distinct stages involved in research resource allocation or investment strategies. One stage involves an initial preordering of research programs based on some judgment of the potential value of the research. This decision may be made during the process of allocating resources to the research system or to the individiual research institute or station. A second stage involves the selection of individual research projects that can advance the work of the preselected program most effectively. This second stage always involves explicitly or implicitly a consideration of cost-benefit and cost-effectiveness criteria. This does not mean that these are the only criteria that are involved in project selection. A research director may, for example, continue to support a modest level of research activity by a relatively unproductive staff member in order to get some return, however modest, from the fixed costs of salary and related benefits.

In a small research institution with a highly personal style of management, allocation decisions may be made in conferences with the individual staff members or in committee meetings with research teams. Project documentation may serve primarily as confirmations of decisions rather than as an imput to the decision-making process. This procedure may be highly effective in a research organization in which the director has the professional background and the intellectual capacity to engage in effective dialogue with the individual researchers or research teams about the methodology and significance of the research effort.

In larger research organizations, using this personal management style is not feasible. Even if using it were feasible at the individual research laboratory, institute, or station level, it would not solve the problem of communication with the higher decision-making levels of a state or national research system. During the 1960s and the 1970s, a great deal of effort was devoted to the development of more-formal systems for the ordering of the information needed for research decision making. Walter L. Fishel has pointed out that these decision information systems "typically have two primary functions: filtration and condensation." Filtration is concerned with separating the relevant from the irrelevant. Condensation is concerned with the reduction of relevant data through analysis or other useful information techniques to the most meaningful form for the research administrator.

In the next section, I present a partial inventory and an assessment of the several formal approaches that have been developed for research planning. The approaches are grouped under scoring models, experimental approaches, and benefit-cost methods.

Scoring models

The earliest and most widely used models involve the scoring or ranking of research areas or projects by panels.[7] The panels may consist of peers, and they may also involve administrators and users. Two scoring models that have been employed in the United States at the state or federal level are described in this section.

The National Program Study. The preparation of *A National Program of Research for Agriculture* by the National Association of State Universities and Land Grant Colleges and the U.S. Department of Agriculture in 1965 and 1966 involved an exceptionally ambitious attempt to utilize scoring methods in the planning of agricultural research. The task force set up to conduct the study was charged with evaluating the strengths and weaknesses of the federal-state research program, with identifying future research priorities, and with recommending the levels of support and manpower required for agricultural research over the following 10 years.

The first step in the study involved the development of the three-way classification system that is now used in the computer-based Current Research Information System (CRIS) for the storage and retrieval of research reports. (See table 11.1.) This system was then used to inventory the financial support and scientist-years devoted to research by the categories in the classification system.

Future research needs were estimated by ranking each of 91 research objectives, or goals, by the 8 criteria listed in table 11.2. Scores were obtained by having panels in each of the commodity or resource areas rate each problem from 1 to 5 according to how well it satisfied each of the 8 criteria and then multiplying the ratings by the weights listed in the table. The scores were then used to project the socially desirable number of scientist-years for each problem area for 1977. The results of this exercise, aggregated into 9 general research goals, are shown in figure 4.1 (p. 80).

Table 11.2. Criteria and Weights Used for Establishing Relative Program Projections in the *National Program of Research for Agricultural Study*

Criterion	Weight
Extent to which the research meets state experiment station, department, and national goals	9
Scope and size considering area, people, and units affected	8
Benefits of research in relation to costs	7
Urgency of research	10
Contribution to knowledge	9
Feasibility of implementation and likelihood of successful completion in a reasonable period of time	5
Likelihood that the research results will be available elsewhere	6
Likelihood of extensive and immediate adoption of results	6
	60

Source: U.S. Department of Agriculture and Association of State Universities and Land Grant Colleges, *A National Program of Research for Agriculture* (Washington, D.C.: U.S. Government Printing Office, 1966), p. 29.

The methodology employed in the national program was conceptually simple but operationally complex. Several participants have indicated that they regarded the use of the scoring method as among the least valid aspects of the planning effort.

The North Carolina model. In 1972 the North Carolina Agricultural Experiment Station initiated a very intensive review of its research program. The immediate goal was to determine the relative emphasis that the North Carolina station should give to the research problem areas identified in the *National Program*.

A joint administration-faculty effort was mounted to conduct an exhaustive review of all the research programs and projects at the station and to explore possible redirections for the future. Twenty task forces, each composed of 5 to 10 research and extension faculty members and, in some cases, state agency personnel, reviewed the station's entire program. These committees recommended quantitative changes in research support and scientific effort. Each recommendation was ranked according to prespecified scoring criteria. Following the completion of the task forces' reports, 18 extramural panels consisting of scientists from other universities and the Cooperative Research Service (USDA) evaluated and rated the task forces' recommendations. The 23 academic departments at North Carolina State University were then asked to evaluate the task forces' and extramural panels' reviews from a disciplinary perspective and also to rank the task forces' recommendations.

The scoring models used in the North Carolina study were based on a revision of the criteria and weights used in the *National Program* study. Separate criteria and weights were developed for each of four major research areas. The initial revisions were developed by the experiment station's administration. They were submitted by mail to members of the Research Planning Advisory Committee to obtain suggestions for revisions of the criteria sets and weights. Revisions, weights and explanations were developed by each member of the committee, were summarized by the administration, and were re-evaluated along an interactive Delphi procedure format.[8] This procedure was repeated twice. The four criteria sets and weights developed in this way are listed in table 11.3.

Each member of the three groups—the task forces, the extramural panels, and the department heads—then independently scored the task forces' recommendations for changes in research problem area resources on a five-point scale without knowing the weight attached to each criterion. Project rankings were then obtained from the scores by using two standards: the average score by all raters and the average score of all raters minus one standard deviation.

The purpose of the latter method was to lower the rank order of research problem areas having the greatest degree of variability among the scores by individual participants. It turned out that there were considerable differences among raters and among groups of raters. The rank-order correlation between department heads' scores and extramural panels' scores was only 0.45. The correlation between the task forces' and the extramural panels' rankings was 0.42 and between the task forces' and the department heads' rankings only 0.24. Furthermore, an analysis of variance suggested that the variations among the groups' scores were not significantly different than the variations among individuals in the same group.

A perspective on the use of scoring methods. The two cases described in this section cannot, of course, fully represent the large number of research evaluation and planning efforts that have made use of scoring methods. Others have been reviewed in the papers by Shumway and by Norton and Davis referred to at the beginning of this section. The two cases do, however, serve to illustrate some of the weaknesses of the scoring method approach. And they also serve to suggest occasions when the use of scoring methods may be appropriate.

One of the serious problems with scoring approaches has been the difficulty of getting the participants to play by the rules. This problem was regarded as a particularly serious one by some of the participants in the *National Program* study. In the North Carolina state study, many participants simply refused to rank some of the research problem areas. Many Delphi studies experience a high "dropout" rate—panelists simply tire of the effort before the last iteration is completed.

A more serious problem with scoring approaches has been the difficulty of designing a reasonably independent and relevant set of criteria or objectives. Considerable progress in this direction was made between the list employed in the *National Program* (table 11.2) and the list employed in the North Carolina state study (table 11.3). But problems remain. How independent, for example, are such criteria as: "extent to which proposed research is consistent with station, regional, and national goals"; and "cost relevance—expected benefits in relation to costs"?

The most serious problems emerge when an attempt is made to combine individual project weights with criterion or objective weights to arrive at a global allocation of research resources at the state or national level. The aggregation of scores is loaded with booby traps. The rules adopted for calculating scores tend to be arbitrary and the shadow prices that are implicit in the weights that emerge from the aggregation process often bear little relationship to the relative weights that would be cast up by market processes or to the shadow prices that might be generated by a rigorous application of the constrictions of utility theory.

But scoring methods can be very useful when employed at a less aggregate level. They are probably most useful when employed in the scoring of individual projects against a single criterion, such as scientific or economic significance. the USDA's competitive research program, which is designed to support basic research in each of four areas—(1) biological nitrogen fixation, (2) photosynthesis, (3) genetic mechanisms for crop improvement, and (4) biological stress in crop plants—represents a useful example. A simple ranking of the projects submitted in each area by a panel of knowledgeable scientists can

Table 11.3. Criteria for Evaluating Research Problem Areas at the North Carolina
Agricultural Experiment Station

Research Area	Criterion	Criterion Weight
A. Biological sciences and technology	1. Urgency—basic information needed to aid in solution to threat or problem.	20
	2. Cost relevance—expected long-term benefits in relation	15
	3. Degree to which similar research is not now being conducted or not likely to be conducted elsewhere (higher scores if inadequate research results expected elsewhere).	15
	4. General importance and potential for contribution to knowledge. Higher scores to be assigned for greater scientific merit and potential for contribution to faculty development and improved academic performance.	50
	Total	100
B. Animals and plants	1. Extent to which proposed research is consistent with station, regional, and national goals in agriculture and forestry. Consider economic value of the crop or animal enterprise and its products to people of North Carolina.	35
	2. Cost relevance—expected benefits in relation to costs.	20
	3. Extent to which similar research of adequate quality is not being conducted on this commodity elsewhere (higher score for RPAs and sub-RPAs for which adequate results are not likely to be available elsewhere), and degree of urgency of need for research results.	20
	4. Potential for contribution to knowledge.	25
	Total	100
C. Environment and natural resources	1. Extent to which proposed research is consistent with station, regional, and national goals in natural resource development and conservation.	35
	2. Cost relevance—expected benefits in relation to costs.	15
	3. Extent to which similar research of adequate quality on this resource is not being conducted elsewhere (higher scores for inadequate research elsewhere) and whether or not there is (1) a threat to natural resource, (2) public pressure, or (3) a critical need for environmental protection.	15
	4. Potential for contribution to knowledge.	20
	5. Extent to which the research will aid in meeting broader public service commitment of the school and university, beyond traditional statutory charge of the experiment station.	15
	Total	100
D. Food-fiber-people-economics	1. Extent to which recommended research is consistent with station, regional, and national goals of promoting and protecting public health and improving family	

Table 11.3—*Continued*

Research Area	Criterion	Criterion Weight
	living; potential for improving quality of life and developing rural communities in North Carolina.	35
	2. Cost relevance—expected benefits in relation to increased costs of research in these areas, resulting from these recommendations.	20
	3. Extent to which similar research of adequate quality is not being conducted elsewhere (higher scores for inadequate research elsewhere) and whether there is (1) public support for research to evaluate the impact of improved agricultural technology. (2) a threat to public health, or (3) a need for information to support new processing industries	20
	4. Potential for contribution to knowledge	25
	Total	100

Source: C. Richard Shumway, "Models and Methods Used to Allocate Resources in Agricultural Research: A Critical Review," in *Resource Allocation and Productivity in National and International Agricultural Research*, Thomas M. Arndt, Dana G. Dalrymple, and Vernon W. Ruttan, eds. (Minneapolis: University of Minnesota Press, 1977), p. 443.

probably be conducted without too much bias. But the initial selection of the four priority areas and the specification of the priority that each should receive in the allocation of financial and scientific effort appear to be a task for which scoring methods are poorly suited.

Experimental approaches

Methods of identifying research priorities have been developed that are based directly on the modeling of physical and biological relationships and that incorporate parameters drawn from the analysis of experimental results and field investigations. These experimental approaches to the establishment of research priorities have typically placed greater emphasis on the value of removing constraints on performance or genetic improvement at the level of the individual field or plot, or even the individual plant or animal, than on the scoring methods discussed above. In this section I examine three such methods: yield constraint models, selection indexes, and plant growth models.

Yield constraint models. Attempts have been made at both the International Center for Tropical Agriculture (CIAT) in Colombia and at the International Rice Research Institute (IRRI) to conduct farm-level investigations by joint agroeconomic research teams in order to obtain data on crop production constraints.[9] In some cases, these farm-level observations have been supplemented by experiments designed to confirm or to refine the measures obtained from the farm-level observations. Analytical methods range from yield response and partial budgeting to relatively sophisticated crop loss or crop production systems models.

The single-commodity approach typically involves two stages. The first is an agroeconomic observation in farmers' fields that describes the production process, identifies factors that limit production and productivity, and attempts to estimate their relative importance. Data are sought on agrobiological factors (cropping systems, cultural practices, soil quality, plant type, pathogens, and others), on economic factors (use and prices of inputs, gross and net revenues), and on institutional factors (credit, tenure, and others). The second step involves the use of agrobiological experiments to provide more accurate measures, for example, of the yield-depressing effects of a soil problem.

The multicommodity, or farm-level, approach involves, in addition, more complex cropping systems trials or experiments that explore interaction effects between, for example, crop varieties and cropping practices. The CIAT has employed a systems engineering methodology in order to analyze the dynamic response of small-farm systems as a function of input and output relationships with the biological, ecological, and institutional environment. The CIAT model also included links to sector-level demand and supply relationships, thus permitting the analysis of aggregate output, productivity, and distributional impacts. The more complex farming systems' simulation efforts at this time can be viewed more as pilot efforts than as operational planning instruments.

Selection indexes. Animal geneticists have used weighting methods to construct composite measures, termed selection indexes, since the mid-1940s. The selection indexes are used to evaluate the incremental contribution of the several traits that enter into breeding programs to the economic value of alternative selection strategies.[10]

The problem that led to the development of selection indexes is that the traits that are influenced by a breeding effort are usually not independent of each other. Important traits are often negatively correlated. Furthermore, genetic traits are discovered by a stochastic search process, and there is a high degree of uncertainty in the transmission of traits to offspring. In breeding decisions, therefore, the breeder is confronted with great uncertainty in attempting to make trade-offs among traits.

The initial selection indexes were constructed by a linear weighting of the traits affected by the breeding program according to the economic value of each trait. In the more sophisticated variants, the index was constructed to reflect (1) the variance and covariance of the measured expression of the trait (phenotypic value), (2) the variance and covariance of the increment in the measured expression of the trait judged by the improvement in the progeny over the expected population value (the breeding value), and (3) the economic value of the trait. Because cost differences were often not fully considered, breeding indexes are more appropriately viewed as partial rather than as total productivity measures.

Recently, there have been a number of efforts to integrate increasingly sophisticated genetic and economic models. The selection process involves a search for the extreme values of specific traits. It also involves the use of quantitative information on genetic and genetic-environmental interactions. The economic value of the traits depends on input and product market price relationships and on the technical relationships linking inputs and output.

Plant growth models. The development of plant growth models has been an expanding area of research since the development of canopy photosynthesis models.[11] These models involve the application of biophysical principles to crop growth and the construction of computer models to simulate growth processes.

One of the more successful of these efforts has been a model (SIMCOT II) that simulates the growth of the cotton plant from emergence to maturity. The effort began in 1968 with attempts to model respiration and flowering. Considerable effort was made to reinterpret earlier experimental data within the modeling framework. By the early 1970s, the work had progressed to the modeling of the supply and use of nitrogen and water. By 1973, efforts were being directed to the modeling of the interaction between insect pests and the cotton plant. By the late 1970s, efforts had progressed to the point where a boll weevil-eradication model could be attempted.

The potential for the use of plant growth models for research design and planning is very promising. The modeling efforts offer the possibility of making more precise identifications of the constraints on crop production. When the results are combined with economic data, the value of removing growth or yield constraints can be estimated. At the present stage of development, it is still necessary to build a great many ad hoc assumptions into such models. It remains difficult to make accurate probability statements about the properties of the models.

A perspective on the use of experimental approaches. The experimental approach, particularly crop loss and plant growth modeling variants, is experiencing rapid methodological development. An important objective of this effort is to provide empirical information on the results of research designed to relax constraints, to improve particular traits, to shift response relationships or to achieve other objectives of applied research efforts. Because these investigations can be conducted as part of a research program, rather than as a diversion from research, the results can be fed directly back into the redesign of research effort at the individual research program level. The information generated from the experimental approaches can also be cast in a form that is useful to research-planning and budget offices.

Benefit-cost methods

The benefit-cost information systems attempt to be more explicit than the scoring models with respect to the inputs into the research process, the outputs expected from the research, and the relationship between the cost streams and the benefit streams.[12] Information of the type generated by the experimental approaches is directly useful in benefit-cost as well as in other forms of impact evaluation.

The Minnesota agricultural research resource allocation and information system. A system for the collection and computer processing of information that could be used either for the subjective evaluation of research activities or for the formal estimation of projected cost-benefit measures was developed and tested by Walter L. Fishel at the University of Minnesota in 1970. The model, which he labeled MARRAIS, involved three major steps: specification, estimation, and analysis. (See figure 11.1.)

The model was designed to provide benefit-cost, benefits-minus-cost, and internal rate of return measures for research projects or programs. Knowledgeable scientists in the field of study related to the proposed research project were surveyed in order to obtain the information needed. The surveys provided estimates of the average annual expenditures, the time requirements, and the scientific and technical feasibility of the research effort. Subjective probability distributions of costs and values were generated for alternative levels of annual expenditures by a Monte Carlo sampling procedure.

MARRAIS is still, more than a decade after its development, clearly one of the most logically thought out and procedurally sophisticated

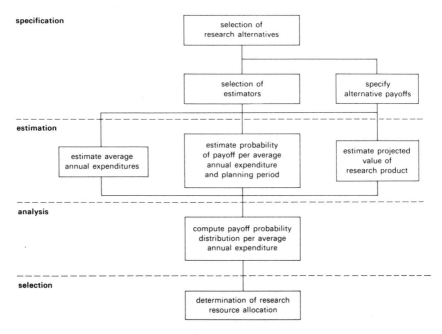

specification

selection of
research alternatives

selection of
estimators

specify
alternative payoffs

estimation

estimate average
annual expenditures

estimate probability
of payoff per average
annual expenditure
and planning period

estimate projected
value of
research product

analysis

compute payoff probability
distribution per average
annual expenditure

selection

determination of research
resource allocation

Figure 11.1. Modular Form of Cost-Benefit Estimation Model. (From Walter
L. Fishel, "The Minnesota Agricultural Research Resource Allocation Information System and Experiment," in Walter L. Fishel, ed., *Resource Allocation in
Agricultural Research* [Minneapolis: University of Minnesota Press, 1971],
p. 354.)

research-planning models available. Its high cost to users has been an
obstacle to its routine application.

Simplified benefit-cost models. In 1977 William Easter and George
Norton developed a simplified MARRAIS-type model to analyze the
U.S. land-grant universities' national budget requests for soybean and
corn production research. Benefit-cost ratios were calculated from
the lower range of estimates provided by scientists on the yield and
cost effects of each research line and on the expected adoption rates
for the new technology.

An important aspect of the analysis was the sensitivity of the benefit-cost ratios to variations in the probabilities of success, the expected yield increases, the product prices, and the length of the lags
between research expenditures and the variability of the results to
the farmers. Distributional effects were examined. These included
the impact on prices of fats and oils, on meat prices, and on gross

farm income. Aggregate gains to consumers and producers resulting from the proposed research were also projected.

A simplified cost-benefit analysis was also used by A. A. Araji, J. R. Sim, and R. L. Gardner to evaluate research and extension programs for several commodities in the western region of the United States and for pest-management programs in several regions of the United States. They also estimated the reduction in productivity that would result from eliminating maintenance research.

A perspective on the use of benefit-cost methods. A major limitation of the more technically sophisticated cost-benefit methods, such as MARRAIS, is that the costs involved in estimation, data storage and maintenance, and economic analysis can easily be underestimated by several magnitudes. A major advantage of even the simpler benefit-cost approaches is that they provide a consistent metric for relating the value of the resources used in research to the benefits that flow from research. The benefit-cost projections can, however, be no better than the judgments about research costs, the probability of research success, and the estimates of research benefits that are generated by the scoring methods or the estimates generated by the experimental approaches on which they are based. This is true regardless of the sophistication of the methodology used to derive the benefit-cost estimates.

LINKING PROJECT SELECTION TO RESEARCH RESOURCE ALLOCATION OBJECTIVES

I have employed, in the last two sections, the convention of focusing first on the considerations involved in the allocation of public research resources to achieve multiple objectives (the objective function) and then on the methods that are available to research administrators and planners to establish priorities among research projects and programs. The problem that remains unresolved is how to link these two bodies of analyses.

There are no fully developed methodologies that are capable of a simultaneous solution to this problem. Several methods have been employed in attempts to achieve convergence. The next section provides a perspective on three of these methods. All three methods represent a variation of growth accounting.

Consistency models

Perhaps the simplest of the several approaches involves a continuous monitoring of the sources of output growth. An example of this approach is an attempt that I made in the mid-1950s to evaluate the implications of projections of resource investment requirements that had been made by the U.S. Department of Agriculture, the President's Water Resources Policy Commission Report, and the President's Materials Policy Commission Report.[13] These reports were concerned with the question of the capacity of American agriculture to meet future food and fiber requirements. The emphasis of the studies was on "the transitory nature of present food surpluses." The Water Resources Policy Commission Report suggested, for example, that the equivalent of 100 million acres of cropland might have to be brought into production to meet 1975 agricultural production requirements.

The approach employed in assessing these projections utilized an equation of the Cobb-Douglas type (linear in logarithms) with a shift factor that captured the effect of productivity growth to examine the consistency between the projected output requirements and the alternative rates of growth of inputs and productivity. Four basic models, with annual rates of productivity growth ranging from zero to 2.4 percent per year were calculated. The several projections, along with the actual changes in output, inputs, and productivity, are presented in table 11.4. The projections implied that continuation of historical rates of productivity growth could be consistent with even a modest decline in land inputs. The realized rate of productivity growth was quite similar to the most rapid rate projected. The input mix projections, under very rapid technical progress, substantially underestimated the rate of decline in labor inputs and the rate of increase in current inputs.

A similar consistency model has been used by G. Edward Schuh to evaluate the 1972 Brazilian agricultural plan.[14] The results indicated that it would not be possible to meet the very high output growth rates projected in the plan within the constraints imposed by the anticipated growth of resource inputs and total productivity growth.

The consistency models have an advantage over models based on partial productivity projections such as land productivity (yield per hectare) or labor productivity (output per worker) in that they force the planners to face up to the problems of resource substitution in selecting the program activities designed to influence the rate of productivity and output growth. The major weakness of the simple consistency models of the type outlined above is that they do not incorporate direct links between research investment and technical change or between technical change in resource substitution and productivity growth.

Table 11.4. Alternative Projections and Realized Farm Output and Factor Input Indexes for 1960 and 1975 (1950 = 100)

	Zero Technical Progress[a]		Slow Technical Progress[b]		Rapid Technical Progress[c]		Very Rapid Technical Progress[d]		Realized Technical Progress[f]
	Low Land Inputs (I)	High Land Inputs (II)	Low Land Inputs (III)	High Land Inputs (IV)	Low Land Inputs (V)	High Land Inputs (VI)	Low Land Inputs (VII)	High Land Inputs (VIII)	
1960 Projections Inputs:									
Labor	88	88	88	88	78	78	78	78	65
Land	96	104	96	104	96	104	96	104	90
Capital[e] (A)	178	172	140	136	149	143	124	121	114
(B)	183	177	145	140	153	147	127	124	
Current[e] (A)	214	207	169	163	178	172	148	145	170
(B)	204	198	161	155	171	164	141	138	
Contributions to Output from:									
Inputs	122	122	112	112	110	110	100	100	96
Technological change	0	0	10	10	12	12	22	22	24
Total output	122	122	122	122	122	122	122	122	118
1975 Projection Inputs:									
Labor	81	81	81	81	67	67	67	67	30
Land	90	110	90	110	90	110	90	110	96
Capital[e] (A)	346	318	199	169	218	201	132	122	133
(B)	378	348	218	185	238	219	144	133	
Current[e] (A)	547	505	317	240	346	318	210	193	472
(B)	491	441	285	234	311	277	189	173	

Table 11.4—Continued

	Zero Technical Progress[a]		Slow Technical Progress[b]		Rapid Technical Progress[c]		Very Rapid Technical Progress[d]		Realized Technical Progress
	Low Land Inputs (I)	High Land Inputs (II)	Low Land Inputs (III)	High Land Inputs (IV)	Low Land Inputs (V)	High Land Inputs (VI)	Low Land Inputs (VII)	High Land Inputs (VIII)	
Contributions to Output From:									
Inputs	160	160	135	135	129	129	100	100	96
Technological change	0	0	25	25	31	31	60	60	54
Total output	160	160	160	160	160	160	160	160	150

Source: Projections: Vernon W. Ruttan, "The Contribution of Technological Progress to Farm Output: 1950-1975," *Review of Economics and Statistics*, 37 (February 1956), pp. 61-69: Realized: Donald D. Durost and Evelyn T. Black, *Changes in Farm Production and Efficiency* (Washington, D.C.: U.S. Department of Agriculture, Economics, Statistics, and Cooperatives Service, Statistical Bulletin No. 612, November 1978).

a. Increased inputs are assumed to account for the entire increase in output.

b. Technological change is assumed to occur at a sufficiently rapid rate to permit an increase in output per unit of input of 1.0 percent per year between 1950 and 1975. This is the 1910-1950 rate calculated on the basis of 1945-1948 prices and techniques.

c. Technological change is assumed to occur at a sufficiently rapid rate to permit an increase in output per unit of input of 1.23 percent per year between 1950 and 1975. This is the 1910-1950 rate calculated on the basis of 1910-1914 prices and techniques.

d. It is assumed that technological change occurs at a sufficiently rapid rate to account for the entire increase in output. This requires an increase in output per unit of input of 2.2 percent per year between 1950 and 1960 and 2.4 percent per year between 1950 and 1975.

e. Estimate A for capital and current inputs is based on the assumption that the ratio of capital to current inputs (C1/C2) will continue to decline at the same percentage rate as during the period 1910-1914 to 1945-1948. Estimate B is based on the assumption that the 1925-1927 to 1949-1950 rate will continue. See text for further discussion of estimates A and B.

f. Calculated for 1948-1953, 1958-1963, and 1973-1977. Capital indexes based on mechanical power and machinery; current inputs based on agricultural chemicals.

Sources of productivity growth models

In a recent report by the U.S. Department of Agriculture, Yao-chi Lu, Philip Cline, and Leroy Quance extended the consistency model approach to incorporate the effect of alternative levels of research investment on productivity growth and, more important, to assess the probable impact of specific technical advances on productivity growth.[15]

The first step in the Lu-Cline-Quance approach was to establish statistically the relationships between the effect of research and extension expenditures on the total productivity index. The effects of research and extension were separated from the short-run effects of weather and the longer-run effects of the level of education of the farm labor force. The statistical analyses permitted a careful identification of the average annual flow of benefits, and hence the impact of research and extension investment. The analysis implied that for the United States as a whole the benefit flow increases gradually for 6 or 7 years and then declines to a negligible level by the 14th year.[16] In a situation characterized by no real increase in the level of research expenditures (in which increases in public expenditures for agricultural research are offset by inflation) an annual increase in productivity of about 1 percent per year is projected. A real increase in research and extension funding of 3 percent per year would push the productivity growth rate to 1.1 percent per year.

The second step in the analysis was to consider the effect of several technological breakthroughs on productivity growth. An extensive literature review identified 12 areas in which technical breakthroughs, such as those that might be analyzed in a constraints model, might occur.[17] These areas were identified by using Delphi methods during interviews with research and extension workers in the Agricultural Research Service, the Cooperative State Research Service, and the Science and Education Administration.

Three areas were considered to have exceptional potential—photosynthesis enhancement and bioregulators in crop production and twinning in livestock production. Subjective probability distributions were constructed for the availability of each of the new technologies, and adoption profiles were developed for the diffusion of the new technologies. Estimates of research costs and productivity impact were made. A number of simulations were run to estimate the impact of the three technologies on the rate of productivity growth. The new technologies would not have a significant impact on production until the 1990s. A medium impact projection suggests that the new technologies could result in an increase in the projected rate of productivity growth from 1.1 to 1.3 percent per year for the period between 1975 and 2000 and an increase of 1.5 percent per year between 2000 and 2025. Even these rates are well below the rates achieved during recent years. (See table 10.1.)

The source of productivity growth approach is clearly an important advance over the older consistency approach. Its major limitation is, from some perspectives, that it involves an assessment of only the efficiency or productivity implications of the new technologies. The value of the approach is that it does provide some indication of the magnitude of the resource savings that could result from focusing a long-term research effort on a major area of research. The productivity measures could then be incorporated in models designed to measure the distributional effects of technical change, such as those discussed in the first section of this chapter.

Trade-off models

It seems clear that the interests of the public that provides the resources and of the legislative bodies that appropriate the funds to support agricultural research extend beyond the goal of efficiency in commodity production. Attempts have been made to develop models that are capable of evaluating the effects of research portfolio choice on a spectrum of variables often classified under such headings as equity and security, as well on the production, consumption, and growth criteria.

The Resource Allocation System for Agricultural Research (RASAR) developed by R. G. Russell in the United Kingdom is an example of a model of this type.[18] The ultimate goal of agricultural research was identified as having nine dimensions in three broad categories: *consumption*, including (1) quality, (2) quantity, and (3) availability; *security*, including (4) human safety, (5) economic defense, (6) food sources security, and (7) conservation; and *equity*, including (8) distribution, and (9) individual rights.

A mathematical programming model for assimilating the complexity of criteria and data into a form that could be used for decision making was developed. The individual units of analysis were research project units small enough to have a single primary objective and large enough to require significant amounts of time and money. Administrative constraints—such as financial and staff limitations, policy constraints such as urgency of the problem and social acceptance, and scientific and technical constraints stemming from the state of knowledge—were built into the model. The system outputs included: (1) *the research program*, the set of projects to receive support; (2) *support levels*, the level and role of support for each project; (3) *program utility*, which summarized the impact of the project on each dimension that entered in to the utility measure; and (4) *program sensitivity*, including a measure of the program's sensitivity to variations in program assessments, an indication of the projects that were barely included or excluded, an indication of whether reformation of a project would change its probability of being included, and an indication of how critical the weight assigned to each criterion is in the selection decision.

The model developed by Russell represents a very sophisticated attempt to link scientific judgments with respect to the opportunities to advance technology with both economic and noneconomic indicators of the value of the research output. Russell pointed out that even in this model a major issue remains unresolved. There is no way to combine the incommensurate measures in the goal dimensions to provide a project utility rating except to use an arbitrary weighting function!

THE LIMITS AND POTENTIAL OF RESEARCH PLANNING METHODOLOGIES

Where does this review of research resource allocation principles and methods lead with respect to decision making about research priorities and the allocation of research resources? How much effort can the director of a national research program, the director of an experiment station, or the leader of a commodity or resource research program afford to allocate to research on research management? These efforts tend to be in direct competition with research projects for resources. The general principle that resources devoted to management must add more to the efficiency of the system than the cost in terms of research projects not funded is difficult to apply in practice.

Clearly, the parity (or congruence) model that was outlined in the first pages of this chapter represents an inadequate response to society's concern about the value of new knowledge and new technology and to the science community's perception of the possibilities for advancing science and technology. But departures from the parity model should be based on informed judgment about the potential impact and value of scientific and technical effort. For example, does the low ratio of research expenditure to market value for soybeans reflect a judgment that advances in soybean productivity are hard to come by or a judgment about the value of soybeans to the American economy, or is it a result of bias in the system by which resources are allocated to research in the United States?

Increasingly powerful methodologies are becoming available to the directors of individual research programs, research institutes, and experiment stations for interpreting scientific, technical, and economic information in a manner that can increase the effectiveness of research effort, whether evaluated in terms of advances in knowledge or technology. In order to have access to these methodologies, resources must be allocated to interdisciplinary experimental design and system-modeling efforts such as those described in the sections on scoring methods, experimental approaches, and benefit-cost methods.

A major advantage of these methodologies is that they can be carried out as an integral part of a research program. Their results become directly available to individual scientists and research teams. Their results can be fed back immediately into research planning and design. They can also provide the information that is needed by research-planning units operating at the central level of a research organization.

But research policy and planning are not simple technical exercises that can be left in the hands of research scientists and managers! Judgments about the priority of public-sector support for agricultural research in relation to other demands for public resources must, in most countries, come out of an intricate bargaining process that goes on between national (or state or provincial) legislative bodies and executive agencies. Judgments about the relative emphasis that should be given to saving labor, saving land, and saving energy—as well as the priority that should be attached to research on the spillover effects of production technologies on the health and welfare of producers and consumers and the impact of agricultural technology on environmental amenities and on the structure of rural communities—must also come out of this same bargaining process.

The development and use of the research-planning and resource-allocation methodologies reviewed in this chapter should not, however, be viewed as being inconsistent with the legitimate role of political decision processes. The political dialogue leading to research resource allocation should be fully informed about the costs and benefits of research resource allocation decisions. Information on the historical and potential impact of technical change on productivity growth and on the distributional impact of technical change can represent a valuable input into these bargaining processes, even if it serves primarily to keep the debate reasonably honest. It is important to know whether utilization research gets used or whether hard tomatoes end up on the fresh vegetable counter or in catsup and pizza sauce. It is important that the self-interests of enthusiasts and promoters not be allowed to obscure the judgments of plant scientists, engineers, and economists about the *advances* in knowledge and technology that will have to be realized if biomass is to become a more efficient source of energy.

What is my final judgment on the methodologies for research resource allocation presented in this chapter? There can be little question that the judicious selection and application of the new methodologies that have been reviewed in this chapter could represent exceedingly powerful tools in the hands of a research director who insisted on making research resource allocation decisions and on understanding how and why he or she made them.[19] They could also be an embarrassment to a weak research director who hoped that research decisions would be revealed by applications of the planning methodologies that are available to his or her staff. These and related analytical methodologies represent potentially powerful aids in the decision-making process. But they can achieve effectiveness only in the hands of a research manager who has the intellectual vigor to grapple with both the substance of the research program and the tools that are available for research decision making.

Let me add one final cautionary note. The methods outlined in the chapter have assumed that the objective of agricultural research can be evaluated primarily in terms of the rate and direction of productivity growth and in terms of how the dividends from productivity growth are partitioned among different groups of producers and between producers and consumers. Over the last several decades, this presumption has been continually challenged. In the final chapter of this book, an attempt is made to consider some of the challenges to articulating and implementing a more comprehensive set of objectives for decision making in agricultural research.

NOTES

1. I am indebted to Jeff Davis, Bobby R. Eddleman, Walter L. Fishel, Robert W. Herdt, Yoav Kislev, George W. Ladd, Yao-chi Lu, Bryon E. Melton, George Norton, Joseph P. Purcell, Richard Sauer, and Richard C. Schumway for their comments on and criticisms of an earlier draft of this chapter.

2. These efforts have given rise to a substantial literature. Among the more useful items are the following: Walter L. Fishel, ed., *Resource Allocation in Agricultural Research* (Minneapolis: University of Minnesota Press, 1971); I. Arnon, *The Planning and Programming of Agricultural Research* (Rome: FAO, 1975); Thomas M. Arndt, Dana G. Dalrymple, and Vernon W. Ruttan, eds., *Resource Allocation and Productivity in National and International Agricultural Research* (Minneapolis: University of Minnesota Press, 1977); W. B. Back, ed., *Technology Assessment: Proceedings of an ERS Workshop, April 20-22, 1976* (Washington,

D.C.: National Economic Analysis Division, Economic Research Service, U.S. Department of Agriculture, AGERS-31, September 1977); George W. Norton, Walter L. Fishel, Arnold A. Paulsen, and W. Bert Sundquist, eds., *Evaluation of Agricultural Research* (St. Paul: University of Minnesota Agricultural Experiment Station, Miscellaneous Publication 8-1981, April 1981). For a useful review of this and related literature, see G. Edward Schuh and Hillo Tollini, "Costs and Benefits of Agricultural Research: State of the Art and Implications" (Washington, D.C.: Consultative Group on International Agricultural Research, October 1978).

3. See James K. Boyce and Robert E. Evenson, *Agricultural Research and Extension Programs* (New York: Agricultural Development Council, 1975), pp. 83-96. In reviewing the evidence on the congruence between commodity research expenditure and commodity value ratios, Boyce and Evenson noted that "the low income countries . . . have lower research-commodity congruence than the more mature research systems" and that "almost every region has moved toward congruence over time" (pp. 95, 96). An alternative label for the "parity model" might be the "congruence model" of research resource allocation.

4. See U.S. Department of Agriculture and Association of State Universities and Land Grant Colleges, *A National Program of Research for Agriculture* (Washington, D.C.: U.S. Government Printing Office, 1966). The classification scheme developed in 1966 has been modified as it has been implemented and used. See Agricultural Research Advisory Committee, *Manual of Classification of Agricultural and Forestry Research: Classifications Used in Current Research Information System* (Washington, D.C.: U.S. Department of Agriculture, January 1978 revision). For a discussion of related efforts among other OECD countries, see G. Wansink, "Co-operation in Current Research Information" (Paris: OECD Directorate for Food, Agriculture, and Fisheries, September 1979).

5. This section depends very heavily on J. D. Ramalho de Castro and G. Edward Schuh, "An Empirical Test of an Economic Model for Establishing Research Priorities: A Brazil Case Study," in *Resource Allocation and Productivity in National and International Agricultural Research*, Thomas M. Arndt, Dana C. Dalrymple, and Vernon W. Ruttan, eds. (Minneapolis: University of Minnesota Press, 1977), pp. 498-525. For a more technical treatment of the gains and losses from technical change, see Yujiro Hayami and Robert W. Herdt, "Market Price Effects of Technical Change on Income Distribution in Semi-subsistence Agriculture," *American Journal of Agricultural Economics*, 59 (May 1977), pp. 245-56; Per Pinstrup-Andersen, Norha Ruiz de Londono, and Edward Hoover, "The Impact of Increasing Food Supply on Human Nutrition: Implications for Priorities in Agricultural Research and Policy," *American Journal of Agricultural Economics*, 58 (May 1976), pp. 131-42.

6. Hans P. Binswanger, *The Economics of Tractors in South Asia* (New York: Agricultural Development Council; and Hyderabad, India: International Crops Research Institute for the Semi-arid Tropics, 1978); John H. Sanders and Vernon W. Ruttan, "Biased Choice of Technology in Brazilian Agriculture," in *Induced Innovation: Technology, Institutions and Development*, Hans P. Binswanger, Vernon W. Ruttan, and others, eds. (Baltimore: Johns Hopkins University Press, 1978), pp. 276-96.

7. This section on scoring models depends very heavily on the following sources: U.S. Department of Agriculture and Association of State Universities and Land Grant Colleges, *A National Program of Research for Agriculture*; J. C. Williamson, Jr., "The Joint Department of Agriculture and State Experiment Stations Study of Research Needs," in *Resource Allocation in Agricultural Research*, Walter L. Fishel, ed. (Minneapolis: University of Minnesota Press, 1971), pp. 289-301; Arnold Paulsen and Donald R. Kaldor, "Evaluation and Planning of Research in the Experiment Station," *American Journal of Agricultural Economics*, 50 (December 1968), pp. 1149-62; John F. Malstede, "Long-range planning at the Iowa Agricultural and Home Economics Experiment Station," in *Resource Allocation*

in Agricultural Research, Walter L. Fishel, ed. (Minneapolis: University of Minnesota Press, 1971), pp. 326-43; C. Richard Schumway and R. J. McCracken, "Use of Scoring Models in Evaluating Research Programs," *American Journal of Agricultural Economics*, 57 (November 1975), pp. 714-18; C. Richard Schumway, "Models and Methods Used to Allocate Resources in Agricultural Research: A Critical Review," in *Resource Allocation and Productivity in National and International Agricultural Research*, Thomas M. Arndt, Dana G. Dalrymple, and Vernon W. Ruttan, eds. (Minneapolis: University of Minnesota Press, 1977), pp. 436-60; Roland R. Robinson, "Administration of Federal Agricultural Research Funds by the Science and Education Administration: Cooperative Research" (Washington, D.C.: USDA, SEA/CR, September 1978); and George Norton and Jeffrey S. Davis, "Review of Methods Used to Evaluate Returns to Agricultural Research," in *Evaluation of Agricultural Research*, George W. Norton, Walter L. Fishel, Arnold A. Paulsen, and W. Bert Sundquist, eds. (St. Paul: University of Minnesota Agricultural Experiment Station, Miscellaneous Publication 8-1981, April 1981), pp. 26-47.

8. "Delphi is a formalized method designed to promote consensus without obscuring variations in evaluation or scoring. It consists of a series of individual interrogations to a group of experts, interspersed with information and opinion feedback. Some questions inquire into the reasons for previously expressed opinions. A collection of such reasons is then presented to each respondent who is invited to reconsider his earlier estimate. Delphi attempts to improve the committee approach by subjecting views of individual experts to each other's criticism in ways that avoid face to face confrontation" (C. Richard Schumway, "Models and Methods to Allocate Resources," p. 448). See also Yao-chi Lu, "Ex Ante Evaluation of the Separate Effects of Research and Extension," in *Evaluation of Agricultural Research*, George W. Norton, Walter L. Fishel, Arnold A. Paulsen, and W. Bert Sundquist, eds. (St. Paul: University of Minnesota Experiment Station, Miscellaneous Publication 8-1981, April 1981), pp. 240-46.

9. The methodology employed at the CIAT has been described by Per Pinstrup-Andersen and David Franklin, "A Systems Approach to Agricultural Research Resource Allocation in Developing Countries," in *Resource Allocation and Productivity in National and International Agricultural Research*, Thomas M. Arndt, Dana G. Dalrymple, and Vernon W. Ruttan, eds. (Minneapolis: University of Minnesota Press, 1977), pp. 416-35; and by John H. Sanders and John K. Lanam, "Definition of the Relevant Constraints for Research Resource Allocation on Crop Breeding Programs," *Agricultural Administration* (forthcoming, 1981). For a description of the IRRI's methodology, see S. K. De Datta, K. A. Gomez, R. W. Herdt, and R. Barker, *A Handbook on the Methodology for an Integrated Experiment-Survey on Rice Yield Constraints* (Los Banos, Laguna, Philippines: International Rice Research Institute, 1979). For a review of recent developments in this field of crop-loss modeling, see W. Clive James and P. S. Teng, "The Quantification of Production Constraints Associated with Plant Diseases," in *Applied Biology*, vol. 4, T. H. Coaker, ed. (London: Academic Press, 1979).

10. This section depends primarily on George W. Ladd and Craig Gibson, "Micro Economics of Technical Change: What's a Better Animal Worth?" *American Journal of Agricultural Economics*, 60 (May 1978), pp. 236-40; Bryan E. Melton, "Basic Breeding Concepts and Relations," in *Applications of Economics in Animal Breeding*, George W. Ladd, ed. (Ames, Iowa: Iowa State University, Department of Economics Staff Paper No. 98, October 1979); Yoav Kislev and Uri Rabiner, "Economic Aspects of Selection in the Dairy Herd in Israel," *Australian Journal of Agricultural Economics*, 23 (August 1979), pp. 128-46.

11. D. N. Baker, J. D. Hosketh, and W. G. Duncan, "Simulation of Growth and Yield in Cotton," *Crop Science*, 12 (1972), pp. 431-39. I have also had the benefit of a memorandum from J. D. Hasketh (USDA Crop Production Systems Research, Mississippi State University) to Walter L. Fishel (USDA Systems and Policy Analysis, Agricultural Research Service), "Status of Simcot II — A Model for Predicting Cotton Growth and Yield," 1977.

12. This section depends primarily on Walter L. Fishel, "The Minnesota Research Resource Allocation Information System and Experiment," in *Resource Allocation in Agricultural Research*, Walter L. Fishel, ed. (Minneapolis: University of Minnesota Press, 1971), pp. 344-81; K. William Easter and George Norton, "Potential Returns from Increased Research Budget for the Land Grant Universities," *Agricultural Economics Research*, 29 (October 1977), pp. 127-33; A. A. Araji, J. R. Sim, and R. L. Gardner, "Returns to Agricultural Research and Extension Programs: An Ex Ante Approach," *American Journal of Agricultural Economics*, 60 (December 1978), pp. 964-68; Norton and Davis, "Review of Methods Used to Evaluate Returns to Agricultural Research."

13. Vernon W. Ruttan, "The Contribution of Technological Progress to Farm Output: 1950-75," *Review of Economics and Statistics*, 38 (February 1956), pp. 61-69. For another evaluation of the same material, see J. D. Black and J. T. Bonnen, *A Balanced United States Agriculture in 1965* (Washington, D.C.: National Planning Association Special Report No. 42, 1956).

14. G. Edward Schuh, "O Potencial de Crescimento da Agricultura Brasileira: Algumas Alternativas e suas Consequências" (Brasília: EAPA/SUPLAN, Ministerio da Agricultura, 1972).

15. Yao-chi Lu, Philip Cline, and Leroy Quance, *Prospects for Productivity Growth in U.S. Agriculture* (Washington, D.C.: U.S. Department of Agriculture, Economics, Statistics, and Cooperative Service, Agricultural Economic Report No. 435, September 1979).

16. The lag analysis is similar to that developed by Robert E. Evenson, "The Contributions of Agricultural Research and Extension to Agricultural Productivity" (Ph.D. dissertation, University of Chicago, 1968). The implications of different approaches to lag analysis have been explored by Jeff Davis in "A Comparison of Alternative Procedures for Calculating the Rate of Return to Agricultural Research Using the Production Function Approach" (St. Paul: University of Minnesota, Department of Agricultural and Applied Economics Staff Paper No. P79-19, May 1979).

17. Sylvan H. Wittwer, "Maximum Production Capacity of Food Crops," *Bioscience*, 24 (April 1974), pp. 216-24. The 12 areas identified as having significant productivity impact potential were (1) enhancement of photosynthetic efficiency, (2) water and fertilizer management, (3) crop pest-control strategies, (4) controlled environment or greenhouse agriculture, (5) multiple and intensive cropping, (6) reduced tillage, (7) bioregulators, (8) new crops, (9) bioprocessing, (10) antitranspirants, (11) development of plants to withstand drought and salinity, and (12) twinning. See Lu, Cline, and Quance, *Prospects for Productivity Growth*, p. 40.

18. D. G. Russell, "Resource Allocation in Agricultural Research Using Socio-economic Evaluation and Mathematical Models," *Canadian Journal of Agricultural Economics*, 23 (July 1975), pp. 29-52.

19. For a less optimistic view, see Edwin Mansfield, "The Evaluation of Industrial Research and Development Projects," in *Evaluation of Agricultural Research*, George W. Norton, Walter L. Fishel, Arnold A. Paulsen, and W. Bert Sundquist, eds. (St. Paul: University of Minnesota Agricultural Experiment Station, Miscellaneous Publication 8-1981), pp. 213-18.

Chapter 12

The Social Sciences
in Agricultural Research

The social sciences have been junior partners in the agricultural research enterprise.[1] Relationships between the natural science-based
disciplines and the social science-based disciplines in the past were
characterized more by interdisciplinary aggression than by interdisciplinary collaboration. But much of the tension has evaporated
in recent years.

The past tensions stemmed from several sources. One was the
emergence of fields such as agricultural economics and rural sociology later than fields such as agricultural chemistry and plant breeding.
This lag resulted in a struggle to establish the legitimacy of the social
science-based disciplines within colleges of agriculture, agricultural
research institutes, and ministries of agriculture.

A second source of tension was the different subject matter of the
disciplines. The subject matter of the natural science-based disciplines
is plants, animals, and land. The subject matter of the social science-
based disciplines is humans and human institutions. These two areas
of interest intersect at the level of the farm, where food and fiber are
produced; at the level of the community, where support services for
agricultural production and rural people are organized; and at the
level of the society or nation, where the terms on which consumers
have access to the products of agriculture and agricultural producers
have access to the products of industry and to social and cultural
amenities are determined.

My objective in this chapter is to clarify the role of the social sciences (anthropology, economics, political science, geography, and
sociology) at several levels in the agricultural research system: in the

agricultural research institute, the college of agriculture, and the ministry of agriculture. Before proceeding to these institutional considerations, however, it will be useful to characterize briefly the nature of social science and the sources of demand for social science knowledge.

THE EMERGENCE OF RURAL SOCIAL SCIENCE

The relative strengths of the several social science disciplines and subdisciplines within the field of agricultural research have changed over time in response to technical and institutional changes in the agricultural sector and the changing role of the agricultural sector in the national economy.[2] The priority given to social science research has also changed in response to advances in the power of social science theory and method. A brief review of the history of agricultural economics in the United States is illustrative.

Prior to 1900, agricultural economics did not exist as a field of specialized study either within general economics departments or in colleges of agriculture, although courses in agricultural economy and the economics of agriculture appeared in college catalogs at the University of Illinois and Cornell University before 1870. The rapid growth of agricultural economics as an academic field between 1900 and the early 1920s reflected the emerging interests of a number of people who had been trained in the agricultural disciplines (such as agronomy, horticulture, animal husbandry, and soil science). Their interest was primarily in factors affecting the costs of production, and in the economics of farm management—particularly in problems such as the economics of enterprise selection, and choice of production methods and the financing and growth of the firm. The growing interest in agricultural economics also reflected the concerns of a number of economists with the problems of agricultural policy, the behavior of agricultural commodity markets, and the economics of land use.

These developments resulted in the organization of the American Farm Management Association in 1910, the organization of the Association of Agricultural Economists in 1916, and the eventual consolidation of the two associations in 1919 under the title American Farm Economic Association. The organization of the two separate associations reflected a difference in perspective between those who entered the field of agricultural economics from the agricultural disciplines of agronomy, horticulture, animal husbandry, and soils and those who entered the field with prior training in economics.

The former were interested primarily in problems of microeconomics, while the latter were interested primarily in problems of macroeconomics and institutional economics. After the merger of the two associations, this difference in perspective continued to manifest itself in terms of discussions regarding the appropriate scope of the field of agricultural economics (was it a separate discipline or an applied field of economics) and the emphasis that should be given to the biological sciences and applied agriculture relative to economic theory and other fields of applied economics in the education of agricultural economists. This dialogue was muted, but not fully resolved, when the association changed its name to the American Agricultural Economics Association in 1964.

A second major event in the development of agricultural economics was the organization of the Bureau of Agricultural Economics in the U.S. Department of Agriculture in 1921 under the direction of Henry C. Taylor. The establishment of the bureau before the department had initiated the major action programs of the 1930s enabled the bureau to develop a tradition and a commitment to research that has been difficult to duplicate in other economics research units within the federal government. The close professional relationship and the continued mobility of agricultural economists between the Bureau of Agricultural Economics (now the Economic Research Service) and the academic departments have been the major sources of strength in the professional development of agricultural economics.

The evolution of agricultural economics since the early 1920s has also been responsive to new ooportunities resulting from developments in economic theory and statistics. Interest in the use of multiple correlation techniques in the analysis of supply, demand, and production relationships was stimulated by the publication of Karl Pearson's early article "The Application of the Method of Correlation to Social and Economic Statistics" and by "Forecasting the Yield and Price of Cotton" by Henry Moore. Moore's work on statistical demand relationships was followed closely by the elaboration of simple and multiple correlation methods by H. A. Wallace, George Snedecor, Mordecai Ezekiel, and L. H. Bean and by further investigations of statistical demand relationships by Holbrook Working, Fred Waugh, Mordecai Ezekiel, Henry Schultz, and others. Elmer Working's classic article on the identification problem, "What Do Statistical Demand Curves Show," was a major theoretical contribution from this same collaboration among economic theorists, statisticians, and agricultural economists. Since World War II, new analytical tools—including the structural equations system pioneered by the

Cowles Commission, the Leontief interindustry analysis, and the closely related methods of linear and nonlinear programming—and better economic time series and survey data have combined to produce a renewal of interest and activity in this area. The work at the U.S. Department of Agriculture by Karl Fox, Richard Foote, Marc Nerlove, and others on supply and demand relationships and commodity models was particularly important in providing new analytical insights, in testing the utility of alternative analytical approaches, and in providing a quantitative basis for evaluating the economic effects of agricultural policy decisions.

Early work in farm management and production economics (that by G. F. Warren at Cornell University and G. A. Pond at the University of Minnesota, for example) typically emphasized accounting and budgeting techniques of analysis. The application of statistical methods in the 1920s led to major innovations in the exploration of agricultural production relationships. W. J. Spillman's studies represented the first major attempt to use statistical techniques in the economic analysis of data from agricultural experiments. Howard R. Tolley, John D. Black, and Mordecai Ezekiel pioneered the use of statistical analysis of production relationships based on survey data collected from individual farms. The first systematic treatment of these developments within the framework of the neoclassical economic theory of the firm appeared with the publication by John D. Black of *Introduction to Production Economics* in 1926. Further progress in the analysis of agricultural production relationships was delayed until after the advances in the theory of the firm made by John R. Hicks and others in the late 1930s. These theoretical developments, when combined with the advances achieved in econometrics and mathematical economics during the 1940s, led to an explosive growth in the number of empirical investigations of agricultural production functions during the 1950s by the "Iowa-Chicago" school of agricultural economics.

A more recent area of intensive interaction between agricultural and general economists has been in the field of agricultural and economic development. As a result of both an intellectual and a policy commitment to the problem of economic growth in low-income, predominantly agricultural countries, general economists have found themselves increasingly concerned with the role of agriculture in national economic growth. Agricultural economists, working in similar circumstances, have found themselves giving more careful attention to the implications of firm- and sector-level analysis for national economic growth than when their analysis was being conducted primarily in Western economies in which the structural transformation from a primarily agrarian to an urban-industrial economy had been largely completed. The fruitfulness of this interaction was acknowledged in 1980 when the Nobel prize in economics was awarded to Theodore W. Schultz and Arthur W. Lewis. The award emphasized the contributions that Schultz and Lewis had made to theory and practice in agricultural development. Schultz's work in the field of economic development represented an extension of his earlier work in agricultural economics. Lewis's contribution reflected his earlier work in the field of international trade and commercial policy.

Several other areas of collaboration between general economics and agricultural economics can be mentioned. Agricultural commodity trade has traditionally occupied an important role in trade theory. Interest in the economic policies of the EEC and in the stabilization of commodity trade between the developed and less-developed countries continues to make international trade a fruitful area for the joint efforts of general economists and agricultural economists. Another area of mutual interest has been in the area of market structure and organization. Much of this work is based on advances in the theory of imperfect competition developed by Joan Robinson and Edward Chamberlin in the 1930s and on the market structure-conduct-performance framework developed by Joseph Bain in the 1950s.

During the last several decades, agricultural economists have also become increasingly involved in the economics of natural resource economics and policy. The evolution of the older field of land economics into the modern field of resource economics illustrates the response of agricultural economists and other economists to changes in the supply of land and raw materials and to changes in the demand for resource amenities. Two factors have been involved in this development. The economics of conservation and land use and of rural land use planning and taxation attracted the interest of a number of economists, particularly those who had been sensitized to land and resource conservation and use issues by training in Europe, into agricultural economics in the 1890s and early 1900s. As a result of the increased concern about the adequacy of the natural resources base to sustain national economic growth in the late 1940s and early 1950s, stimulated by the President's Water Resources and Materials Policy Commission reports, the field of land economics expanded to include other natural resource areas. The rising demand for environmental amenities in the 1960s resulted in a greater emphasis on the problems of water resource development and the economics of environmental management. In the 1970s, rising energy prices resulted in a further expansion of the demand for knowledge in the field of resource economics. These developments were also characterized by fruitful collaboration among agricultural economists and general economists in the interrelated fields of public finance, location, and regional economics.

In addition to its close relationships with the field of applied biology and with general economics and statistics, agricultural economics was closely linked to rural sociology during the formative years of the two fields. Many departments were organized as departments of agricultural economics and rural sociology. In spite of the close administrative links between the two fields, their contributions to each other have been quite limited. However, economists' and sociologists' interest in problems of urban and rural poverty and in the diffusion and adoption of technical change is leading to collaboration between the two fields. Nevertheless, the relationship between economics and sociology, and between agricultural economics and rural sociology, remains uneasy. Why this should be remains a puzzle to me. One possibility is that there is a much stronger normative orientation—a stronger commitment to social reform—in rural sociology than in agricultural economics. It is interesting to note that the same tension does not seem to exist between agricultural economics and anthropology. Anthropologists, for example, have established highly complementary working relationships with both the technical agricultural disciplines and the economics staffs at several of the international agricultural research institutes.

Excessive fragmentation along geographic and subdisciplinary lines and excessive parochialism reflected in the state or regional orientation of much of experiment station research have frequently been identified as major factors limiting the effectiveness of agricultural economics research. The criticism of fragmentation and parochialism is valid. Yet, this very parochialism and fragmentation of agricultural economics have also represented a source of strength and a basis for many of the field's contributions. Its parochialism has contributed to the interest of agricultural economists in focusing their attention on the economic problems of individual farm production and marketing firms. Its fragmentation has contributed to the interest of agricultural economists in examining specific commodity demand, supply, and production relationships. Agricultural economists have studied the economics of beekeeping in the Sacramento Valley (J. K. Galbraith), the effects of price supports on the dry bean industry in Michigan (Dale Hathaway), the effect of the soil bank program in southeastern Pennsylvania (George Brandow and James Houck), the impact of government programs on the potato industry in the Red

River Valley (Roger Gray, James Sorenson, and Willard Cochrane), and the impact of regulations restricting pesticide and animal drug use on crop and animal production (J. C. Headly and Gerald Carlson).

The fragmentation of agricultural economics along subdisciplinary lines may also have accounted for the ease with which it has expanded from its initial emphasis on problems of production economics and farm management to encompass marketing of agricultural commodities and factor inputs; commodity supply demand and trade relationships and policy; land, natural resource, and regional economics; and problems of agricultural development and economic growth.

Close association with the experimental and statistical methodology employed in applied biology made agricultural economists particularly receptive to methodological developments leading to greater precision in the quantification of economic and technical relationships, in the empirical testing of hypotheses and generalizations, and in the provision of quantitative measures of the effects of alternative private- and public-sector decisions. The dynamic changes that have taken place in the technology of agricultural production and in the institutions that link the agricultural and the nonagricultural sectors have forced agricultural economists and other rural social scientists to incorporate an acute sense of history into their attempts to understand the behavior of rural society, rural economy, and rural policy.

THE DEMAND FOR SOCIAL
SCIENCE KNOWLEDGE

What has led to the expanded role of agricultural economics, rural sociology, and in some places political science and cultural anthropology as components of agricultural research programs and agricultural education institutions? How does society value the investment it has made in the production of rural social science knowledge in agriculture?

Over the last several decades, agricultural economists and other social scientists have provided research managers with increasingly powerful tools for valuing applied research in the biological and physical sciences.[3] One is generally familiar with the calculations showing 30 to 60 percent rates of annual return on investments in agricultural technology research in countries like the United States and Japan and even higher rates of return on individual commodities such as hybrid corn in the United States, cotton in Brazil, and wheat in Mexico. (See chapter 10.) More recently, a good deal of effort has been devoted to adapting the models used to estimate historical rates of return to investment in research for use in estimating the contribution of research to future productivity growth. (See chapter 11.)

The basic concept on which the evaluation of the returns to agricultural production research rests is that the demand for knowledge is derived from the demand for technical change in commodity production. Once the output of research was clearly conceptualized as an input into the process of technical change in commodity production, processing, and distribution, this link made it possible to develop models to measure the *ex post* returns to research. It then became possible to make *ex ante* estimates of the relative contribution of alternative uses of research resources and to attempt to specify rules that research managers might follow in the allocation of research resources.

The same effort has not yet been devoted to the development of formal methodologies for the valuation of economic (and social science) research. Social scientists have only begun to adequately conceptualize the contribution of knowledge in the social sciences to institutional innovation and performance.[4]

The first step in an attempt to value new knowledge in economics is to specify the sources of demand for that knowledge. It is clear that the demand for knowledge in economics is not derived primarily from either private or public demand for technical change in commodity production. The demand for knowledge in economics and in the other social sciences—as well as in related professions such as law, business, and social service—is derived primarily from a demand for institutional change. Stated another way, changes in the demand for knowledge in economics are primarily a function of changes in demand for institutional innovation and for efficiency in institutional performance.

Shifts in the demand for institutional innovation or improvements in institutional performance may arise from a wide variety of sources. The Marxian tradition has emphasized the importance of technical change as a source of demand for institutional change. Douglass North and Robert Thomas attempted to explain the economic growth of Western Europe between 900 and 1700 primarily in terms of innovation in the institutions that governed property rights.[5] A major source of institutional innovation was, in their view, the rising pressure of population against increasingly scarce resource endowments. Theodore W. Schultz, focusing on more recent economic history, identified the rising economic value of labor during the process of economic development as a primary source of institutional innovation. North and Thomas would apparently have agreed with Schultz that "it is hard to imagine any secular economic movement that would have more profound influence in altering institutions than would the movement of wages relative to that of rents."[6]

The theory of induced technical change was discussed in chapter 2. The North-Thomas and Schultz models of the demand for institutional innovation open up the possibility of a theory of induced institutional innovation that is capable of generating testable hypotheses regarding alternative paths of institutional innovation. The theory of induced institutional innovation also raises the possibility of a more precise identification of the link between the demand for improvement in institutional performance and the demand for knowledge in economics and in the social sciences generally. Advances in knowledge in the social sciences in response to the demand for more effective institutions offer an opportunity to reduce the costs of institutional innovation, just as advances in knowledge in the biological sciences and agricultural technology have reduced the costs of technical innovation in agriculture.

When one accepts the notion that the demand for knowledge in economics, and in the social sciences generally, is derived from the demand for improved institutional performance, it then becomes necessary to consider the responsiveness of the supply of institutional innovation to the same factors that influence the demand for institutional innovation. Is the supply of institutional innovation relatively elastic? Or is society typically faced with a situation wherein the demand for institutional innovation shifts against a relatively inelastic supply curve?

It seems reasonable to hypothesize a close analogy between the supply of institutional innovation and the supply of technical innovation. Advances in knowledge in the social sciences (and in related professions such as law, business, planning, and social service) reduce the cost of institutional innovation just as advances in knowledge in the natural sciences and engineering reduce the cost of technical innovation.

This is not to argue that the institutional innovation is entirely dependent on formal research leading to new knowledge in the social sciences and professions. Technical innovation did not wait until research in the natural sciences and technology became institutionalized. (See chapter 13.) Similarly, institutional innovation may occur as a result of external contact or internal stress. The objectives of institutionalization of social science research include greater efficiency in the allocation of social science research resources to speed the production of new knowledge that is designed to be used as inputs into those areas of institutional innovation on which society places a relatively high priority and to apply the new knowledge to bring about a more precise linkage between the institutional innovations that are implemented and the objectives of institutional innovation.

One of the most dramatic examples of the effect of new knowledge in the social sciences on institutional innovation and efficiency has been the knowledge of macroeconomic relationships identified with the Keynesian revolution. No effort has been made to estimate the economic gains generated by the new knowledge in enabling the developed economies of the West to operate at close to full employment since World War II. There are, however, the estimates made by Arthur M. Okun of the contributions to U.S. economic growth of the reductions in the personal and corporate income tax under the Revenue Act of 1964. Tax cuts in corporate and personal taxes amounting to the $13 billion in 1964 and 1965 were made with the explicit objective of reducing the gap between actual and potential GNP. The tax cuts were recommended by the Council of Economic Advisors on the basis of quantitative projections based on estimated disposable income, consumption, investment, and inventory relationships. Okun's ex post estimates indicate that the tax reduction contributed $15 billion during the first two years after the tax cut and an ultimate $36 billion to the growth of U.S. GNP. In retrospect, these estimated gains should be partially discounted to the extent that they contributed to the inflation since the late 1960s.

In a similar manner, it should be possible to compute the returns to broad lines of research in agricultural economics. For example, research leading to the quantification of commodity supply and demand relationships is expected to lead to more efficient functioning of supply management, food procurement, and food distribution programs; research on the social and psychological factors affecting the diffusion of new technology is expected to lead to more effective commodity production campaigns; and research on the effects of land tenure or group farming arrangements is expected to lead to institutional innovations leading to greater equity in access to rural resources and greater productivity in the utilization of resources in rural areas.

The demand for knowledge that can contribute to improvements in the effectiveness of the institutions that serve rural areas appears to have risen sharply in both developed and developing countries in the 1960s and 1970s. In the past, technical constraints on production have generally represented a more serious barrier to agricultural and rural development in poor countries than institutional constraints have. As some of the technical constraints have been released, institutional constraints have again emerged as increasingly significant barriers to the realization of higher levels of productivity of both human and physical resources in rural areas. This places research in the rural social sciences higher on the agenda of research priorities than it was in the past.

The premise that the demand for knowledge in the rural social sciences is derived from the demand for more effective rural institutions places a special burden on the organizations that fund or manage research programs in the rural social sciences. Resources directed to research on issues that are not of real social or economic significance have a relatively low priority. Furthermore, research that is of potential significance has no clear payoff unless the results are communicated in such a way that the new knowledge enters into decision-making processes and institutional design.

In the next sections, I consider the role of social science research in the agricultural research institute, the agricultural college or university, and the agriculture ministry.

SOCIAL SCIENCE RESEARCH IN THE AGRICULTURAL RESEARCH INSTITUTE

What should be the role of social science research in the agricultural research institute? I have in mind here the freestanding or autonomous institute whose primary objective is the production of new scientific knowledge or new technology directed toward improving productivity in crop or animal production, processing, or distribution. The Rothamsted Agricultural Experiment Station, the Rubber Research Institute of Malaysia, the International Rice Research Institute, and the USDA Peoria Research Laboratory are examples. (I discuss separately, in the next section, those agricultural research centers and experiment stations that operate within a college or university setting.)

What are, or should be, the sources of demand for social science knowledge within the agricultural research institute? This is a question that research directors have not found easy to answer for themselves, their staffs, or their funding sources. My own experience is relevant at this point. In 1963, when I joined the staff of the International Rice Research Institute, I became the first economist in what has become the CGIAR-sponsored complex of international agricultural research institutes. (See chapter 5.) When I arrived at the IRRI, I was shown to an office in the very attractive new institute complex.

The office was conveniently located near the library. It had a brass plate in the door with the label Agricultural Economics. In the weeks that followed, however, neither the director nor the associate director of the IRRI conveyed to me a very clear idea of why they needed an agricultural economist or what contribution they expected from the economics unit at the IRRI. This may have been a wise strategy. It permitted me, and the agricultural economists and other social scientists who followed me at the IRRI, to evolve an effective role during a process of intensive interaction with institute management and other institute staff as our knowledge of the farming systems and the institutional environment in Southeast Asia developed.

Three roles

As I reflect on my own experience at the IRRI and on the experience of other social scientists at national and international agricultural research institutes, it seems to me that there are three roles that the social scientists have occupied at agricultural research institutes.[7] In some cases, the social scientists have been isolated, or have isolated themselves, from the major thrust of the institute's program and have pursued personal or professional objectives. This role has not been effective for either the individual social scientist or the institute. It has typically led to the alienation of the social science staff members from their colleagues in other fields and to tension between the social science staff and the institute's administration. When this has occurred, it has often discouraged the institute's management from attempting to build an effective social science staff.

In what has been a more typical situation, social scientists have directed their attention to the evaluation and dissemination of the new knowledge and technology developed by the institute—the new practices, the new crop varieties, and the improved equipment. The rural sociologists have done diffusion and communication studies and have staffed the information offices. The agricultural economists have done farm management and have served as outreach and extension specialists. Institute managers and staff have often held the view that the role of the social scientist is to communicate all the wonderful products of the research institute to the national extension services and the funding agencies. Where a social science staff has been cast in a purely service-oriented role, however, low staff morale and difficulty in retaining an effective social science capacity have tended to result.

In some cases, the social scientist has been cast in the role of program analyst with the responsibility of working with the director's office on issues of program analysis and planning. Social science skills are highly relevant to the process of research management and research resource allocation. The location of an institute's social science capacity in the director's office with a definition of the social science role primarily as program analysis, however, places the social science staff in a very vulnerable position. The social science staff is seen by scientists in the natural science-based disciplines as rendering judgments about their contributions to the institute's program.

An effective agricultural research institute needs to have access to the three areas of social science research outlined above. The efficient allocation of research resources is the first priority of the institute's management. Failure to bring social science knowledge to bear on this problem is wasteful in the extreme. The screening of new knowledge and new technology for cultural and economic viability is essential for the credibility of the outreach activities of the institute. (See chapter 11.) A modest level of basic research in areas of theory and method that are of direct relevance to activities in the other two areas is essential if the social science staff is to continue to be able to bring the best social science capacity to bear on social science research relevant to the institute's mission. In a well-organized research institute, there should be the same symbiotic integration among the agricultural economics and the agronomy and plant-breeding programs, for example, that there is between entomology or plant pathology and plant breeding or between plant breeding and agronomy.

Yield constraints research programs

The yield constraints research program at the IRRI is one of the best examples of effective complementarity with which I am familiar.[8] Between 1973 and 1977, IRRI agronomists, economists, and statisticians developed and implemented a multidisciplinary research project to study rice yield constraints. Collaborative yield constraint projects were developed with colleagues throughout Asia. The initial impetus for the project was the observation that, although a substantial percentage of Asia's rice land was planted to modern varieties, average yields on farms that adopted the modern varieties remained far below those demonstrated to be possible by research scientists. Far from reaching the researchers' yields of 6 to 8 metric tons per hectare, good farmers rarely exceeded yields of 3 to 4 metric tons per hectare, while most farmers obtained yields between 2 and 3 metric tons per hectare. (See figure 12.1.) The IRRI constraints project staff pointed out that this situation suggested two questions: Why have some farmers still not accepted the modern varieties? and why are many farmers unable to achieve the potential high yields of modern rice technology? They noted that the first question had received considerable research attention but that the second question had not been addressed effectively by researchers.

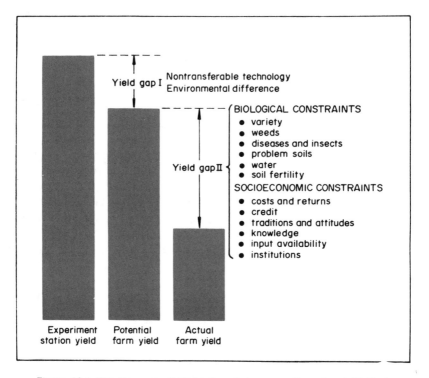

Figure 12.1 The Concept of Yield Gaps between an Experiment Station's Rice Yield, the Potential Farm Yield, and the Actual Farm Yield. (From Kwanchai A. Gomez, Robert W. Herdt, Randolph Barker, and Surajit K. DeKatta, "A Methodology for Identifying Constrants to High Rice Yields on Farmers' Fields," in *Farm Level Constraints to High Rice Yields in Asia: 1974-77*, [Los Banos, Laguna, Philippines: International Rice Research Institute, 1979], p. 30.)

A major objective of the yield constraints project has been to divide this difference between farm-level and experiment-station yields among three major constraint categories: physical, biological, and institutional constraints.

The physical constraints imposed by the environment in which the farmer produces rice include such factors as soil, climate, and water control. The individual farmer has relatively little control over such factors, but they can be modified by appropriate public investment programs (in irrigation and drainage, for example).

The biological constraints imposed by the effectiveness of biological processes and technology include such factors as the availability of cost-effective insect- and weed-control techniques and the availability of high-yielding varieties for particular ecological situations (such as deep-water or rain-fed rice). Some of these constraints can be removed by directing scientific and technical effort to the solution of the problems that give rise to the constraints.

The institutional constraints imposed by the economic, social, and cultural environment in which the farmer lives and works include such factors as farm size and tenure, the structure of input and product markets and the prices that prevail in these markets; and the effectiveness of information channels and the capacity of the farmers to understand and utilize technical and economic information. Some of these constraints can be modified by implementing appropriate market policies and extension programs. Others may require more fundamental institutional changes in property rights and political representation.

The constraints project made use of data obtained from farm surveys, experiments in farmers' fields, and experiments conducted under highly controlled experiment-station conditions. Figure 12.1 illustrates the result of an attempt using data generated by the project to partition the differences between average rice yields in the Philippines of less than 2 metric tons per hectare in the mid-1970s and the approximately 8 tons per hectare that could have been realized with a high-level input under irrigated conditions in the dry season under experimental conditions. The broad categories among which the constraints are partitioned represent aggregations of the several subcategories.

The significance of the constraints project is that it serves to identify those limitations on rice yields that are amenable to reduction through the scientific and technological research at the International Rice Research Institute and the national research institutes and experiment stations. The relative importance of a particular factor, such as insect control, in accounting for the difference between farm-level yields and potential yields gives some indication of the gain that could be obtained by removing the constraint.

The constraints project also identified that part of the yield gap that was not amenable to reduction by advances in knowledge of crop production or practices or by the development of new rice varieties or by research-based control of pests and disease. It also serves to identify the size of the yield gap that is amenable to the implementation of market and resource development policies.

My use of the constraint model to illustrate a useful pattern of collaboration between social scientists and other scientists within a research institute is not intended to imply that there are not other very fruitful patterns of collaboration. Issues of farming intensity, which the rice-yield constraints project has avoided, are being analyzed by multidisciplinary teams working within the framework of cropping and farming systems research programs at several international agricultural research institutes and national programs.[9] These programs are also providing a flow of information that identifies research priorities and that can be fed into outreach and extension programs. They are also providing methodological challenges that are leading to advances in both theory and method in the several collaborating disciplines. The director of the International Potato Center (CIP) has indicated that social science research has led to a focusing of engineering research on village-level processing and storage and to an attempt to develop true potato seed that could replace tuber propagation for both commercial and subsistence production, particularly in the hot-humid tropics.

In closing this section, let me again emphasize a point that has been made earlier. The role of the social scientist in the agricultural research institute is both enlarged and constrained by the mission of the institute. It is enlarged by the opportunity to become productively engaged in a research enterprise that can focus professional effort of a broad spectrum of disciplines on the development of new technology. It is constrained by the necessity of structuring personal and professional objectives in a manner consistent with the institute's mission. The viability and value of the social science program requires synergism with, rather than autonomy from, the rest of the institute's programs.

SOCIAL SCIENCE RESEARCH IN THE COLLEGE OF AGRICULTURE

The role of the social sciences in an agricultural college or university or in a college of agriculture within a larger state or national university derives directly from the mission of the college itself. The mission of the University of Minnesota's Institute of Agriculture—which includes the College of Agriculture, the College of Forestry, the College of Home Economics, the Agricultural Experiment Station, and the Cooperative Extension Service—has been stated as follows.

The traditional mission of the Institute of Agriculture has been the development and transmission of knowledge in those areas of technical and institutional change which are relevant to advances in agricultural productivity, more effective use of natural resources, and improvement in the quality of life in rural areas—primarily in Minnesota. In recent years this mission has been interpreted more broadly in terms of national and international responsibilities.

The mission and responsibility of a social science department within a university setting are much more diverse than those of a social science unit within an autonomous research institute. They also differ significantly from those of a social science department located within a college of arts and sciences. At the University of Minnesota, for example, the Department of Agricultural and Applied Economics has a responsibility for teaching the undergraduate- and graduate-level applied economics courses dealing with the agricultural economy and rural development. It is responsible for the economics research functions of the Minnesota Agricultural Experiment Station. It is also responsible for off-campus educational programs in the area of agricultural economics and rural development that are organized through the Extension Service. Most staff members are responsible for teaching activities in the College of Agriculture and for research at the Agricultural Experiment Station. Some may have joint responsibility for on-campus teaching activities in the College of Agriculture and for off-campus educational programs conducted through the Extension Service. And some may engage in all three activities. A professor who specializes in the field of agricultural policy might, for example, have 25 percent of his or her time budgeted to teaching, 50 percent to research, and 25 percent to extension work. Another might have 50 percent budgeted to extension work in dairy marketing, 20 percent to teaching, and 30 percent to research.

The pattern of close articulation of teaching, research, and extension work, which exists in most state universities in the United States, tends to be more widespread than in most other countries. Yet, there are few university departments of agricultural economics that are not responsible for some mix of teaching, research, and off-campus education or service.

The pattern that I have described for agricultural economics departments is also characteristic of many rural sociology departments. There is, however, a somewhat greater tendency for rural sociology to be organized as a section within a sociology department than for it to be set up as an independent department. Anthropology departments usually include a relatively high proportion of staff members whose professional interests focus on rural communities. Students of agricultural history, agricultural geography, and agricultural politics are often found in departments of history, geography, and political science.

One way of characterizing the work of economists whose appointments are in a department of agricultural economics or of sociologists whose appointments are in a department of rural sociology is to contrast their role with that of economists or sociologists with appointments in a department of economics or sociology in a college of liberal arts (CLA). In a CLA department, classroom teaching represents the primary institutional commitment of the academic staff. Research is supported or encouraged primarily through reduction in teaching loads. Only limited institutional funds are available to the university for the support of research. Research support must be obtained primarily in the form of grant or contract funds from sources outside of the university. Contribution to the advance of disciplinary knowledge is given heavy weight in promotion criteria. Applied macroeconomic research with significant national policy significance is typically weighted more heavily than applied microeconomics research. Service, usually in the form of consultancies and membership on policy advisory committees, is viewed as a byproduct of teaching and research. It is valued as a useful contribution to the public image of the university, but it is rarely a programmed activity. An important implication of the difference in the college of agriculture and the college of liberal arts missions is that the social scientist located in a CLA department is likely to feel a primary commitment to a discipline rather than to the college or the university. It is the scientist's contribution to the discipline that assures his or her professional mobility and minimizes the financial and bureaucratic constraints exercised by the college or the university.

The social scientist with an appointment in a college of agriculture is subject to many of the same constraints and incentives that are operative in a college of arts and sciences. There are, however, additional incentives and constraints that arise from the institutional (state and federal) funding of research and service (extension). Institutional funding of agricultural research has the effect of placing a high priority on research that will contribute to state economic development. Research on market relationships or on the public policy alternatives for commodities that are important to the state or on local resource and regional development problems is valued more highly than it would be if state funding were not available to the institution. Public service activities involving highly structured educational efforts are organized through the extension service. Incentives in the college of agriculture are typically structured to encourage a state or regional orientation of research and extension efforts. There is, however, always a tension between the incentive and reward structures that are relevant for a college of agriculture and those that are accepted as relevant for those parts of the university that have only limited state support for research and teaching programs.

It is also useful to contrast the role of social science research in a college of agriculture with that in a freestanding research institute. One major difference stems from the much greater autonomy of academic departments in the university environment. It is much more costly in terms of physical and intellectual effort to organize multidisciplinary research efforts within, say, the University of Minnesota's College of Agriculture than it would be at the International Rice Research Institute. It would be difficult, for example, to identify many multidisciplinary projects at a U.S. college of agriculture that involve as close interaction between the social science and the technical agricultural disciplines as the interaction involved in the IRRI constraints and cropping systems projects.

I do not intend to imply that such interdisciplinary efforts do not occur in a university environment. The article by Earl Swanson referred to earlier documents a large number of collaborative efforts. My argument is that the structure of organization and incentives within the university tend to discourage rather than encourage such collaboration. When it does occur, however, it usually takes place within a somewhat looser coordinating structure, often at the level of what Swanson referred to as "editorial integration" or collaboration. It is often organized through a series of interrelated subprojects, closely related graduate thesis research, or other relatively independent contributions.

A particularly fruitful example of collaboration among agronomists and economists occurred during the mid-1950s in connection with efforts to develop more effective analytical methods for the design, analysis, and economic interpretation of agronomic-economic research. The research program, funded by the Tennessee Valley Authority, involved collaboration among economists and agronomists from Iowa State University, the University of Kentucky, Michigan State University, North Carolina State University, Purdue University, and the Tennessee Valley Authority. The research objective was to develop designs for fertilizer-response experiments that would provide the information needed for the economic analysis of the experimental results. It also involved the testing of alternative response functions and their implications for experimental design.[10]

It is not uncommon, of course, for economists to employ the results of experimental work in crop or animal science or in engineering in their own research. Neither is it uncommon for scientists in other disciplines to employ economic data or simple budgeting or programming techniques in interpreting the results of their research. Effective multidisciplinary collaboration among social science and technical agricultural departments involving collaboration at the research design stage or in advancing theory and method remains an exception to usual practice.

Even within agricultural economics, research involving group efforts tends to be informal or loosely structured rather than tightly organized. One of the more successful efforts was the series of studies of supply and demand relationships for agricultural commodities organized by the Interregional Committee on Agricultural Policy. A major product of this research effort was the definitive report by George Brandow entitled *Interrelations among Demand for Farm Products and Implications for Control of Market Supply.*[11] The study provided the empirical foundation for estimates of the impact of the agricultural commodity programs that were introduced during the 1960s.

A more typical example is the evolutionary pattern of development in theory and method and the gradual cumulation of research results illustrated by the large number of studies of rates of return to agricultural research reported in table 10.3. The first rate of return estimates for hybrid corn and sorghum were made by Zvi Giliches in his 1956 doctoral thesis at the University of Chicago. This initial study was followed by estimates of the rate of return to agricultural research on a sector-wide basis in the early 1960s. The work by Griliches was continued by several of his students in the middle and late 1960s. Willis Petersen estimated the rates of return to poultry research, Robert Evenson developed new methodology for estimating the rates of return for the U.S. agricultural sector as a whole, and Ardito Barletta estimated the rates of return to research on wheat and maize in Mexico. Since the early 1970s, there has been a virtual explosion of studies that have further advanced the methodology for estimating rates of return, that have covered a number of additional commodities, and that have extended the work on rate of return to the agricultural sector in a number of other countries. Attempts have also been made to adapt the methodology employed for the rate of return studies to use in research planning.

A similar example of the pattern of collaboration that typifies university-centered research can be drawn from the research on the diffusion of agricultural technology by rural sociologists, geographers, and agricultural economists.[12] The sociological tradition of research on diffusion was initiated with the study by Bruce Ryan and Neale Gross at Iowa State College on the diffusion of hybrid seed corn in two Iowa communities during the early 1940s. This earlier research, and much of the research that followed, attempted to explain the time sequence of adoption in a particular community in terms of the personality characteristics of individual farmers and the pattern of interaction among farmers. The diffusion research tradition in geography, which began with Torsten Hagerstrand's studies in Sweden during the 1940s and early 1950s, has focused primarily on spatial rather than temporal diffusion. Research within the geographic tradition has tended to focus on patterns of interaction among individuals and on the role of the hierarchical structure of neighborhoods and regional centers and subcenters. The economics tradition of diffusion research dates from Griliche's work in the mid-1950s, which placed major emphasis on the technical and economic factors they condition the differential profitability of innovation over both time and space. During the late 1950s and early 1960s, an intense debate was carried out primarily within rural sociology on the relevance of the "economic" and "sociological" diffusion models. Interaction among the sociologists and the geographers occurred somewhat later and at a less intense level.

What has been the result of the evolution of diffusion theory and research method? As of yet, the three schools of diffusion research have not resolved the issues that have divided them since each tradition became aware of the existence of the other traditions. There is not yet a unified model of technology diffusion. Yet, I cannot help but believe that the issues that divide the three traditions, some of which appear to be terminological rather than conceptual, could be resolved if the leading theorists from each tradition, or even several of the best practitioners, could seriously confront each other's research over a period of a few months.

In the examples just cited, I have stressed some of the factors that limit the effectiveness of research in the rural social sciences within the framework of the agricultural experiment station or research center that is located within a university environment. The most serious of these limitations appears to be the fragmentation of effort that seems inherent in the academic pattern of organization. This pattern is reinforced, rather than countered, by a research project system that has been criticized (in the Pound report) for encouraging small budgets, fractional allocation of staff time, and short time perspectives.

One implication of the fragmentation of social science research within the academic context is that the experiment station leadership, in contrast to the international agricultural research center leadership, has rarely been able or willing to utilize social science research capacity in its own decision-making processes. Social science research capacity is only rarely organized to provide a direct input into research resource allocation at the agricultural experiment station administrative level. This is in part a reflection of the weaker role of the experiment station management, relative to the departments, in an academic environment than in a research institute environment. The failure of such activities to emerge as a core element in experiment station research programs has also, in my judgment, reflected a lack of imagination and intellectual leadership at the experiment-station level. Yet, both experiment station biological scientists and social scientists often do participate in multidisciplinary efforts directed toward the assessment of research priorities when such efforts are organized by external agencies (such as the National Research Council of the National Academy of Sciences or the Council on Agricultural Science and Technology) or by almost any outside contractor that insists on such collaboration and is willing to pay for it.

A second constraint is related to the sociopolitical environment that conditions both the support and the focus of social science research within the experiment station. The experiment station and extension linkages, which have been responsible for the close articulation between research and extension activities and state-level development, at times have limited the efforts of social scientists at the state experiment stations. In recent years, these constraints appear to be more in the nature of self-limitations imposed by a commitment to applied work directed to state or local problems rather than of external constraints imposed by state-level funding. The staff members in rural social science departments conduct their research in an environment in which there is always tension between (1) the demands by the department's clientele, both within and outside the university, for new knowledge that is directly relevant to the operation of private firms and public programs or to policy formulation and (2) the need to maintain professional capacity and contribute to disciplinary progress. (See chapter 3.) These tensions are not easily resolved.

In stressing the limitations that the academic environment places on the capacity to focus research effort, even when such effort is organized under the auspices of the agricultural experiment station, one should not lose sight of the factors that make a major research-oriented university a highly favorable research location as compared to either a freestanding research institute or a ministry of agriculture. Perhaps the most important is the intense interaction that occurs among graduate students, junior faculty, and senior faculty within the framework of graduate-training activities. Graduate-level teaching forces senior researchers to relate their own work to an expanding range of research issues and methodologies. Graduate students are probably more severe critics of the quality of research effort than colleagues are. The opportunity does exist, in spite of the constraints suggested earlier, for productive interaction between basic and applied research (or theoretical and empirical investigations) and across disciplines. The interaction between teaching research and service (or extension) does provide an opportunity for flexible career development. The result is that one rarely finds within the academic environment of a major research university the intellectual stagnation that the Pound report identified in several of the USDA regional utilization laboratories.

SOCIAL SCIENCE RESEARCH IN THE
MINISTRY OF AGRICULTURE

A modern ministry of agriculture has four primary functions: (1) the conduct and coordination of agricultural research; (2) the gathering, dissemination, and interpretation of agricultural statistics; (3) the management of agricultural development programs; and (4) the operation of the nation's agricultural commodity and food programs. With these four responsibilities, the minister of agriculture, working with a support staff, is a central figure in the formulation of agricultural policy. The minister's office or (in the United States) the secretary's office becomes a center for dialogue within the national administration, between the administration and the national legislature or parliament, and between the government and the several political constituencies with interests in agricultural development, natural resources, commodity, and food policies. In many governments, however, these responsibilities are fragmented among, for example, a ministry of natural resources, a ministry of agriculture, and a national food board. Within the administration, important coordinating functions are often exercised by a council of economic advisers, a national planning commission, a budget office, or a ministry of finance.

The role social science plays within a ministry of agriculture depends both on historical tradition and the capacity of a particular minister to utilize social science knowledge. There are, however, essentially two functions involved regardless of tradition or personality.

One is a staff function that involves organizing the information about programs and policies that the minister needs in order to interact effectively at the policy level with the rest of the government, with the legislature, and with the several agricultural and food constituencies and, in nations in which agricultural trade is important, with other governments. A ministry of agriculture that has an inadequate staff in the social sciences leaves a minister exposed to shifts in the political winds and the flow of social and economic currents.

Social scientists in a ministry of agriculture also have a very important analysis and information function. It is important that the dialogue within the government and between the government, its several constituencies, and the public be conducted in an environment in which there can be reasonable agreement about the social and economic impact of policy or program alternatives. This permits the dialogue to center on the desirability of the impact rather than to degenerate around arguments of "my facts versus your facts." When I worked as an agricultural economist at the President's Council of Economic Advisors in the early 1960s, for example, there was a very intensive effort made to achieve agreement among the council, the Bureau of the Budget, and the staff economics group in the Department of Agriculture on the commodity production, farm income, consumer price, and budget implications of the alternative commodity programs that were under consideration. Policy debate could then proceed without disagreement on the impacts of alternative programs. The ministry of agriculture in most modern societies also recognizes its obligation to make information on a whole range of economic and social indicators (such as prices paid, prices received, production costs, and the income, education, and health of rural people) available to its rural, farm, and agribusiness constituencies.

Both the staff function and the information function require that rather substantial resources be devoted to statistical services and economic and social analysis. In the USDA, for example, there are more than a thousand economists and statisticians employed to serve the staff, analysis, and information needs of the agency. Smaller governments cannot, of course, have access to a comparable level of capacity. The Ministry of Agriculture in the Philippines employs approximately a hundred economists and statisticians with graduate training. In Nepal, there were in the late 1970s fewer than 20 economists with training at the master's level or above, of which 12 were employed by the government and 6 by government-related institutions.

In countries such as Nepal or even the Philippines, there can be much less specialization on the part of economic analysts and planners.

On a global level, only two institutions, the USDA and the FAO, have the capacity to assemble and analyze global information on agricultural commodity production, markets, and trade. The agricultural ministries in smaller countries do need to have the capacity to interpret in terms of their own needs the statistical information and the analytical reports that are available from the FAO, the USDA, the International Food Policy Research Institute (IFPRI), the international lending and donor agencies, and other national and international agencies. The social science units in the agricultural ministries of both the large and small countries also need to have the capacity to draw on the results of the research carried out at the agricultural research institutes and universities. The USDA, for example, often cooperates with university econometricians and commodity analysts for the development or improvement of analytical models (for the soybean or feed grain subsector, for example). Effective cooperation of this type requires a high level of analytical capacity at the agency level. When this capacity is not developed or when it is allowed to atrophy, the agency loses its capacity to monitor the quality of contract research. An active in-house research program is necessary for an agency to have the capacity to effectively contract for and use research performed outside the agency.

The limited professional resources that are available in the rural social sciences in most poor countries require that rather careful consideration must be given to the establishment of priorities in institutionalizing capacity for economic information and analysis in agricultural ministries. Dr. Raj Krishna, an agricultural economist who served as a member of the Planning Commission of the Government of India, identified six priority areas for policy-oriented research by agricultural economists: (1) input-output coefficients, (2) return-cost studies, (3) demand and outlook studies, (4) project evaluation, (5) distribution and unemployment, and (6) agricultural and general growth.

The list presented by Krishna is consistent with the needs of most developing countries. It must be kept in mind, however, that India is more adequately endowed with professional resources in agricultural economics and the other rural social sciences than any other developing economy and more than most developed countries. The need to economize in the use of the professional resources available for information, analysis, and planning forces to the surface another set of questions concerning the methodology that should be brought to bear in the analysis of agricultural development and commodity policies. One approach to this issue has been to argue that the problems of agricultural development in poor countries are so complex that only the most powerful programming and econometric sector-analysis methodology are appropriate. A second perspective has been to insist that the statistical information systems and the professional resources available for analysis and interpretation are so limited that only the simpler social accounting and statistical methodologies should be employed. These arguments have been discussed in a series of sector analysis workshops and have been summarized in several valuable papers by Erick Thorbecke.[13]

My own perspective on this discussion is based on the induced innovation hypothesis that has been employed in the analyses of priorities for research leading to technical change. The path a particular nation should follow in institutionalizing capacity for social science research must take into consideration its current endowments of professional resources, the capacity of its educational system to add to the quantity and quality of professional resources, and the demand for knowledge in the rural social sciences implied by the opportunities and constraints on agricultural development and by the social and political system that a particular nation has inherited. A society dominated by a powerful land-owning elite will reflect quite different demands and will impose quite different constraints on the development of social science capacity than a society in which agricultural production is organized in a small-scale peasant structure. A society in which the educational system is capable of producing a large number of middle-level professionals in the rural social sciences will have quite different options available for the development of analytical capacity than a society that must send its young professionals abroad for advanced training or depend in part on expatriate advisers to staff its research and planning units.

Attempts to develop information and analysis capacity in a ministry of agriculture should also be sensitive to the limitations of the institutional transfer syndrome. Institution building is an evolutionary process. Success in institutionalizing information and analysis capacity depends on the internalization of a body of craft skills and traditions as well as access to technical analytical capacities. This implies that there is no way to avoid the initial steps of developing effective statistical services, of institutionalizing the skills in social and economic accounting that emerged in the developed countries during the 1930s, and of developing the capacity to conduct statistical analyses of commodity market relationships that emerged in the 1940s and 1950s. As these capacities become institutionalized, acquiring modern data-processing equipment and modern sector-analysis capacity leading to greater precision in analysis and planning is a natural development.

Even under the most favorable circumstances, the development of an effective social science capacity in a ministry of agriculture is not an easy task. Maintaining a high level of capacity over time is even more difficult. In part this stems from the close link between social science analysis and the agricultural policy process in the ministries. Within a ministry, there is no way to avoid considerable tension between the needs of short-run analysis of current policy issues and the development of analytical capacity. (See chapter 3.) Neither is there a way to obscure the political implications of the analysis that is conducted in the ministry. Information on regional and personal income distribution or the implications of particular rural development or commodity policies can generate political support or opposition to the policies and programs to which the government (or administration) is committed. Even the refinement of a commodity model can enhance the capacity of the program's administrators to administer a commodity program—and more effective administrators can generate gains or losses for producers, middlemen, and consumers.

As a result, there are often substantial lacunae in social science staffing. Ministries have typically been successful in institutionalizing the capacity for statistical services and economic analysis more than for the other social sciences. As ministries of agriculure have expanded their concern with rural development, their lack of an effective capacity in anthropology and sociology has often been a serious constraint on effective policy formulation and program design. In the late 1940s, the USDA was forced to dismantle its emerging capacity for sociological research on community development because of the Congress's dissatisfaction with a series of sociological studies on the adjustment problems of returning servicemen in racially segregated rural communities in the South.[14] The U.S. Congress's continued opposition has prevented the rebuilding of a more effective social science capacity to work in problems of rural development.

Even in economics, the development and maintenance of capacity in a number of important areas has periodically been jeopardized by controversy over the appropriate role and organization of economics in the USDA. A central Bureau of Agricultural Economics (BAE) was established in 1922. Under a series of strong leaders, the bureau developed an exceptional capacity, in comparison to other federal agencies, in the fields of survey research and economic analysis. It had a strong professional orientation, and its work was highly regarded by the economics profession.

In 1938 the BAE was reorganized and made responsible for "general agricultural program planning and economic research for the secretary and the Department as a whole." The program-planning functions, particularly land-use planning, came under strong attack from other agencies within the department, from the Congress, and from the American Farm Bureau Federation.

In the early 1950s, economics work in the Department of Agriculture was again reorganized and decentralized. One reason for the reorganization was that the new Republican secretary regarded the BAE as being ideologically committed to the commodity programs of the previous Democratic administration. A second reason was a judgment that economists could make more effective contributions if they were located administratively within appropriate operating units in the department rather than in a central bureau.

In the early 1960s, economic work was again reorganized under a central bureau structure. This time, however, the staff's function was more clearly separated from the information and analysis functions than in the old BAE. The Staff Economics Group was established in the office of the secretary under a director of economics. The information and analysis responsibilities were organized under the Statistical Reporting Service (SRS) and the Economic Research Service (ERS). The new organization was not, however, able to fully insulate the ERS from the burden of policy analysis or from the political forces that were critical of the policy directions being conducted in the secretary's office. There is a general impression, in the agricultural economics and economics professions outside of the USDA, that the quality of the staff and the quality of analysis deteriorated after the mid-1960s.

In 1978 a new reorganization with a stated commitment to modernizing the economic capacity at the USDA and to increasing its effectiveness in policy analysis was undertaken. However, the new leadership did not reverse, but rather reinforced, the ascendancy of research managers who had little research experience and little appreciation for the problems of research staff development and nurture. Particularly serious was the failure to maintain a clear distinction between the staff and analytical functions. The result appears to be a further erosion of the analytical capacity that is needed in order to maintain the effectiveness of the staff function. Many skilled analysts who had served their professional apprenticeships at the USDA moved to other appointments in government and the universities just as they were reaching the stage at which they might have provided effective professional leadership in the department.

Efforts to develop and to maintain social science research capacity in the USDA continue to be subject to considerable stress. This includes tension between (1) the professional criteria employed in research program design and project selection and responsiveness to the constituency information and analysis demands and (2) the political and bureaucratic desire to use social science analysis to legitimize program activity and the demand of the social science analyst to achieve exposure of research results to professional colleagues and nonofficial participants in the policy dialogue.[15]

Similar and often considerably more severe stress has characterized efforts to build a social science capacity in the agricultural ministries of many developing countries. Recent attempts to institutionalize sector-analysis capacity in agricultural ministry economics and planning units in countries such as South Korea, Thailand, Mexico, the Philippines, and Tunisia represent useful cases. There has been continuing tension between the allocation of professional resources to the long-run objective of building sector-analysis capacity and the short-run objective of bringing available information to bear on the policy-decision-making process.

It seems clear that the continuing tension in ministry social science research units cannot be resolved by the simple expedient of reorganization. The competing demands for (1) expansion of analytical capacity, (2) responsiveness to the latent demand for social science knowledge by the social and economic constituencies that are not adequately represented in the political marketplace, such as the rural poor, and (3) the biases in demand for knowledge by politically powerful clientele are too strong to be neutralized. In his 1976 presidential address to the American Agricultural Economics Association, Kenneth R. Farrell, then deputy administrator of the USDA's Economic Research Service, suggested that one approach to overcoming the political constraints on policy research in the USDA and the parochialism of state experiment station social science research would be to establish an autonomous national food and agricultural policy research institute.[16] He argued that the institute should be funded largely through private sources. If, however, the relationship between social structure and the demand for social science knowledge suggested in chapter 3 is valid, even the institutional innovation suggested by Farrell would be inadeuqate. Effective demand for social science knowledge in those areas that are not now represented by effective political constituencies depends on changes in the structure of economic and political power.

A PERSPECTIVE ON THE DEMAND FOR
SOCIAL SCIENCE KNOWLEDGE

There were in the early 1980s few agricultural research programs in which the social science disciplines were not represented. The ability of the social sciences to "colonize" agricultural research institutions has, however, been highly uneven. Agricultural economics is the only field that is fully institutionalized across the broad spectrum of agricultural research institutes, university-related agricultural experiment stations, and agricultural ministries.

What accounts for the limited development of professional capacity in the other social science disciplines in agricultural research institutions? Lester Thurow has argued, in the case of economics, that the "part of the imperial success of economics is due to the fact that the profession has a client relationship with society. Economists have not been generally instrumental in shaping society's agenda, but they have been willing to work on that agenda—whatever it is."[17]

The social science research agenda in the field of agricultural development and policy has clearly been heavily weighted in terms of economic criteria. The other social science disciplines have been more critical than economics of the agenda that society—often represented by a legislator or a minister chosen from one of the more conservative rural constituencies—has drawn up for social science research. It has been exceedingly difficult to resolve the problem of how to respond to the biased demands for social science knowledge that are channeled through existing institutions and at the same time attempt to reform the process by which the existing agenda has been established.

Don Paarlberg, in a series of important papers, argued that in the United States a set of new priorities in which community, environmental, and equity considerations are more heavily represented has been placed on the policy agenda.[18] The World Bank, U.S. Agency for International Development, and a number of other international aid agencies have been attempting to force issues such as integrated rural development and programs designed to meet basic needs higher on the policy agenda in the development assistance programs for which they provide support.

A stronger demand for social science knowledge of the type that psychologists, sociologists, political scientists, anthropologists, and historians are more able than economists to supply may be emerging. A question that the social science disciplines must ask is whether

they will respond to this rising demand or whether their past alienation from the policy process will limit their capacity to respond.

I have suggested in this chapter that society is able to realize the
gains from investment in a social science research capacity to work
on problems of agricultural and rural development in terms of more
rapid institutional innovation and improved institutional performance. These gains are not realized without substantial cost. The development of a social science capacity capable of producing a continuous stream of new knowledge directed toward institutional
innovation and performance imposes severe stress on a number of
institutions. There will be stress within the several social science
disciplines over the allocation of professional resources between attempting to understand basic behavioral relationships and using social
science knowledge to speed innovation and improve performance.
There will be stress among the several social science communities and
disciplines over the priorities in expanding professional capacity and
over the priorities in disciplinary and multidisciplinary research activities. There will be tension between the social science- and the natural
science-based disciplines and professions over the extent to which
each will be guided by its own choice of priorities or by the findings
of related disciplines. Finally, the results of the new knowledge that
flows from social science research will produce tension among the
political system, the broader society that it represents, and the institutions that are responsible for the conduct of social science research
—research institutes, experiment stations, universities, and ministries
—over the value, the legitimacy, and the implications of the new
knowledge that emerges from social science research.

NOTES

1. I am indebted to Randolph Barker, Robert Chandler, Lowell Hardin, David M.
Freeman, Alan B. Paul, Richard L. Sawyer, G. Edward Schuh, Eldon D. Smith, Seleshi
Sisaye, Bernard H. Sontag, William F. Whyte, and the members of the University of Minnesota Agricultural Development Workshop for their helpful comments on an earlier draft
of this chapter.

2. This section depends heavily on Vernon W. Ruttan, "Agricultural Economics," in
Economics, Nancy D. Ruggles, ed. (Englewood Cliffs, N.J.: Prentice-Hall, 1970), pp. 144-
51; and "The Social Sciences in Agricultural Research," *Canadian Journal of Agricultural
Economics*, Proceedings of the Canadian Agricultural Economics Society annual meeting,
August, 1980, pp. 1-13. For a more complete review of the historical development of agricultural economics in the United States, see H. C. Taylor and A. D. Taylor, *The Story of Agricultural Economics* (Ames, Iowa: Iowa State University Press, 1952). For a history of the
development of agricultural economics in Europe, see Joseph Nou, *Studies in the Development of Agricultural Economics in Europe* (Uppsala: Almquist and Wicksells, 1967). For a
history of the development of rural sociology, see Lowry Nelson, *Rural Sociology: Its*

Origins and Growth in the United States (Minneapolis: University of Minnesota Press, 1969). See also Robert A. Scott and Arnold R. Shore, *Why Sociology Does Not Apply: A Study of the Use of Sociology in Public Policy* (New York: Elsevier, 1979). For a review and assessment of the more recent developments in agricultural economics, see the four-volume review of agricultural economics literature edited by Lee R. Martin, *A Survey of Agricultural Economics Literature: Traditional Fields of Agricultural Economics, 1940s to 1970s*, vol. 1; *Quantitative Methods in Agricultural Economics, 1940s to 1970s*, vol. 2; *Economics of Welfare, Rural Development and Natural Resources in Agriculture, 1940s to 1970s*, vol. 3; *Agriculture in Economic Development*, vol. 4 (Minneapolis: University of Minnesota Press, 1977, 1978, 1979, forthcoming). For an excellent and brief critical review of the development of agricultural economics from a philosophy of science perspective, see Yang-Boo Choe, "Toward an Idea of Agricultural Economics: A Critique on the Idea of the Applied Economics of Agriculture," *Journal of Rural Development*, 1 (November, 1978), pp. 1-21.

3. This section depends rather heavily on Vernon W. Ruttan, "Induced Institutional Change," in Hans P. Binswanger and Vernon W. Ruttan, *Induced Innovation: Technology, Institutions and Development* (Baltimore: Johns Hopkins University Press, 1978), pp. 327-57, and on an unpublished paper by Vernon W. Ruttan and R. J. Hildreth, "The Legitimacy of Agricultural Economics and the Demand for Knowledge in Agricultural Economics" (Serdang: University Pertanian Malaysia, Faculti Economi Sember Dan Perniagaantani, Staff Paper No. 1975-14, 1975).

4. For a discussion of an initial attempt to estimate returns to economic research, see George W. Norton and G. Edward Schuh, "Evaluating Returns to Social Science Research: Issues and Possible Methods," in *Evaluation of Agricultural Research*, George W. Norton, Walter L. Fishel, Arnold A. Paulsen, and W. Bert Sundquist, eds. (St. Paul: University of Minnesota Agricultural Experiment Station, Miscellaneous Publication 8-1981, April 1981), pp. 247-61. For a discussion of an initial attempt to develop a research administration methodology for agricultural economics research, see Richard W. Cartwright, "Research Management in a Department of Agricultural Economics," (Ph.D. dissertation, Purdue University, 1971).

5. Douglass C. North and Robert Paul Thomas, *The Rise of the Western World* (London: Cambridge University Press, 1973).

6. Theodore W. Schultz "Institutions and the Rising Economic Value of Man," *American Journal of Agricultural Economics*, 50 (December 1968), p. 1120.

7. For other perspectives on the role of the agricultural economist in an interdisciplinary setting, see Earl R. Swanson, "Working with Other Disciplines," *American Journal of Agricultural Economics*, 61 (December 1979), pp. 849-59; B. H. Sontag and K. K Klein, "Prospects and Problems in Interdisciplinary Research at Canadian Research Stations," *Canadian Journal of Agricultural Economics*, Proceedings of the annual meeting of the Canadian Agricultural Economics Society, August 1977, pp. 53-62.

8. The reader is referred to several reports on the yield constraints project: International Rice Research Institute, *Constraints to High Yields on Asian Rice Farms: An Interim Report* (Los Banos, Laguna, Philippines: IRRI, 1977); R. W. Herdt, and T. H. Wickham, "Exploring the Gap between Potential and Actual Rice Yields in the Philippines," *Food Research Institute Studies*, 14(2) (1975), pp. 163-81; S. K. DeDatta, K. A. Gomez, R. W. Herdt, and R. Barker, *A Handbook on the Methodology for an Integrated Experiment—Survey on Rice Yield Constraints* (Los Banos, Laguna, Philippines: IRRI, 1978); Randolph Barker, "Adoption and Production Impact of New Rice Technology—The Yield Constraints Problem," in *Farm-Level Constraints to High Rice Yields in Asia: 1974-77* (Los Banos, Laguna, Philippines: IRRI, 1979), pp. 1-26.

9. For a perspective on farming systems research, see Consultative Group on International

Agricultural Research, *The Review of Farming System Research at the International Agricultural Research Centers: CIAT, IITA, ICRISAT and IRRI* (Rome: CGIAR/TAC Secretariat, April 1978). See also James G. Ryan, *Farming System Research in the Economics Program* (Hyderabad, India: ICRISAT, November 1977).

10. The results of several conferences reporting the results of the collaborative effort are reported in E. L. Baum, Earl O. Heady, and John Blackmore, eds., *Methodological Procedure in the Economic Analysis of Fertilizer Use Data* (Ames, Iowa: Iowa State College Press, 1956); E. L. Baum, Earl O. Heady, John T. Pesek, and Clifford G. Hildreth, *Economic and Technical Analysis of Fertilizer Innovations and Resource Use* (Ames, Iowa: Iowa State College Press, 1957).

11. G. E. Brandow, *Interrelations among Demands for Farm Products and Implications for Control of Market Supply* (University Park, Pa.: Pennsylvania State University Agricultural Experiment Station Bulletin No. 680, August 1961).

12. For an early review of the several traditions of diffusion research, see Elihu Katz, Herbert Hamilton, and Martin L. Levin, "Traditions of Research on the Diffusion of Innovations," *American Sociological Review*, 28 (April 1963), pp. 237-52. For a more recent review, see R. K. Lindner and P. G. Pardey, "The Micro Process of Adoption—A Model," a paper presented to the Australian and New Zealand Association for the Advancement of Science, Auckland, January 1979.

13. Erick Thorbecke, "Sector Analysis and Models of Agriculture in Developing Countries" (Ithaca, N.Y.: Cornell University, Department of Economics, 1978). See also Willard W. Cochrane, *Agricultural Planning: Economic Concepts, Administrative Procedures and Political Processes* (New York: Praeger, 1974), pp. 82-89.

14. Charles M. Hardin, *Freedom in Agricultural Education* (Chicago: University of Chicago Press, 1955, and New York: Ano Press, 1976), pp. 166-68. The Hardin book contains very useful history and perspective on the political pressures that were brought to bear on social science research both in the U.S. Department of Agriculture and in the colleges of agriculture between the mid-1920s and the early 1950s.

15. For additional examples and analyses of the tension involved in institutionalizing social science research capacity, see Irving Louis Horowitz and James Everett Katz, *Social Science and Public Policy in the United States* (New York: Praeger, 1975). Horowitz and Katz's observation—that social science research capacity has been employed most effectively in those public policy and program areas in which a broadly based consensus exists on social values and policy objectives and that, where substituted dissensus exists, it has been difficult for social science research to avoid the polarities of interest-group and ideological politics—is consistent with the social science research experience at the USDA.

16. See Kenneth R. Farrell, "Public Policy, the Public Interest and Agricultural Economics," *American Journal of Agricultural Economics*, 58 (December 1976), pp. 785-94.

17. Lester C. Thurow, "Economics 1977," *Daedalus*, 106 (Fall 1977), p. 80.

18. Don Paarlberg, "A New Agenda for Agriculture," *Policy Studies Journal*, 6 (Summer 1978), pp. 504-6; *Farm and Food Policy: Issues of the 1980's* (Lincoln, Neb.: University of Nebraska Press, 1980), pp. 59-64.

Chapter 13

Responsibility
and Agricultural Research

The productivity of modern agriculture is the result of a remarkable fusion of technology and science.[1] In the West, this fusion was built on ideological foundations that, from the early Middle Ages, have valued both the improvement of material well-being and the advancement of knowledge.

This fusion did not come easily. The advances in tillage equipment and cropping practices in Western Europe from the Middle Ages until well into the 19th century evolved entirely from husbandry practice and mechanical insight. "Science was traditionally aristocratic, speculative, intellectual in intent; technology was lower-class, empirical, action oriented."[2] This cultural distinction persisted in the folklore regarding the priority of basic science over applied science long after the interdependence of science and technology eliminated the functional and operational value of the distinction.[3]

The power that the fusion of theoretical and empirical inquiry has given to the advancement of knowledge and technology since the middle of the 19th century has dramatically increased their impact on the integrity of traditional institutions and on natural environments. Agricultural scientists have been dedicated to the liberation of humans from the limitations of the natural world. It is not unrealistic to argue that agronomists, along with engineers and health scientists, have been the true revolutionaries in the 20th century.[4]

Within the rural communities, the production surpluses generated by the new technology have changed the way that income is shared between the owners of land and the suppliers of labor—among landowners, tenants, and landless workers. The growth of production has also changed the way the income flows are partitioned among agricultural producers and other sectors of society. Within the nonagricultural sector, it has set in motion competition between the public sector and the private sector for control over the new income streams. Within the public sector, it has generated tension between the forward-looking development bureaucracies and the backward-looking control bureaucracies.

In those parts of the world where the constraints on the natural fertility of the land have been released and the power of technology has been harnessed, the old servile relationship between those who labor but own no land and those who own land but do not labor has been broken. Farmers, unlike peasants, feel no need to tip their hats and render a *servo vostro* or *un bacio la mano* to *padrones*. The successful exploitation of productive agricultural technology has been encouraged where there has been an identity of interest between the suppliers of labor and the suppliers of land. Historically, this identity has been most effectively achieved under a system of agriculture in which the two functions are largely merged, as in an owner-operator system, or in a system in which the property rights of the tenant have been highly developed.

Both conservative intellectuals burdened with a romantic view of the communities of the past and radical ideologues mesmerized by utopian plans for the future have been suspicious of the prosperity that technical change brings to rural communities. Progressive farmers are viewed as kulaks rather than yeomen. The policy goals and class interests of both the conservatives and the radicals are more effectively advanced by viewing rural life through the screen of a subculture of peasantry.

THE AGRICULTURAL SCIENTIST AS HERO AND VILLAIN

It has not been difficult to discover heroic qualities in the pioneers who have carried the banners for the agricultural revolution. One can recall many examples:

- Liebig battling to establish the theory of the mineral nutrition of plants and Mendel patiently distilling the elementary laws of genetics from the color of peas in his monastery's garden.
- Harry Ferguson, the self-taught mechanic, applying basic physical principles to the integrated design of tractors and tractor equipment.
- Donald Jones finding it necessary to escape from the orthodoxy of

the corn-breeding program at the Illinois Agricultural Experiment Station to the obscurity of Connecticut in order to have the freedom to explore the potential value of hybrid vigor.

- The intellectual and physical commitment of N. J. Vavilov, the great plant pathologist-geneticist-wheat breeder, protecting the integrity of the Institute of Plant Breeding against the ideological opportunism of T. D. Lysenko.

In the United States, one recalls those who struggled over the freedom to conduct agricultural research in spite of pressures from vested interests—the Iowa margarine incident in the early 1940s, the low-nicotine tobacco case in Kentucky during the early 1950s, and the struggle to limit the impact of pesticides on the natural environment and human health in the 1960s. It is also difficult to avoid listing the science entrepreneurs who, each in their own time, pioneered the development of the great colonial research institutes, organized and nurtured the development of new national research systems, and created the international agricultural research system in the same heroic mold.[5]

But agricultural scientists have been reluctant revolutionaries! They have wanted to revolutionize technology but have preferred to neglect the revolutionary impact of technology on society. They have often believed that it would be possible to revolutionize agricultural technology without changing rural institutions. Because of their beliefs, they have often failed to recognize the link between the technical changes in which they took pride and the institutional changes that they either failed to perceive or feared to accept. As a result, they have often reacted with shock and anger when confronted with charges of responsibility for institutional changes—in labor relationships, tenure relationships, and commodity market behavior—that were induced by technical change. (See chapter 3.)

During the 1960s and 1970s, a new skepticism emerged about the benefits of advances in science and technology.[6] A view emerged that the potential power created by the fusion of science and technology—reflected in the cataclysm of war, the degradation of the environment, and the psychological cost of social change—is obviously dangerous to the modern world and to the future of the human species. The result has been to question seriously the significance for human welfare of scientific progress, technical change, and economic growth.

Agricultural science has not escaped these questions. Some interpret the mechanization of land preparation and harvesting as a source of poverty in rural areas rather than as a response to rising wage rates.

The milling of grain by the use of wind and water power was counted as progress in 12th century Europe. But today's critics view the substitution of rice mills for hand pounding as destructive of opportunities for work in 20th-century Java. There are those who regard the use of yield-increasing fertilizers that increase food production as poisoning the soil rather than as removing the pressure of agricultural production on marginal lands and fragile environments. The new income streams that flow from more productive farms are viewed as destructive of the integrity of rural communities rather than as enabling rural people to participate in a society in which the gap between rural and urban income, life-styles, and culture has narrowed.

The changing perspective on the role of science in society also has led to a change in the way society views the scientist, the engineer, and the agronomist. They have been demoted from cultural heroes to cultural villains. They are assumed to be motivated only by their own curiosity and class interests. They are seen as leading the world into a cul-de-sac from which there will be no escape. The breeding of high-yielding wheat varieties that replace traditional varieties is interpreted as destructive of the genetic resources necessary for responding to future ecological changes.

What should the agricultural scientist or science administrator make of these charges? Can they be dismissed as the mistaken or malicious rhetoric of romantics, populists, and ideologues?[7] How does one engage in fruitful dialogue about the role of science in society in an atmosphere that is so politically and emotionally charged?

A first step is to recognize that similar economic and social forces have generated both the drive for technical change—designed to lead to advances in the productive capacity of plants, animals, machines, and people—and the drive for institutional change—designed to achieve more effective management of the direction of scientific and technical effort and capacity. (See chapter 2.) The increased scarcity of natural resources (of land, water, and energy) continues to create a demand for technologies that are capable of generating higher levels of output per worker, per hectare, and per kilocalorie. The rising value that a society places on the health of workers and consumers and on the environmental amenities such as clean water, clean air, and clean streets has led and continues to lead to a demand for effective social controls over the development and use of agricultural technology.

These demands go beyond considerations of economic feasibility, viability, and impact of the type discussed in chapter 11. Issues of ethical and aesthetic sensibility are also involved.

RESPONSIBILITY FOR RESEARCH RESULTS

This enhanced sensitivity to the moral and aesthetic as well as to the economic implications of technical change imposes an expanded responsibility on both public and private decision-making processes. There is a demand for a greater responsibility in the way scientific and technological advances are put to use. Should government respond to this demand by changing the institutions that induce the generation of new knowledge and new technology? Should government assume a stronger role in the adoption and use of new technology? Should it attempt to encourage greater aesthetic and moral sensitivity on the part of scientists, engineers, agronomists, and science administrators? What can be expected from such efforts?

An attempt was made to address these issues by the National Academy of Sciences Committee on Science and Public Policy in 1969. The committee argued that it is useful to "distinguish clearly between *technologies* or *technological systems* — codified ways of deliberately manipulating the environment to achieve some material objective and the economic arrangements through which such technologies become available and are subject to social control — arrangements that have been described as "supporting systems."[8] The committee noted that the automobile and the highway network make a technology or a technological system and that the rules of accident law, automobile insurance schemes, and traffic police officers are components of the corresponding supporting system. In the committee's judgment, greater sensitivity to secondary and tertiary impact would, in most cases, lead not to different technologies but to different support systems.

The committee's distinction between technological systems and supporting systems is similar to the distinction made in this book between technologies and institutions and, in a dynamic setting, between technical change and institutional change. However, the web of interrelationships among technical and institutional systems is more complex than that implied by the committee. (See chapter 3.) It includes the inducement or generation as well as the adoption and use of new technologies. For example, a small-farm agricultural system in which resources are distributed with a reasonable degree of equity, as in Taiwan, will induce a distinctly different set of technical innovations than a system organized along hacienda-minifundio lines, as in many Latin America countries.

The difficulties that face governments in attempting to respond to the public demand for greater moral responsibiltiy in the generation and use of new technology can be illustrated more concretely by referring to specific examples. The current controversy over the employment displacement effects of the tomato harvester, which was discussed in chapter 8, serves as one useful illustration. Another illustration of an even more complex set of ethical considerations, is the case of research on tobacco improvement. The two cases were chosen, not because either case is the most significant one that could be selected, but because they illustrate in a dramatic way principles that are much more pervasive.

Technical change and employment displacement: The tomato harvester case

The introduction of the machine harvesting of tomatoes has been accompanied by an especially vigorous debate. It has been viewed as the product of a uniquely effective collaboration between mechanical engineers and plant scientists. And it has been attacked for its effect in displacing farm workers and small producers.[9]

In 1978 a suit on behalf of the California Agrarian Action Project and a group of farm workers was filed against the University of California Regents. The suit charged that the regents had allowed agribusiness corporations and their own economic interests to influence their decisions to spend public tax funds on developing agricultural machines. The relief sought by the plaintiffs included an order to compel the university to use the funds it receives from its machinery patents to help farm workers displaced by those machines.

In December 1979, U.S. Secretary of Agriculture Bob Bergland announced in Fresno (California) that he intended to stop USDA funding for research that might put farm laborers out of work.[10] The dean of the University of California's College of Agricultural and Environmental Sciences at Davis criticized Bergland for attempting to impose restrictions on the freedom of academic research. In the summer of 1980, Secretary Bergland established the Agricultural Mechanization Task Force to examine in greater depth the policies of the USDA with respect to mechanization research.

Clearly, the farm workers displaced by labor-saving machinery deserve a reasonable degree of protection from unemployment. This is a legitimate claim on the new income streams—the productivity dividends—resulting from the adoption of the new technology. But who among the displaced workers deserves protection? Do the displaced workers who immediately found other employment have a legitimate claim on the new income stream? What about the workers who found other employment but at lower wage rates? And what about the tomato growers in Indiana and New Jersey who lost part of their market due to the lower cost of tomatoes grown in California? Who should pay the compensation? Should it be the inventors and manufacturers of the labor-displacing equipment? Should it be the farmers who captured the initial gains from lower costs or the processors who expanded their production as a result of their ability to buy tomatoes at a lower cost? Or should it be the consumers who ultimately gained as competitive forces transferred the lower costs of production on to consumers?

The answers are implicit in the questions. The gains of productivity growth are diffused broadly and the costs should be borne broadly. Society has an obligation to provide generalized protection through a comprehensive system of severance payments, unemployment insurance, and retraining programs for workers who are displaced for any reason—by labor-saving technology, competition from imports, business failures, or fluctuations in aggregate economic activity. In a wealthy society like the United States, a worker should not have to prove specific displacement—that he or she was displaced by a tomato harvester or a Toyota—in order to be eligible for such protection.

The first line of defense against the impact of displacement is an economy in which productivity is growing and employment is expanding. Society has little obligation to compensate the worker who can readily find alternative employment. The second line of defense is a program of severance payments and unemployment insurance that is effective for all workers, those who are forced to seek seasonal or casual employment. A society that provides generalized protection will be in a stronger position to realize the gains from technical change and to diffuse these gains broadly than a society that insists on specific or categorical protection.[11] The failure to develop institutions capable of protecting farm workers from the effects of seasonal unemployment and technological displacement has resulted in the transfer of an excessive burden of displacement costs on farm workers. This, in turn, has induced a legal and political response that, if effective, could slow technical change and limit the gains from productivity growth. A more appropriate set of labor-market institutions would facilitate greater equity in sharing the gains from technology.

In a society in which employment opportunities are expanding rapidly and protection from unemployment is adequately institutionalized, neither the individual researcher nor the director of the research team involved in the development of a tomato harvester or a lettuce harvester needs to be excessively burdened by the moral implications of trade-offs between the economic and social costs and the benefits of mechanization. Public policy has relieved them of that burden. But who should bear the burden of responsibility in a society that is too poor to protect workers from technical displacement? Or in a wealthy society that forces the burden onto its poorest citizens?

Efficiency in the production of a health hazard:
The case of tobacco

Tobacco is a commodity that has been the subject of moral debate and political intervention since it first became a commercial export from colonial America. It was introduced to Europe in the 1550s and was officially condemned as a health hazard in England by James I in 1604. Firm medical evidence came much later. An association between cancer of the lips and mouth and tobacco use was reported by several observers during the 18th century. In the 1950s and 1960s, conclusive evidence was produced of the association between cigarette smoking and lung cancer, coronary artery disease, chronic bronchitis, and emphysema. The narcotic effects of smoking and part of the source of the health hazard are due to the nicotine and related alkaloids in tobacco smoke.[12]

What are the moral responsibilities of agricultural researchers and research administrators regarding a crop that not only induces chemical dependency but also kills people—that has a high probability of shortening the lives of those who consume products that are made from it.

Under these circumstances, one would think that efforts to develop tobacco varieties with low nicotine content would have the support of both farmers and consumers. Yet, a successful effort made in the early 1950s by Prof. W. D. Valleau of the University of Kentucky to develop low-nicotine varieties of tobacco was bitterly attacked by Kentucky farmers because of the new varieties' potential competition with burley tobacco.[13]

The price of Burley tobacco was maintained by a system of acreage allotments. The Burley Growers' Association attempted to prevent further research on low-nicotine tobacco and to prevent the marketing of the low-nicotine varieties. A bill was introduced in the Kentucky legislature to limit experimentation and to prevent the growing of any type of tobacco that would jeopardize the profitable production of burley. The bill was passed by the state Senate but defeated in the House after being vigorously opposed by the dean of the University of Kentucky's College of Agriculture and the Louisville

Courier-Journal. Thus, Professor Valleau's research produced a land-mark case in the history of freedom to do research, but it did not result in lower nicotine in cigarette tobacco.[14]

In retrospect, one has little difficulty in supporting the objectives of Professor Valleau's research to reduce the nicotine content of burley-type tobaccos. Even a marginal contribution to the reduction of the chemical dependency and health hazards associated with cigarette smoking would seem to be desirable. But what about the issue that underlies this judgment?

Should public funds be used to do research to reduce the costs and improve the productivity of a product that induces chemical dependency or shortens life expectancy? What are the moral responsibilities of the members of the Kentucky and North Carolina legislatures that appropriate funds for tobacco research? What are the moral responsibilities of the directors of the agricultural experiment stations in these states? And what about individual scientsts who devote their lives to understanding the physiology and the nutrition of the tobacco plant? Is the farmer who grows the tobacco absolved from responsibility by the fact that there is a market demand for tobacco? Are the legislators and the experiment station directors absolved by the fact that tobacco has been one of the more profitable crops available to small farmers in the depressed areas of Kentucky and North Carolina? Are scientists relieved of responsibility by an appeal to the freedom to do research?

These concerns also bear on the moral responsibilities of research administrators, scientists, engineers, and agronomists who are employed by the tobacco companies. Some tobacco breeders contend that the tobacco companies have encouraged the development of tobacco varieties that, while containing lower levels of suspected carcinogens, retain a sufficiently high level of nicotine to maintain the desirable taste and habit-forming effects of tobacco smoke. What are the moral implications for the tobacco breeder, whether employed by a private firm or a public research institution, of responding to market criteria when the market is most effectively enhanced by inducing chemical dependency?

As evidence has accumulated on the health hazards of smoking, there has been in the United States a greater tendency to insist that the legislators who appropriate the funds, the science administrators who allocate funds, and the individual researchers who carry out the work in both public- and private-sector research institutions must assume a greater responsibility for the impact of their actions on the health of those who use tobacco. What inferences can be drawn about moral responsibility from the behavior of a society in which the government spends billions of dollars on medical and health care made necessary by smoking and millions of dollars on research on tobacco-related disease and on campaigns to discourage smoking while at the same time it spends large amounts on tobacco improvement and programs to stabilize tobacco prices?

As in the case of the tomato harvester, there are institutional changes that would relieve research administrators and scientists of the moral dilemma posed by tobacco research. If a public consensus were to result in making the sale of tobacco products illegal in the United States, it would be doubtful that the directors of the Kentucky and North Carolina state agricultural experiment stations would allocate any more resources to tobacco improvement than they now allocate to marijuana research. There has not yet been sufficient convergence of opinion to take the steps that would be needed to limit the amounts of dependency-forming and carcinogenic substances in ciagarettes. An attempt to move toward complete prohibition would require a careful balancing of the desirable effects on individual health against the undesirable effects of prohibition associated with attempts at enforcement.[15]

TOWARD SOME GUIDELINES TO MORAL RESPONSIBILITY

The tomato and tobacco research cases pose extremely difficult moral problems for agricultural researchers and research managers. The centuries-long struggle in Western society to free scientific inquiry from the constraints of the church make it unlikely that the answers to issues of moral responsibility for new knowledge and new technology will be sought from traditional religious sources.

Where, then, can the scientist or science administrator look for guidance on issues of moral responsibility? One possibility is a philosophy of inquiry that recognizes the objective status of both positive and normative knowledge.[16] The philosophy of scientific inquiry to which most scientists subscribe, either explicitly or implicitly, imposes only two criteria as a test of *objective knowledge: correspondence* and *coherence*. The test for correspondence requires that knowledge be continually tested against experience and observation. The test for coherence requires that the scientific explanation meet the test of *logic*—that it be explicable in terms of the general knowledge of scientific principles. In order to meet the test of correspondence, it is important that a scientific statement be stated clearly. A concept that cannot be stated in a clear, unambiguous manner does not lend itself to rigorous testing. The requirement of clarity is important because all scientific knowledge is tentative and hence subject to rigorous testing and continuous revision.

This view of scientific method, known as logical positivism, has been of great significance in leading to the quantification of scientific knowledge. Biometricians, econometricians, and others are able to clearly distinguish between the logical structure of their concepts, which can be tested for coherence, and the empirical content of their statements, which can be tested for correspondence. One limitation of logical positivism is that it tends to ignore normative knowledge, knowledge about what is good or bad. Indeed, it is a fundamental principle of logical positivism that there is no empirical, objective, or true knowledge of the normative.

The recent social criticism of science can be interpreted as a challenge to this view. It has also been challenged by some philosophers of science who draw an analogy between the tentative "dialectical" nature of scientific knowledge that must be continually tested for correspondence with empirical observations and the tentative nature of normative knowledge. They argue that normative experience, such as the goodness of a healthy body or the badness of injustice, implies that normative knowledge can also be tested against the criteria of coherence, correspondence, and clarity. Like positive knowledge, normative knowledge is tentative and must be continually tested.

Acceptance of comparable objectivity of positive and normative knowledge does not, however, lead directly to prescriptions about right or wrong behavior. It is not always wrong to do what is bad—when it is the least bad that can be done under the circumstances. Neither is it always right to do what is good—when something even better can be accomplished with the same or less effort or resources. Thus, the knowledge that cigarette smoking has bad effects does not automatically imply a decision that cigarette smoking should be prohibited. A decision to prohibit cigarette manufacture and trade would involve a weighing of the bad effects of smoking (chemical dependency, damaged health, and shortened longevity) against the bad effects that might be induced by the prohibition (such as the corruption of the legal system by illegal trade in tobacco and the loss of personal freedom).

Any decision or rule that transforms knowledge about what is good or bad into a prescription about what should or must be done involves an arbitrary exercise of power because both positive and normative knowledge are imperfect—both being subject to further exploration and revision. Furthermore, decisions about what should be done or what is the right thing to do (to fund a particular research project or to embody new knowledge in new technology) imply the use of both normative and positive knowledge. Such decisions involve both positive and normative knowledge about what the consequences will be, for example, for agricultural production, for human health, for the incomes of hired laborers and farm operators, for the cost of food to consumers, and for the economic and political status of scientists, bureaucrats, and politicians.

Only modest progress has been made in achieving a tested normative knowledge that can serve as a basis for workable prescriptions. Yet, one can perceive in the public discussion of the tomato harvester and tobacco research cases two principles that appear to have fairly broad applications in interpreting a wide range of individual and group behavior in response to issues of moral responsibility.[17]

The first is that a risk or loss that occurs incrementally is associated with less personal concern and induces weaker public response than a risk or loss that occurs in more discrete units. Most smokers seem willing to accept the statistical evidence that cigarette smoking reduces average life expectancy, but almost no one believes that smoking one more cigarette or waiting one more week to stop smoking will have a effect on his or her own life expectancy. In contrast, the loss of a job as a result of the mechanization of a harvest operation or the discplacement of domestic employment by imports is a discrete and often a very painful event. It often generates a substantial political response even when the number of individuals affected is relatively small.

The second principle is that there is a strongly held value that government has a clear responsibility to protect citizens against damage or loss imposed on them by either the purposeful or unintended actions of others. In contrast, there appears to be a strongly held perception that the government has only limited responsibility for protecting citizens against the damage that they do to themselves. The smoker who is willing to acknowledge the effect of his or her habit on life expectancy may also insist that smoking is a matter of personal choice and may be willing to defend tobacco research that will provide him or her with less expensive or better cigarettes.

Although the two principles appear to correspond to a great deal of personal and group behavior, they have not been subjected to rigorous tests of correspondence or coherence. An implication of the emerging philosophy of inquiry perspective is that the continuous testing and evaluation of the values implicit in individual and group behavior could lead toward the normative knowledge needed in evolving a body of workable prescriptions in the field of science policy.

TECHNOLOGIES, INSTITUTIONS, AND REFORMS

There is no way to avoid responsibility for the consequences of the distributional effects of the mechanical tomato harvester, the health effects of improved tobacco, or the environmental effects of persistent pesticides. Even the most unregenerate adherents to traditional forms of market relationships acknowledge that the principle of caveat emptor (let the buyer beware) cannot be accepted as an appropriate guide to market practice. The fusion of science and technology that has been so powerful in releasing humans from the constraints of the natural world have made traditional concepts of absolute freedom, including the freedom to conduct research, obsolete. When the freedom to know can no longer be distinguished from the freedom to do, the nature of that freedom changes.[18]

There can be no retreat from the fusion of science and technology or from the responsibility implied by this fusion. Attempts to define some activities as "pure" or "basic" science on the basis of disinterest in technological implications must be treated as little more than rationalization. Such attempts carry little credibility when accompanied by requests for public support for basic research on the grounds that advances in technology are constrained by advances in basic research.

Recognition that support for even the most basic research must be evaluated on grounds other than its empirical validity or the aesthetics of its own internal logic is not the end of the discussion. The fact that scientists, engineers, and agronomists cannot evade responsibility for the economic and social consequences of their investigations does not resolve the question of what responsibilities an enlightened community should insist that they bear. In the preceding section, I insisted on society's right to hold scientists, engineers, and agronomists responsible for the consequences of the technical and institutional changes set in motion by their research. In this section, I argue that it is in society's interest to let the burden of responsibility rest lightly. I also argue that social control over the effects of technical change can be more efficiently exercised at the adoption stage than at the research or development stage.

When credit is claimed for the productivity growth generated by advances in agricultural technology, responsibility for the effects on the distribution of income among suppliers of labor, land, capital, and industrial inputs cannot be avoided. Neither can responsibility be evaded for the impact on environmental amenities or the health of workers and consumers. But it is in society's interest to let these burdens rest rather lightly on the shoulders of individual researchers and research managers. If society insists that it be assured that advancee in agricultural technology carry minimum risk—that agricultural scientists abandon their roles as reluctant revolutionaries—society must accept the risk of killing the goose that has laid the golden eggs.

I insist that in allocating resources to agricultural research a first consideration must be whether the agricultural science community can visualize opportunities to advance knowledge or technology. The second consideration must be, given the research is conducted, whether or not there will be an economic demand for the knowledge or the technology that is generated by the research. (See chapter 11.)

Both of these criteria are consistent with the motivations of the agricultural science community and its commitment to the power of the interaction among the methods of science and technology. The first consideration meets the criterion of professional integrity. The second meets the need to avoid alienation or, more positively, the need to feel that the results of the exercise of professional skill and insight have meaning for society.

The objective of research management

I argued in chapter 11 that society cannot expect research managers and scientists to commit themselves to the realization of scientific or technical objectives that are unrealistic in terms of the state of scientific and technical knowledge. For example, it was unrealistic in the 1950s to expect that utilization and marketing research could make a significant contribution to the solution of agricultural surplus problems in the United States. The allocation of excessive research resources to these areas led to both a waste of research resources and an erosion in the credibility of the research enterprise.

It is equally wasteful for society to ask agricultural research managers and scientists to adopt objectives that are not evident in the economic or political marketplace. It is unrealistic, for example, to insist that the California Agricultural Experiment Station direct its mechanization or biological research to the needs of the 160-acre farm unless the state of California or the federal government is prepared to support the structural policies necessary to reverse the trend toward large-scale agriculture. A research system cannot be asked to produce knowledge and technology that will not be used without eroding the intellectual integrity and ultimately the scientific capacity of the research system.

It could be argued, against the position just stated, that policymakers should ask research managers and scientists to "discover" society's true objectives (a social welfare function) prior to the time that they are "revealed" in the political or economic marketplace. Such an argument would imply that the research manager should have on his or her staff persons with the capacity not only to analyze the incidence of the benefits and burdens of the technical changes anticipated from a research program but, in addition, to develop a set of weights (shadow prices) that reflect the value society places on the welfare of each individual or group that may be benefited or burdened by the results of the research. The incidence estimates and the welfare weights could then be combined in making research resource allocation decisions. This view suggests that research directors should allocate resources on the basis of a social welfare function prior to the time that it is revealed by either the economic or the political system!

What alternative do I have to suggest? Clearly, a research director must have access to the analytical capacity to gauge the potential incidence of benefits and burdens in order to enter into effective dialogue with the political system about research budgets and priorities. The research director who does not have access to, or fails to use, such capacity stands naked before both critics and supporters. Research leading to a better understanding of the discrepancies or the disequilibrium in the economic, political, and social weighting system is essential. But the objective of such research should not be to provide research directors with weighting systems for internal research resource allocation. The objective should be to contribute to a political dialogue that will result in institutional changes leading to the convergence of the several weighting systems. As these weighting systems converged, research directors would not be forced to choose among alternative responses to an arbitrary or inconsistent set of economic, political, and social weights.[19]

Let me illustrate. John Sanders demonstrated for Brazil and Hans Binswanger for India that biased signals in the form of differential excahnge rates and subsidies encouraged more rapid mechanization of farming operations than would have occurred if the decisions had been guided by more "efficient" prices.[20] One effect was to bias the flow of benefits to larger landowners and to impose an excessive burden on hired laborers. The market bias also resulted in the excessive allocation of research resources in the direction of mechanization. The effect of "inefficient" rice prices in biasing the allocation of resources in agricultural research in Japan was noted in chapter 4.

Research managers do have a clear responsibility to inform society of the impact of pricing systems and tax structure on the choice of technology by farmers; on the incidence of technical change; on the distribution of income among laborers, landowners, and consumers; on the structure of farming and rural communities; and on the health and safety of producers and consumers. They also have a responsibility to enter into the intellectual and political dialogues that are necessary if society is to achieve a more effective convergence between "market" and "shadow" prices and between the "individual" and "revealed" preferences of its citizens. If market and shadow prices for inputs and products can be made to converge, research directors can be given clear signals for the allocation of research resources. When market and efficiency prices diverge, it will be almost impossible to induce research planners to allocate their resources in a manner consistent with the shadow prices. If political processes can lead to greater consistency between revealed preferences and individual values, individual scientists and research managers might no longer be confronted by a situation in which cigarette smoking is branded as dangerous to health and at the same time public resources are appropriated for research on tobacco production.

Agricultural technology as an instrument of reform

A second reason why it is unwise for society to insist that the potential moral and aesthetic consequences of technical change weigh heavily on agricultural research resource allocation is that technical change is such a blunt instrument of reform. A nation's agricultural research budget can be a powerful instrument for expanding its capacity to produce food and fiber, but it is a relatively weak instrument for changing income distribution in rural areas.

In thinking about the incidence of benefits and burdens of technical change, it is important to distinguish between *embodiment, augmentation,* and *incidence.*[21] Technology embodied in one factor of production has the capacity to augment the productivity of other factors, while the incidence of benefits and burdens may be experienced by other factors or even in other sectors. (See chapter 12.) Embodiment is a characteristic of the technology itself. Augmentation and incidence are influenced by the institutional environment into which the technology is introduced.

For example, research embodied in a highly divisible technology, such as improved seed, available to both small-scale and large-scale farmers may have different impacts on income distribution depending on the economic or institutional environment into which it is introduced. If introduced into an environment in which the supply of labor is more elastic than the supply of land, it will augment the returns per unit of land more than per units of labor, even though it may greatly expand the employment of labor per unit of land. If introduced into an environment in which the supply of labor is inelastic, it will augment the returns per unit of labor more than per unit of land, even though there is little increase in employment per unit of land.

The incidence of benefits is also affected by institutional endowments and changes. If the divisible technology is introduced into an economy in which agriculture is carried out under a share-tenure system, the incidence of benefits will be different than if introduced in a system in which farming is carried out primarily by owner-cultivator families. If mechanical technology is introduced into a society in which capital investment is encouraged through interest and tax rate subsidies, it will have a different impact than in a society in which monetary and fiscal policies are more neutral with respect to the use of capital and labor.

The research system often seems to be criticized for its failure to respond to distributional considerations, not because it is a powerful instrument for bringing about changes in the distribution of income, but because it is more vulnerable to pressure than institutions with a broader base of political support. There is a great danger that, in attempting to respond to such pressure, the research system will not become a source of equitable growth but will lose its capacity to contribute to growth at all!

These considerations lead me to the conclusion that the primary consideration in the allocation of resources to agricultural research should be cast in terms of criteria that are relevant to the adoption and diffusion of the technology. These are the same criteria that will result in the highest rates of return to the research when evaluated in economic terms—given the existing economic and institutional structure of the agricultural sector. When changes in the social objective function are expressed in terms of modification in the economic and political structure, research administrators can respond, by employing the same criteria, to the clear expression of a modified social objective function.

Institutional innovation as an instrument of reform

The induced innovation perspective, which I have used as an organizational theme in this book, suggests that technical change can be a powerful source of institutional change. Institutional innovation can, in turn, exert a powerful impact on the rate and direction of technical change. The contribution of technical change to institutional reform is indirect. Its incidence is uncertain and difficult to anticipate. In contrast, institutional innovations can be designed both to generate technical change in a manner consistent with resource endowments and product demand and to bias the incidence of benefits and burdens in a manner consistent with social policy. Institutional innovation is both a more powerful and a more reliable instrument of reform than technical change.

Among the most effective instruments for directing the gains from technical change are institutional changes related to the ownership of resources. The land-to-the-tiller type land reforms as employed in Japan, South Korea, and Taiwan, for example, were a powerful instrument for achieving greater equality in the distribution of income in rural areas. The green revolution's seed-fertilizer technology, when introduced into an environment characterized by reasonable equality in land ownership, as in South Korea, contributed to greater equality in income distribution. This same technology introduced into an environment characterized by great inequality in the distribution of assets, in the Pakistan Punjab, for example, has reinforced this inequality. The education of rural people is also a powerful source of institutional reform. The institutional reforms in land tenure and market organization in Denmark were clearly facilitated by the existence of a literate rural population.

It was argued in chapter 12 that the demand for knowledge in the social science and related professions, including the rural social sciences, is derived from the demand for institutional innovation and more effective institutional performance. If the agricultural research system fails to respond to the actual or latent demand for knowledge in the social sciences relevant to institutional design and performance, it will find itself increasingly confronted with demands that the scientific and technical effort be directed toward the achievement of reform rather than productivity.

THE TECHNOLOGY ASSESSMENT MOVEMENT

Since the late 1960s, the term "technology assessment" has been increasingly introduced into discussions about the distributional, environmental, and aesthetic consequences of research and development.[22] The concept was first introduced in the mid-1960s when Congressman Emilio Q. Daddario (Democrat, Connecticut) proposed the creation of the Office of Technology Assessment, which would provide the U.S. Congress with an "early warning" of the potential impacts of developing technologies and help legislators understand the full implications of decisions and policies related to technology. Daddario's proposal was viewed by some as a response to the anti-technology movement and by others as an attempt to evade the thrust of the movement. There was also confusion over its content. Was it a new and more powerful methodology designed to overcome the limitations of narrower or more partial approaches to problems of technology generation and choice? Or was it a social movement induced by the neglect of the moral and aesthetic impacts of technical change? After considerable debate, Daddario's proposal to establish the congressional Office of Technology Assessment was finally implemented in 1973, and by the late 1970s technology assessment had become a regular or at least an intermittent component of the programs of most federal agencies.

Technology assessment, as practiced by the agencies responsible for technology policy, has had great difficulty living up to its promises. Compromises have been necessary between the objectives of comprehensive assessment and rigorous analysis. It has been necessary to choose between issues of current concern and issues of future significance. Focusing on issues of current concern has resulted in a primary emphasis on problems of technology choice and in limited attention to the longer-run issues of priority in the generation of technology for the future.

Some initiative has been taken to apply a technology-assessment perspective to agricultural research planning and the establishment of agricultural priorities. The Office of Technology Assessment has assessed the alternatives for organizing and financing basic research to increase food production and the effects of regulation on research in the pesticide industry, and it has listed research and development priorities for U.S. food productivity as a continuing issue of major concern.[23] Modest initiatives have been made in the U.S. Department of Agriculture. These include assessments of the impact of minimum tillage, four-wheel-drive tractors, and energy from biomass. At this time, however, the role of technology assessment and the capacity to perform technology assessments at the U.S. Department of Agriculture remains uncertain.

The technology-assessment movement—the attempt to integrate technical, economic, social, aesthetic, and moral considerations—has not yet lived up to its promise. It has experienced difficulty bringing new methodology to bear on problems of evaluation and assessment. It has had difficulty finding an intellectually satisfying way of bringing aesthetic and moral issues to bear on problems of technology choice and generation. These concerns continue to find expression through the courts and through political action rather than through the methodology and institutions established for technology assessment. In spite of its limited accomplishments, however, the technology-assessment movement is leading to the institutionalization of a more effective capacity to anticipate and to evaluate the effects of technical change.

AGRICULTURAL RESEARCH AND THE FUTURE

What should society expect from agricultural science in the future? And what does agricultural science have a right to expect from society if it is to meet society's expectations?

Let me comment first on what society should expect from agricultural science. First, society should insist that agricultural science maintain its commitment to expanding the productive capacity of the resources used in agricultural production. These include the original endowments of nature: the soil, water, and sunlight; the agents that humans have domesticated or adopted for their own purposes: the plants and animals and the organic and mineral sources of energy; the agents humans have invented: the machines and chemicals; and the people engaged in agricultural production. It is essential for the future of humankind that by the end of this century the capacity to maintain this commitment is established in every part of the world. During the last two decades, the world has become increasingly dependent on the productive capacity of North American agriculture. This dependence poses danger both to the developing world and to North America. Effective agricultural science communities and institutions that are capable of producing the knowledge and the technology to reverse the trends of the last several decades must be established. Agricultural science in North America must remain strong enough and cosmopolitan enough to contribute to and learn from the emerging global agricultural science community.

Second, society should insist that agricultural science embrace a broader agenda that includes a concern for the effects of agricultural technology on the health and safety of agricultural producers, a concern for the nutrition and health of consumers, a concern for the impact of agricultural practices on the aesthetic qualities of both natural and artificial environments, and a concern for the quality of life in rural communities and that considers the implications of current technical choices on the options that will be available in the future. These concerns are not new to agricultural science, but they have often been viewed as peripheral or diversionary to the main task of agricultural science by the agencies responsible for the financial support of agricultural research. It is important for the future of agricultural science that these concerns be fully embraced. It is also important that the capacity to work on these problems outside of the traditional agricultural science establishment be maintained so that an effective dialogue can be achieved both within the research community and in the realm of public policy.

What should the agricultural science community expect from society? First, agricultural science should expect that society will gradually acquire a more sophisticated perception of the contribution of agricultural technology to the balance between humans and the natural world. The romantic view that agricultural science is engaged in a continuous assault on nature is mistaken. Society must come to understand that agricultural science can succeed in expanding productive capacity only as it reveals and cooperates with the laws of nature.

Western society is the inheritor of a tradition that views material concern as a defect in human nature. This inheritance leads to a romantic view of humanity's relationship to the natural world. It also leads to a view that technology alienates humans from both the natural world and the natural community. But I cannot believe that a Taiwanese farmer who harvests a yield of 6 metric tons of rice from his 1 hectare as a result of planting higher-yielding varieties, using chemical fertilizers to complement his organic sources of fertility, and applying herbicides to control the weeds in his field feels a greater alienation than his father, who realized less than 2 tons of rice from the same field. Scientists, engineers, and agronomists have the right to expect the philosophers of society to achieve greater insight into humanity's relationship with technology and nature. It is time to recognize that the invention, adaptation, and use of knowledge to achieve material ends does not "reduce" experience but rather expands it.

Second, it is time for the general science community to begin to follow the lead of the agricultural science community in embracing the fusion of science and technology rather than continuing to hide behind the indefensible intellectual and class barriers that have been retained to protect its privilege and its ego from contamination by engineering, agronomy, medicine, and the social sciences. This change will become increasingly important in the future as the close of the fossil fuel frontier joins with the close of the land frontier to drive technical change along a path that entails the emergence of a much larger role for biological and information technology.

The 1970s was a period of declining productivity growth in the United States and several other advanced economies. In a number of developed economies, these dangerous trends were more apparent in the industrial than in the agricultural sector. Rates of return to agricultural research have remained high. (See chapter 10.) The evidence suggests that the institutional linkages that have provided effective articulation between science, technology, and agriculture have continued to be productive sources of economic growth in both developed and developing countries. There is much that can be learned from this experience by those who are not blinded by outmoded status symbols or cultural constraints.

NOTES

1. I would like to express my appreciation to Maury E. Bredahl, Francis C. Byrnes, Jeff Davis, David Ervin, Charles M. Hardin, J. C. Headly, Glenn L. Johnson, Keith Huston, Charles P. Lutz, Don Paarlberg, Gordon C. Rausser, Eldon D. Smith, and David Zilberman for the comments they made on an earlier draft of this chapter. My perspective on the role of science and technology in society has been strongly influenced by the work of Lynn White, Jr., and Samuel C. Florman. See particularly Lynn White, Jr., *Machina Ex Deo: Essays in the Dynamism of Western Culture* (Cambridge, Mass: MIT Press, 1968); Samuel C. Florman, *The Existential Pleasures of Engineering* (New York: St. Martin's Press, 1976).

2. White, *Machina Ex Deo*, p. 79.

3. This same point has been emphasized by Isaac Asimov, "Pure and Impure: The Interplay of Science and Technology," *Saturday Review*, 6 (June 9, 1979), pp. 22-24, 29. See also Kenneth E. Boulding, *The Impact of the Social Sciences* (New Brunswick, N.J.: Rutgers University Press, 1966); and N. Bruce Hannay and Robert E. McGinn, "The Anatomy of Modern Technology: Prolegomenon to an Improved Public Policy for the Social Management of Technology," *Daedalus*, 109 (Winter 1980), pp. 25-53. Hannay and McGinn pointed out that "modern technology . . . is itself obtained with the aid of sophisticated products of modern technology—electron microscopes, computers, spectrographs, and so on. 'Modern science as applied technology' is no less true and no more misleading than 'modern technology as applied science'" (p. 39).

4. Boysie E. Day, "The Morality of Agronomy," in *Agronomy in Today's Society*, J. W. Pendleton, ed. (Madison, Wis.: American Society of Agronomy, 1978), pp. 19-27. I follow Day's convention of using the term "agronomists" to refer to the community of production-oriented agricultural scientists. In the United States, the term "agronomy" has the more narrow connotation of field crop production and management.

5. For a view of the agricultural scientist as a hero, see E. C. Stakman, Richard Bradfield, and Paul C. Mangelsdorf, *Campaigns against hunger* (Cambridge, Mass.: Harvard University Press, 1967). For an earlier and even more heroic view, see Paul de Kruif, *Hunger Fighters* (New York: Harcourt Brace, 1928).

6. For a useful historical perspective, see Edward Shils, "Faith, Utility and the Legitimacy of Science," *Daedalus*, (Summer 1974), pp. 1-15.

7. For an example of romantic criticism, see Wendell Berry, *The Unsettling of America: Culture and Agriculture* (New York: Avon Books, 1977). For a populist perspective, see Jim Hightower, *Hard Tomatoes, Hard Times* (Cambridge, Mass.: Schenkman, 1973). For an ideological perspective, see Francis Moore Lappe and Joseph Collins (with Gary Fowler), *Food First: Beyond the Myth of Scarcity* (Boston: Houghton Mifflin, 1977). See also E. G. Valliantos, *Fear in the Countryside: The Control of Agricultural Resources in Poor Countries* (Cambridge, Mass.: Ballinger, 1977); Susan George, *How the Other Half Dies: The Real Reasons for World Hunger* (Montclair, N.J.: Allanheld, Osman, 1977). For reviews of this literature, see Nick Eberstad, "Malthusians, Marxists, and Missionaries," *Society*, 17 (September/October 1980), pp. 29-35; and Charles M. Hardin, "Feeding the World: Conflicting Views on Policy," *Agricultural History*, 53 (October 1979), pp. 787-95.

8. National Academy of Sciences, *Technology: Process of Assessment and Choice* (Washington, D.C.: U.S. Government Printing Office, July 1969), p. 16.

9. See Wayne D. Rasmussen, "Advances in American Agriculture: The Mechanical Tomato Harvest as a Case Study," *Technology and Culture*, 9 (October 1968), pp. 531-43; Andrew Schmitz and David Seckler, "Mechanized Agriculture and Social Welfare: The Case of the Tomato Harvester," *American Journal of Agricultural Economics*, 52 (November 1970), pp. 569-77; W. H. Friedland and A. E. Barton, "Destalking the Wily Tomato: A Case Study in Social Consequences in California Agricultural Research" (Davis, Calif.: University of California, Department of Applied Behavioral Research, Monograph No. 2, 1975); Richard E. Just, Andrew Schmitz, and David Zilberman, "Technological Change in Agriculture," *Science*, 206 (December 14, 1979), pp. 1277-80.

10. Eliot Marshall, "Bergland Opposed on Farm Machinery Policy," *Science*, 208 (May 9, 1980), pp. 578-80.

11. For a useful discussion of the constraints on the feasiblity of a general policy to provide specific protection, see E. C. Pasour, Jr., "Economic Growth and Agriculture: An Evaluation of the Compensation Principle," *American Journal of Agricultural Economics*, 55 (November 1973), pp. 611-16. For a more positive view, see Gordon C. Rausser, Alain de Janvry, Andrew Schmitz, and David Zilberman, "Principle Issues in the Evaluation of Public Research in Agriculture," in George W. Norton, Walter L. Fishel, Arnold A. Paulson, and W. Burt Sundquist, *Evaluation of Agricultural Research* (St. Paul: University of Minnesota Agricultural Experiment Station, Miscellaneous Publication 8-1981, April 1981), pp. 262-80.

12. For a definitive review of the evidence, see U.S. Public Health Service, *Smoking and Health: A Report of the Surgeon General* (Washington, D.C.: U.S. Department of Health, Education, and Welfare, Public Health Service, HEW Publication No. [PHS] 79-50006, 1980).

13. Charles M. Hardin, *Freedom in Agricultural Education* (New York: Arno Press, 1976 [reprinted from the 1955 edition published by the University of Chicago Press]), pp. 56-61.

14. Tobacco breeders have now gone well beyond Professor Valleau's limited objectives. At present, it is possible to manipulate the nicotine content within a very broad range without significantly altering the yield potential. The ability to manipulate the chemical characteristics of tobacco is probably more advanced than the understanding of the health implications of the chemical characteristics. Over the next decade or so, technical and institutional changes in tobacco harvesting and marketing are likely to result in substantial labor displacement along lines similar to the case of the tomato harvester. Eldon D. Smith, personal communication, June 20, 1980.

15. I would not argue that the prohibition of the production, processing, and use of

tobacco would be any more effective than the attempts that were made during the 1920s in the United States and Finalnd to prohibit the consumption of alcoholic beverages. It is of interest to note that in 1980 Malaysia imposed rather severe restrictions on the advertising of alcoholic beverages and cigarettes. The restrictions on tobacco advertising has been less severe than those on alcoholic beverage advertising primarily because tobacco is produced by large numbers of small farmers.

16. For a discussion of these issues within the context of energy policy, see the appendix "Literature and Philosophic Ideas Underlying Model Validation and Verification," in Glenn L. Johnson and Judith L. Brown, *An Evaluation of the Normative and Prescriptive Content of the Department of Energy Mid-Team Energy Forecasting System (MEFS) and the Texas National Energy Modeling Project (TNEMP)* (Austin, Texas: Energy and Natural Resources Advisory Council, 1980).

17. Chauncy Starr and Chris Whipple, "Risk of Risk Decisions," *Science*, 208 (June 6, 1980), pp. 1114-19.

18. This is consistent with the view expressed by Hans Mohr, "The Ethics of Science," *Interdisciplinary Science Reviews*, 4 (March 1979), pp. 45-53. Mohr noted that "freedom of inquiry . . . does not necessarily imply freedom in the choice of any particular goal; it implies, however, that the results of scientific inquiry may not be influenced by any factor extrinsic to science" (p. 48).

19. The position expressed here is similar to that presented by Peter O. Steiner, "The Public Sector and the Public Interest," in *The Analysis and Evaluation of Public Expenditures: The PPB System*, Robert H. Haveman, ed., for the Subcommittee on Economy in Government of the Joint Economic Committee, Congress of the United States (Washington, D.C.: U.S. Government Printing Office, 1969), pp. 13-14.

20. John H. Sanders and Vernon W. Ruttan, "Biased Choice of Technology in Brazilian Agriculture," in *Induced Innovation: Technology, Institutions and Development*, Vernon W. Ruttan, Hans P. Binswanger, and others, eds. (Baltimore: Johns Hopkins University Press, 1978), pp. 276-96; Hans P. Binswanger, *The Economics of Tractors in South Asia* (New York: Agricultural Development Council; and Hyderabad, India: International Crops Research Institute for the Semi-Arid Tropics, 1978).

21. Hans P. Binswanger, "A Note on Embodiment, Factor Quality and Factor Augmentation," in Hans P. Binswanger, Vernon W. Ruttan, and others, eds. *Induced Innovation: Technology, Institutions and Development* (Baltimore: Johns Hopkins University Press, 1978, pp. 159-63.

22. This section depends very heavily on the following surveys: National Academy of Sciences, *Technology: Process of Assessment and Choice*; Joseph F. Coates, "The Role of Formal Models in Technology Assessment," *Technological Forecasting and Social Change*, 9 (1975), pp. 139-89; W. B. Back, *Technology Assessment: Proceeding of an ERS Workshop* (Washington, D.C.: Economic Research Service, USDA, AGERS-31, September 1977); and Vary T. Coates, *Technology Assessment in Federal Agencies: 1971-76* (Washington, D.C.: George Washington University, Program of Policy Studies on Science and Technology, March 1979). For discussions of some of the limitations of technology assessment, see Robert T. Holt, "Technology Assessment and Technology Inducement Mechanism," *American Journal of Political Science*, 21 (May 1977), pp. 283-301; Lynn White, Jr., "Technology Assessment from the Stance of a Medieval Historian," *Technological Forecasting and Social Change*, 6 (1971), pp. 359-69; C. Marchetti, "A Postmortem Technology Assessment of the Spinning Wheel: The Last Thousand Years," *Technological Forecasting and Social Change*, 13 (1979), pp. 91-93.

23. Office of Technology Assessment, *Organizing and Financing Basic Research to Increase Food Production* (Washington, D.C.: U.S. Government Printing Office, June 1977); Office of Technology Assessment, *Pest Management Strategies in Crop Production*, vol. 1 (Washington, D.C.: U.S. Government Printing Office, October 1979).

Indexes

Index of Subjects

Administration of research: role of administrators in, 35, 219-21, 270; in United Kingdom, 70; in India, 95. *See also* Management, research

Agricultural development. *See* Development, agricultural

Agricultural Development Council, 10

Agricultural economics: emergence of, 24; development of in United States, 299-304; demand for, 305-7; staff responsibilities in, 314; in India, 322. *See also* Social-science research

Agricultural growth, 3-4, 17-20. *See also* Productivity, agricultural; Productivity growth

Agricultural Research Council, United Kingdom, 68-70, 71

Agricultural research council model, 107

Agricultural Research Institute (ARI), 183

Agricultural Research Service, 164, 290

Agricultural workers, in France, 24. *See also* Labor

Aid agencies, role of in national research systems, 125, 126-27

Allocation of resources. *See* Resource allocation

American Agricultural Economics Association, 9, 300

American Farm Bureau Federation, 325

American Farm Economic Association, 299-300

American Farm Management Association, 299-300

American Seed Trade Association, 195

Anthropology, contributions of to agricultural research, 142

Argentina, agricultural productivity growth in, 238

Australia: use of mechanical power in, 28; Commonwealth Scientific and Industrial Research Organization (CSIRO), 106, 110; agricultural productivity growth in, 238

Autonomous research institute model of research, 107

BAE. *See* Bureau of Agricultural Economics

Barbados cane-research station, 59

Bell Laboratories, 57

Benefit-cost methods of research planning, 284-86

Benefits of research: as tool in claim to resources, 267-68; analysis of, 290, 345-46; embodiment, augmentation, and incidence of, 346-47

Biological constraints on optimum yield, 312

Biological technology: as substitution for land, 28, 29; in Japan, 29; costs as inducement for, 31-34. *See also* Technology

Blackman report, 103

Blackwelder Company, 191

Botanic Gardens at Heneratgoda, 101

Brazil: research model for developing countries in, 66; Ministry of Agriculture, 96, 99, 100; history of agricultural research in, 96-98; Brazilian Public Corporation for Agricultural Research (EMBRAPA), 97-100; EMBRAPINHAS, 99; centralization in, 110-11; capacity in, 126; research resource allocation in, 272-73, 274; evaluation of agricultural plan of, 287; rate of return in, 304

Brazilian Public Corporation for Agricultural Research (EMBRAPA), 97-100

"Breeders' rights," 195. See also Plant variety protection

Breeding programs, animal, 282-83

Budget, research: maintenance research in, 59-60; in United States, 80; of EMBRAPA, 99; administration of, 220-21

Bureau of Agricultural Economics (BAE), 78, 300, 324-25

California, University of, mechanization research at, 187-88, 191, 274, 336

California Agrarian Action Project, 336

Canada: mechanization in, 28, 272-73; International Development Research Centre (IDRC), 106, 142; Research Division of Agriculture, 169; agricultural productivity growth in, 238

Caribbean Agricultural Research and Development Institute (CARDI), 174

Carnegie Institution, 57, 229

CATIE. See Inter-American Institute of Tropical Agriculture.

Center of origin, location of research near, 162, 163-64

Central Rice Research, Indonesia, 165-66

Centralization of research facilities, 110-11, 168

CGIAR. See Consultative Group on International Agricultural Research

Chemical pest control. See Pesticides

Chemicals, agricultural, 208

Chemistry, agricultural: emergence of in United Kingdom, 68; Liebig's contribution to, 72

CIAT. See International Center for Tropical Agriculture

CIMMYT. See International Center for the Improvement of Maize and Wheat

CIP. See International Potato Center

Closed economy, 270-71

Colonization, agricultural, 21

Commodity research: in United States, 78; in India, 91, 93; in Malaysia, 98, 106; benefits of to producers, 274-75; demand studies in, 303-4; on production, 305

Commodity trade, 302

Common Catalogs of seed varieties in Europe, 197

Commonwealth Scientific and Industrial Research Organization (CSIRO), Australia, 106, 110

Competitive grants program of USDA, 225, 228, 230, 279, 281. See also Project research grant system

Congruence model, of research resource allocation. See Parity model

Connecticut Agricultural Experiment Station, 57, 77

Conservation model, of agricultural development, 22-23

Consistency model, of growth accounting, 286-89

Constraints on production, 282, 307-8, 312

Consultative Group on International Agricultural Research (CGIAR): centers organized by, 50, 126, 128-29, 131; structure of, 122-23; Technical Advisory Committee (TAC) of, 123, 126, 128, 136, 141-42, 147-48; mentioned, 125, 133, 138, 215, 230; role and accomplishments of, 126, 128-29, 131, 137; direction of, 138; performance of, 143, 229

Consumers: benefits of research to, 257, 270-72; support of agricultural research by, 257-58

Cooperative State Research Service (CSRS), 290

Corn, India's improvement projects in, 93

Costs: of payoffs, 63; of research, 90, 124, 177-78; of obtaining research funds, 217-18. See also Expenditures on research

Cotton, simulated growth model of, 283

Cotton harvester, 187

Council on Agricultural Science and Technology, 319

Counterpart model, of development assistance, 117

Cowles Commission, 301

CRIS. *See* Current Research Information System

Crop research and development: in diffusion model, 24-25; in United States, 78, 192-96; in Japan, 86; in international institutes, 134-36; effects of legislation on, 196-98; by public and private sectors, 198-99; plant growth models in planning of, 283-84

Cropping systems, 22

CSIRO. *See* Commonwealth Scientific and Industrial Research Organization

CSRS. *See* Cooperative State Research Service

Cucumber harvester, 191

Cumulative synthesis approach to innovation, 51-56

Current Research Information System (CRIS), 266-67, 276

Customer-contractor principle, in funding, 70

DDT, 201, 203

Decentralization of research: in United States, 168, 249-51; effects of on productivity, 249-258

Denmark: productivity in, 238, 273; institutional reforms in, 348

Developed countries: productivity in, 19-20, 38, 39-40; diffusion in, 25; assistance to developing countries by, 117-19; emphasis on research benefits in, 267, 270

Developing countries: rates of productivity in, 19-20; diffusion in, 25; research in, 36, 37, 144; research needs of, 39, 130-31, 199, 322-23; professional assistance to, 117-19; rice production in, 142; emphasis on research benefits in, 267, 270

Development, agricultural: strategies in, 20; frontier model of, 21-22; conservation model of, 22-23; urban-industrial impact model of, 23-24; diffusion model of, 24-25; high-payoff input model of, 25-27; induced innovation model of, 27-36

Development assistance, 117-27

Diffusion of agricultural technology, 24-25, 62, 63, 153, 156, 317-18

Directors, research, 7-8, 131, 233. *See also* Administration of research

Disequilibrium in world agriculture, 36-41

Displacement of workers, society's response to, 336-38. *See also* Labor

Distribution expansion, 58-59

Economics research, 305, 306. *See also* Agricultural economics

Education: in development models, 26; as factor in human capital, 37, 39, 40; relationship of to research, 110, 182; of farmers, 249. *See also* Extension; Training

EMBRAPA. *See* Brazilian Public Corporation

Energy supply, effect of on technical change, 33, 272-73

Entomology, and pesticide policy, 203, 204

Environmental issues, in pesticide policy, 203-4, 207

Environmental Protection Agency, 205, 206-7, 209

Environmental Studies Board, 204

Ethics in technology and research, 335-36, 338-40, 340-43

Europe: agricultural development in settlements of, 21; pattern of technical change in, 33; plant variety protection in, 195, 196-97

European Economic Community (EEC), 141, 197, 302

Evaluation of research. *See* Review of research

Expenditures on research: by private sector, 183-86; value produced by, 252, 264-66; classification of, 268-69. *See also* Costs

Experiment stations: induced change in, 35; function of, 45-46; in Germany, 73, 74; in Japan, 85; in national systems, 108; in United States, 250. *See also* Institutes, research

Exports, effects of on commodity prices, 271-72

Extension: in United States, 77, 82; conflicts between research and, 110; effects of, 249, 290

Factor endowments, 28, 84

FAO. *See* Food and Agriculture Organization, United Nations

Farm management, 24, 301

Farmers: roles of in high-payoff input model, 25, 26; in Japan, 85; education of, 249; adoption of new technology by, 257

Farming research, in international institutes, 134-36

Federal Environmental Pesticides Control Act, 205, 206
Federal Pesticide Act of *1978*, 209
Fertilizer: research on, 6, 78; inducements for use of, 31, 32-35
Financial support. *See* Funding of research
Food and Agriculture Organization (FAO), United Nations: function of, 116; mentioned, 117, 126; role in CGIAR, 122, 123, 128, 129; capacity of to process information, 322
Food crisis, 17, 18
Food Machinery and Chemical Corporation (FMC), 191
Food policy, 141-42
Ford Foundation, 119, 122, 131, 142, 215, 230
Foundations, private, 229-30. *See also* Ford Foundation; Funding; Rockefeller Foundation
France, constraints on growth in, 24
Frontier model, of agricultural development, 21-22
Funding of research: in United Kingdom, 68-69, 70; customer-contractor principle in, 70; in Germany, 75; in United States, 77, 82, 183-92, 225-27, 251, 254, 255-56; formula method of, 83, 255-56; in India, 91; in Malaysia, 102, 103; responsibility for, 111-12; in institutions, 131, 177, 215; effects of on research staff and efficiency, 216-24; costs of obtaining, 218; effects of Hatch Act on, 228; strategies for, 231-33. *See also* Project research grant system

Genetic resources, collection of, 133-34, 163, 198
Germany: productivity in, 19, 238; as early model for research systems, 66, 71-72, 73, 83, 84-85; Berlin Academy of Sciences, 73; Friedrich-Wilhelm Universität, 73; Humboldt Universität, 73; Saxon agricultural experiment station, 73, 74; structure of agricultural research in, 73-74, 75; Federal Ministry of Food, Agriculture, and Forestry, 74, 75; Federal Research Center for Agriculture, 75
Goals of agricultural research, 80, 81, 150. *See also* Objectives of agricultural research
Grants: seeking of, 218, 219, 222, 223; from USDA, 252-27

Green Revolution, 17, 138, 139, 348

Harvesting equipment: as example of induced technical innovation, 30-31; and history of tomato harvester, 188, 189-90; displacement of labor by, 336-38
Hatch Act, 9, 77, 83, 228, 250-51
Hawaii, research by private sector in, 119
Heinz Company, H. J., 191
Herbicide development, 59. *See also* Pesticides
Hevea species of rubber, 101, 103
High-payoff input model, of agricultural development, 25-27
Holland, productivity in, 187
Human capital, 37-39, 40, 72
Husbandry, crop and livestock, 22, 24
Hussy reapers, 30

IADS. *See* International Agricultural Development
ICAR. *See* India
ICRISAT. *See* International Crop Research Institute for the Semi-Arid Tropics
IDRC. *See* International Development Research Center
IFPRI. *See* International Food Policy
IITA. *See* International Institute of Tropical Agriculture
Income, farm and nonfarm, 257-58
India: Coimbatore agricultural research station, 89-90; Bacteriological Research Laboratory, 91; Imperial Council on Agricultural Research (ICAR), 91; Ministry of Food and Agriculture, 91, 92; Agricultural Research Institute, 91, 92, 93; Agricultural Research Review Team, 92; Indian Council of Agricultural Research (ICAR), 92-93; 94-95; Central Rice Research Institute, 93; Agricultural Research Service, 93, 95; All-India Coordinated Research Programs, 93, 95; G. B. Pant University of Agriculture and Technology, 94; Review Committee on Agricultural Universities, 94; Udaipur University, 94; Uttar Pradesh Agricultural University, 94; Ministry of Agriculture and Education, 95; mechanization in, 274; agricultural economics in, 322. *See also* Indian research system
Indian Council of Agricultural Research (ICAR), 92-93, 94-95

Indian research system; productivity of, 37, 95; colonial origins of, 66, 89-90; crop improvement projects of, 89-90, 93; organization of, 91-92, 94; management of, 95, 126; centralization of, 111

Indonesia, location of research facilities in, 165-66

Induced innovation model, of agricultural development: emergence of, 26-27; technical innovation in, 29-35, 30-31; institutional innovation in, 35-36, 42; public resources in, 41-43; role of social sciences in, 42. *See also* Innovation; Institutional innovation

Industrialization, impact of on agricultural development, 23-24

Industries, agricultural input of, 183-86. *See also* Private sector

Information: as output of research laboratory, 46-47, 50; institutionalizing capacity in, 323

Innovation: mechanistic process approach to, 51, 54; cumulative synthesis approach to, 51-56; as result of linkages between science and technology, 56-58; role of society in, 60-64; early examples of, 108. *See also* Induced innovation model, of agricultural development; Institutional innovation

Inputs to agricultural production: indigenous, 18; in conservation model of agricultural development, 22-23; in high-payoff input model of agricultural development, 25-26; as part of growth measure, 238-39; complementary, 253; indexes for projecting, 288-89

Insecticides, 200-201. *See also* Pesticides

Institute pour la Recherche dans Agronomie Tropicale (IRAT), 128

Institutes, research: natural history of, 7, 132; need for change and growth in, 35, 42, 47, 230; public, 41; outputs of, 45-46, 50, 63, 155-57; funding of, 47, 222-24; linkages in, 48, 61-64; internal structure of, 50. *See also* Experiment stations; International institutes

Institutional change, demands of technology for, 41-42

Institutional innovation: in induced innovation model, 35-36, 42; in India, 95; effects of on other research, 210-11, 305, 306-7; as an instrument of reform, 347-48. *See also* Innovation

Institutional research support system, 215, 216-21, 229-31

Institutionalization of agricultural research: in United States, 76-77; in Japan, 84-86

Integrated research, extension, and education model, of agricultural research, 107

Inter-American Institute of Tropical Agriculture (CATIE), 140, 174

Interdisciplinary research, 8-9, 152

Interest groups, benefits of innovation to, 61-62

International Agricultural Development Advisory Service (IADS), 128, 175, 177

International Board for Plant Genetic Resources, 133

International Center for the Improvement of Maize and Wheat (CIMMYT): mentioned, 86, 123, 125, 130, 131; establishment of, 119, 122; research at, 120, 142-43; training programs of, 154

International Center for Tropical Agriculture (CIAT): establishment of, 122; mentioned, 122, 129, 131, 164; research at, 134, 139, 281-82

International Convention for the Protection of New Varieties of Plants, 195, 196

International Crop Research Institute for the Semi-Arid Tropics (ICRISAT): social science research at, 12; mentioned, 133, 134, 136, 137, 142

International crop research networks, 175

International Development Research Center (IDRC), Canada, 106, 142

International Food Policy Research Institute (IFPRI), 141-42, 322

International Institute of Tropical Agriculture (IITA): establishment of, 122; mentioned, 128-37 *passim*; relationships of with universities, 155

International institutes: problems of, 50, 126, 129, 132-33; as development models, 117, 119; establishment of, 118, 120-21; linkages at, 129-30, 153, contributions of, 133-43 *passim*; role of social science research at, 141-42. *See also* Institutes, research; International research systems

International Potato Center (CIP), 130, 153, 313

International research systems, 123, 124, 127, 142-44, 172-73. *See also* International

institutes
International Rice Research Institute (IRRI):
direction of, 8; social science research at,
12, 141, 308-9; as model institute, 54,
119-21; programs of, 93, 136, 139, 154-
55, 281-82, 310-13; mentioned, 106,
123-33 *passim*, 172; establishment of,
119-22; impact of, 142-43; allocations
of, 151-52
International Service for National Agricul-
tural Research (ISNAR), 128-29, 175,
177
International Union, 195
Interregional Committee on Agricultural
Policy, 317
Investments in agricultural research, 28, 39,
248-49
Iowa State University, 316
IRAT. *See* Institute pour la Recherche
"Iron triangle," of research policy in United
States, 63-64, 203
IRRI. *See* International Rice Research
Institute
ISNAR. *See* International Service for Na-
tional Agricultural Research

Japan: technology in, 10-11, 29, 31, 32, 33,
86, 87, 273; productivity in, 19, 238-46
passim, 257, 304; land constraints in, 28;
history and organization of agricultural
research in, 83-87; failure of mechanized
agriculture in, 84; Naito Shinjaku Agricultural Station, 84; University of Tokyo
College of Agriculture, 84; Komaba Agricultural School, 84, 85; Experimental
Farm for Staple Cereals and Vegetables,
85; Assigned Experiment System, 86;
Konosu Experimental Farm, 86, 87; National Agricultural Experiment Station,
86, 87; Ministry of Agriculture and Forestry, 86, 87, 88; Agriculture, Forestry,
and Fisheries Research Council, 87;
Hokkaido National Agricultural Experiment Station, 87; Hokuriku National
Agricultural Experiment Station, 87;
Kyushu National Agricultural Experiment
Station, 87; National Institute for Agricultural Sciences, 87; research priorities
in, 87-89; research system of compared
to that of United States, 88

Joint Council on Food and Agricultural Sciences, 262

Kennedy Committee report, 204
Kentucky, University of, 316, 338-39
Knowledge: supply and demand for, 149;
normative and positive, 341-43

Labor: productivity of, 18-19, 19-20, 30,
37-41; in conservation model of agricultural development, 22, 23; effects of
technology on, 23-24, 28, 29-30, 187,
188, 191, 274, 336-38; effects of supply
of on technology, 31-33; underutilization of, 41; role of in allocation decisions, 272-73
Laboratories, agricultural research, 73. *See
also* Experiment stations; Institutes,
research
Labor-intensive technologies, 138
Land: productivity of, 3, 19, 20; substitution of technology for, 28-29; effects of
supply of on technology, 31-32, 32-35;
role of in allocation decisions, 272-73;
reforms in ownership of, 348
Land economics, 302-3
Land-grant universities, 76, 77
Lawes Agricultural Trust, 68
Leadership in research, 48-49, 83, 131-32
Linkages between science and technology,
56-59
Livestock research, 168-70
Location of research facilities, 161-66

Machinery, 33, 136-38. *See also* Mechanical
technology
McIntire-Stennis Cooperative Forestry Research, 228
Maintenance research, 59-60
Malaysia: research system of, 66, 173; role
of rubber in, 100-102; Department of
Agriculture, 101-2; Malaysian Rubber
Producers' Research Association, 103;
Malaysian Rubber Research and Development Board (MRRDB), 104-5; Malaysian Pineapple Industry Board, 105; diversification of agricultural research in,
105-6; Malaysian Agricultural Research
and Development Institute (MARDI),
105-7; Oil Palm Research Institute of

Malaysia (OPRIM), 106; School of Agriculture, 106. *See also* Rubber Research Institute of Malaysia

Management, research, 4, 17, 99, 143-44, 218, 344-46. *See also* Administration of research

MARDI. *See* Malaysia

Market structure, 41, 302

MARRAIS. *See* Minnesota Agricultural Research Resource Allocation and Information System

Mechanical technology: as substitute for labor, 28, 29-30, 31-33, 274; role of private sector in, 137; role of public sector in, 137, 186-92

Mexico: wheat development in, 25, 119; and Rockefeller Foundation, 119, 122; Ministry of Agriculture, 119, 123, 230; social science research in, 304, 326

Michigan State University, 191, 316

Ministry(ies) of agriculture: as model for organization of research, 107; social science research in, 320-26

Minnesota, research in, 168, 284-85

Minnesota, University of: Minnesota Agricultural Experiment Station, 194, 314; Institute of Agriculture, 313-14; Department of Agricultural and Applied Economics, 314

Minnesota Agricultural Research Resource Allocation and Information System (MARRAIS), 284-85

Minnesota Crop Improvement Association, 194

Mita Farm Machinery Manufacturing Plant, 84

MRRDB. *See* Malaysia

Nairobi, University of, 176

National Academy of Sciences, 204, 250, 319, 335

National Association of State Universities and Land Grant Colleges, 276-77

National Cattlemen's Association, 169

National Dairy Forage Center, 170

National Environmental Policy Act, 205

National Fertilizer Development Cener, 78

National Pork Producers' Council, 169

National Research Council, 204, 228

National research systems: role of payoff

matrix in, 62; developmental stages of, 107-9; in developing countries, 125-26, 128-29, 144, 154

National Science Foundation, 183, 216, 230

National Wool Growers' Association, 169

Natural resource economics, 302-3

Nebraska, University of, 169

Negative payoffs, 63

Nepal, 321

New Zealand, agricultural productivity growth in, 238

Nigeria, location of research facilities in, 163

Norin varieties, 86, 87

North Carolina: North Carolina Experiment Station, 277; model of research planning in, 277-78, 279, 280-81; North Carolina State University, 278, 316

Objectives of agricultural research, 149, 150-51, 272

Office of Technology Assessment, 349

Open economy, 271-72

OPRIM. *See* Malaysia

Organization of agricultural research systems, 171-73

Output: of research, 63, 151, 217, 220-21; in agricultural production, 238-39, 288-89

Output per hectare. *See* Land, productivity of

Output per worker. *See* Labor, productivity of

Outreach programs, 155-57. *See also* Extension

Pakistan, 348

Parity model, of research resource allocation, 265-67, 292

Patents: for food processes and products, 184; for plants, 194-96

Payoff matrix in technical innovation, 61, 62-63. *See also* Returns to investment in research

Personnel, management of, 220. *See also* Management, research; Staff, research

Pest management programs, 210

Pesticides, 58, 202, 203-10

Pharmaceutical Society Museum in London, 100

Philippines: rice research in, 25, 119-22; productivity in, 29, 30; mentioned, 126, 326; University of the Philippines College of Agriculture (UPCA), 129, 155; project on yield constraints, 310-13; Ministry of Agriculture, 321

Physical constraints on optimum yield, 311-12

Planning, research: methodologies of, 9-10, 263, 292-94, 349; capacity of national systems in, 109; in small countries, 176-77; issues in, 262-63; scoring model of, 276-81; experimental approach to, 281-84; benefit-cost methods of, 284-86

Plant growth models, 283-84

Plant improvement and development, 58, 102-3, 192-96. See also Crop research and development

Plant Patent Act of 1930, 195

Plant variety protection, 194-99

Plant Variety Protection Act, 195-96, 198, 199

Political influences on research, 61, 62, 112, 125, 164-65, 191, 293-94, 345-46

Postharvest technology, 137

Pound report, 318, 320

PPBS. See Program planning

President's Council of Economic Advisors, 307, 321

President's Materials Policy Commission Report, 287

President's Water Resources Policy Commission Report, 287

Price support program for rice, in Japan, 88-89

Private sector: in Japan, 88; funding responsibilities of, 111-12; mechanization research by, 137, 191; role of in research, 175, 183-86, 210-11; role of in varietal development, 192, 194, 198-99; pesticide research by, 206-7, 209

Producers, benefits of research to, 270-72

Product development, 186

Product markets, as stimulus for agricultural development, 23

Product price, in relation to cost of research, 90

Production economics, 301

Productivity, agricultural: definition of, 18; in frontier model of agricultural development, 21-22; in urban-industrial impact model of agricultural development, 23-24; comparisons of in different countries, 27-28, 37, 40; in United States in 19th century, 76; decline of as a result of pesticide policy, 207; transfer of gains from, 257-58. See also Labor, productivity of; Land, productivity of; Productivity growth

Productivity, research, 11-12, 63, 126, 229-30; in India, 95; influence of scale and location of research facilities on, 167, 170, 249; decline in, 173; in pesticides research, 207; of scientists, 218-19

Productivity growth: definition of, 18, 238; in diffusion model, 24; contributions of research to, 142-43, 241-49; in Japan, 238, 239, 241, 242-43, 246; measures of, 238-39, 286-92; in United States, 238, 239-41, 247-49, 257; gains from 270-73. See also Productivity, agricultural

Program Planning and Budgeting System (PPBS), 9

Project research grant system, 215-25 passim. See also Competitive grants program of USDA

Project selection, 286-92

Public sector: support for research by in Germany, 72, 73; funding of research by, 111-12, 182-84, 252; role of in research, 137, 186, 210-11; research on mechanization by, 186-92; as source of new varieties, 192, 198-99; role of in pesticide research, 209-10; value of research by, 254-55, 258

Purdue University, 7, 316

Quantification of research benefits and costs, 157-58

RASAR. See Resource Allocation System

Rebuttable presumption against registration (RPAR), 206

Regional institutes, 174

Registration time for agricultural chemicals, 208

Regression analysis, 246

Research entrepreneurship strategy for funding, 231-32

Research institutes. See Institutes, research

Research Planning Advisory Committee, 278

Research output. See Output

Resource allocation: within research facilities, 7-8, 172, 275-76; systems of, 9, 151-52; planning methodologies of, 10, 262-64, 292-94; in meeting societal needs, 37, 42; role of benefits in, 62, 267, 270-75; in United States, 79-80, 254, 258-59; distortion of in Japan, 88-89; by American pesticides manufacturers, 208; and grant seeking, 229; parity model of, 264-67, 292; role of resource endowments in, 272-73; in Brazil, 272-73, 274; Minnesota system of, 284-85; linkage of to project selection, 286-92; role of social scientist in, 310; criteria for, 344, 347. See also Planning, research

Resource Allocation System for Agricultural Research (RASAR), 291

Resource endowments, 32, 37-39, 40

Resources for the Future, 215, 230

Returns to investment in research, 37, 142-43, 228-29, 241-43, 254-55, 256; in high-payoff input model, 25-26; importance of diffusion process to, 155-56; social versus private, 182, 183, 199; in mechanization research, 192; from institutional support system, 229; measures of, 252-53, 304-5, 317

Review of research: by external agent, 147-48; objectives of, 149-50; program areas in, 151-57; preparation for, 157, 158; organization of, 159

Rice blast nursery, 54-56

Rice research: and development of high-yielding varieties, 25, 26, 142; in Japan, 31, 86, 87; in India, 93; yield constraint study on, 310-13

Rockefeller Foundation: role of in Indian research, 93; Office of Special Studies established by, 119; support of research in Mexico by, 119, 122, 230; support of research in Philippines by, 119-22; funding of international institutes, 122, 128, 131; mentioned, 123, 142, 175

Rothamsted Agricultural Experiment Station, 68, 108

Rothschild reforms, 70, 71, 224

Royal Botanical Gardens at Kew, 100

RPAR. See Rebuttable presumption

RRIM. See Rubber Research Institute

Rubber, as object of research in Malaysia, 101-5

Rubber Growers' Association, 102

Rubber Research Institute of Malaysia (Malaya) (RRIM): mentioned, 66, 100, 119; organization of, 102-5; review of, 103; returns from research in, 229

Rural development, 42

Rural sociology, 24, 303, 314-15. See also Agricultural economics; Social science research

Saxon agricultural experiment station, 73, 74

Scale of agricultural research facilities, 161, 166-68, 170

Science: role of, 56-59, 334; agricultural, 73, 350-52

Science-oriented research, 247-49, 256

Scientific method in ethical decisions, 340-43

Scientist-years in research, 81, 82, 87-88, 277

Scoring model, of research planning, 276-81

Scotland, Agricultural Chemistry Association in, 68, 108

SEARCA, See Southeast Asian Regional Center

Seed certification, 192, 194, 196-97

Seed multiplication and distribution, 192-93, 194

Selection indexes in research planning, 282-83

Singapore Botanic Gardens, 101

Small countries: needs of, 173-78 passim; importance of applied fields in, 174; private sector research in, 175, cost of research in, 177-78

Small farms: research for, 4-6, 138, 140, 274-75; adoption of technology by, 139

Smith-Lever Act, 250

Social controls on technology, 200, 343-44

Social science research: institutionalization of, 12-13; importance of to institutional innovation, 42, 306-7; at international institutes, 141-42; intersection of with natural sciences, 298; on institutional constraints, 307-8; role of social scientist in, 309-10, 313, 315-16; collaborative efforts in, 310, 313, 316-18, 322; in colleges and universities, 313-20; constraints on, 316-19, 324-28 passim; role of in ministry of agriculture, 320-26;

in poor countries, 322-23. *See also* Agricultural economics; Rural sociology
Social Science Research Council, 215
Socialization of agricultural research, 41
Socioeconomic structure, role of in technological change, 61, 62
Sociopolitical environment, 319
South Korea, 326, 348
Southeast Asian Regional Center for Graduate Study and Research in Agriculture (SEARCA), 137
Spillover of research, 247, 249, 255-56, 257-58
Staff, research: development of, 47, 49, 98-99, 176-77; in Brazil, 98-99; of RRIM, 105; in Mexico, 119; strategy for, 122; outreach problems of, 156-57; of facilities in small countries, 175; role of in funding, 232, 233
State agricultural research system: organization of in United States, 77, 78, 172; funding of, 83, 225, 226, 228; in Brazil, 99; returns from research by, 247, 248
Structural rigidity in research systems, 171-73
Subcommittee on Agriculture of the Senate Appropriations Committee, 164-65
Substitution view of public research on mechanization, 187
Sugarcane breeding, 58-59, 60, 89-90
Sukamandi, rice research institute at, 165-66
Supply and demand: for innovation, 61, 62; for machinery, 137; for products, 257; for social-science research, 304-8
Supply-side argument for location of research facilities, 162-63
Supporting research, 152-53
Supporting systems of technology, 335
Surplus, agricultural, 83-84

Taiwan, 28, 187, 348
Technical change: in agricultural development, 18, 27, 36; utilization of, 40-41; responsibilities of, 335; effects of institutional innovation on, 347-48. *See also* Technology
Technical inputs, 37-39, 40. *See also* Inputs to agricultural production
Technology: change in, 10-11, 33, 41-42; management of, 17; diffusion of, 24-25; in factor substitution, 28, 29; linkage of

with science, 56-59, 331, 333, 343; in United States, 76; in Japan, 85; transfer of, 175-76; impact of, 290-91, 331-32, 333-34, 346-47; systems of, 335; assessment of, 348-50; relationship of with nature, 351. *See also* Biological technology; Technical change
Technology-oriented research, 247-49, 256
Tennessee Valley Authority (TVA), 4-6, 78, 316
Thailand, 326
Tobacco, research on, 338-40, 341
Tomato harvester, 188, 189-90, 336-38
Tractor horsepower per male worker, 33, 34
Trade-off models of growth accounting, 291-92
Training: programs in, 118, 153, 154-55; of research staff, 153-54, 176
Transcendentalist approach to innovation, 51, 53
Tunisia, 326
TVA. *See* Tennessee Valley Authority

United Kingdom: agricultural revolution in, 22; as early model for research, 66, 84; history of agricultural research in, 67-70, 71, 72; Department of Scientific and Industrial Research, 68; Development and Road Improvement Act of *1901*, 68; funding of research in, 68-69, 70; Agricultural Research Council, 68-70, 71; Council for Scientific Policy, 69; Department of Education and Science, 69; Ministry of Agriculture, 69; Agricultural Development and Advisory Service (ADAS), 70; Central Policy Review Staff, 70; Ministry of Agriculture, Fisheries, and Food (MAFF), 70, 71; research reform in, 70, 71; as example of autonomous model, 107; Rothamsted Agricultural Experiment Station, 108; review schedule of research in, 147; criticism of agricultural research in, 224
United Nations Development Program, 122
United Planters' Association, 101
United States: as model for research, 66; history of agricultural research in, 76-77; organization of research system in, 77, 254, 258; reforms in research system in, 79; limitations of research system in, 80; criticism of agricultural research in,

80, 110, 181-82, 224, 225; Agricultural Act of *1977*, 82; future of agricultural research system in, 82-83; research system of compared to Japan's, 84, 88; International Development and Food Assistance Act of *1975*, 118; characteristics of public agricultural research in, 249-51; Food and Agricultural Act of *1977*. *See also* United States Deparment of Agriculture

United States Agency for International Development (USAID), 92, 93, 97, 136, 166, 215, 327

United States Department of Agriculture (USDA): organization of, 76-77, 78, 171-72; Bureau of Animal Industry, 77; Bureau of Biological Survey, 78; Bureau of Entomology, 78; Bureau of Plant Industry, 78; Bureau of Soils, 78; Office of Experiment Stations, 78; Weather Bureau, 78; Bureau of Agricultural Economics, 78, 300, 324-25; Agricultural Research Service of Science and Education Administration (USDA/SEA/AR), 78, 169, 170, 254, 255; Research Review Committee, 79; evaluation of, 80, 82; Joint Council on Agricultural Research, 82; National Agricultural Research and Extension Users Advisory Board, 82, 262; staff of, 110, 250-51, 321; National Seed Storage Laboratory, 133; review by, 147; Agricultural Research Service, 164, 290; Meat Animal Research Center (MARC), 168-70; research locations of, 170-71, 254; policy on mechanization research, 188; Science and Education Administration, 191; role of in plant breeding and protection, 196, 198; pest control policies of, 203, 206, 207; grants received by, 222; Competitive Research Grants Office, 225; funding by, 228, 230, 279, 281, 336; returns from research by, 247-48; data collection at, 266, 322; Current Research Information System (CRIS), 266-67; planning by, 276-77, 290, 349; Cooperative Research Service, 278; mentioned, 287, 290-91; social science research at, 301, 324-26; Economics Research Service (ERS), 325; Staff Economics Group, 325; Statistical Reporting Service (SRS), 325; Agricultural Mechanization Task Force, 336

Universities: agricultural in India, 92, 94; agricultural in Brazil, 99; support for research at, 224, 230-31; social science research at, 313-20

University contract model, 117, 118, 123-24

UPCA. *See* Philippines

Urban-industrial impact model, of agricultural development, 23-24

Vavilov Institute, Soviet Union, 133

West African Rice Development Association (WARDA), 126, 128, 129, 174, 175

Wheat research: and varieties development, 25, 26, 142, 255; in Japan, 86; in Mexico, 119; network of, 127

Wisconsin, University of, 184

World Bank: mentioned, 10, 122, 165, 166, 327; support of for reform in Brazil, 97; funds for Malaysia from, 106; use of counterpart model by, 117-18; role of in CGIAR, 122-23

World Food and Nutrition Study, 183

World food production, 17-18

Yield constraints, research on, 281-82, 310-13

Yield-increasing technologies, 200

Index of Names

Araji, A. A., 286

Bain, Joseph, 302
Bakewell, Robert, 67
Barletta, Ardito, 317
Bean, L. H., 300
Bergland, Bob, 188, 191, 336
Binswanger, Hans P., 167, 345
Black, John D., 301
Boussingault, J. B., 108
Brandow, George, 303, 317
Bredahl, M., 256

Carlson, Gerald, 304
Carson, Rachel, 80, 201
Chamberlin, Edward, 302
Cline, Philip, 290
Cochrane, Willard, 304
Cock, James, 162

Daddario, Emilio Q., 348-49
Davis, Jeffrey, 279
Deere, John, 108
De Janvry, Alain, 61
Dupree, A. Hunter, 77

Easter, William, 285
Einstein, Albert, 225
Evenson, Robert, 58, 126, 167, 247, 256, 317

Ezekiel, Mordecai, 300, 301

Farrell, Kenneth R., 326
Ferguson, Harry, 332
Fishel, Walter L., 276, 284
Foote, Richard, 301
Fox, Karl, 301
Fulton, Robert, 51

Galbraith, J. K., 303
Gardner, R. L., 286
Gilbert, Sir Henry, 68
Gray, Roger, 304
Griliches, Zvi, 163, 253, 317, 318
Gross, Neale, 318

Hadwiger, Don, 164
Hagerstrand, Torsten, 318
Hathaway, Dale, 303
Hayami, Yujiro, 10
Headly, J. C., 304
Hicks, John R., 301
Hightower, Jim, 80
Houck, James, 303
Humboldt, Wilhelm von, 73

Idachaba, Francis Sulemanu, 163

Janvry, Alain de, 61
Jefferson, Thomas, 108

Jennings, Peter, 162
Jones, Donald, 57, 332

Kamien, Morton, 167
Kislev, Yoav, 58
Knipling, Edward, 203
Krishna, Raj, 322

Lawes, Sir John Bennet, 68, 108
Leontief, Wassily, 301
Lewis, Arthur W., 302
Liebig, Justus von, 68, 72-73, 74, 85, 108,
    250, 332
Lima, Luis Fernando Cirne, 97
Lu, Yao-chi, 290

McCormick, Cyrus, 108
Malstead, Illona, 183
Mansfield, Edwin, 229
Marconi, Guglielmo, 51
Mendel, Gregor, 332
Moore, Henry, 300
Moseman, Albert H., 133, 163
Mosher, Arthur, 158

Nerlove, Marc, 301
North, Douglass, 305, 306
Norton, George, 279, 285

Okun, Arthur M., 307
Ou, S. H., 55

Paarlberg, Don, 327
Pearson, Karl, 300
Peterson, Willis L., 256, 317
Pond, G. A., 301
Pound, Glenn S., 80, 318, 320

Quance, Leroy, 290

Ridley, H. N., 101
Robinson, Joan, 302
Rothschild, Lord, 70, 71, 224
Russell, R. G., 291, 292
Ruttan, Vernon, 256
Ryan, Bruce, 318

Sanders, John, 345

Schmitz, Andrew, 188, 253
Schmookler, Jacob, 167
Schuh, G. Edward, 287
Schull, George H., 57, 229
Schultz, Henry, 300
Schultz, Theodore, 123, 162, 302, 305, 306
Schwartz, Nancy, 167
Scriven, M., 150
Seckler, David, 188, 253
Shockley, William, 57
Shumway, C. Richard, 279
Sim, J. R., 286
Snedecor, George, 300
Sorenson, James, 304
Spillman, W. J., 301
Swanson, B. E., 154, 155
Swanson, Earl, 316

Tang, A., 246
Taylor, Henry C., 300
Thaer, Albrecht, 73
Thomas, Robert, 305, 306
Thorbecke, Erick, 323
Thunen, Johann Heinrich von, 23, 108
Thurow, Lester, 327
Tolley, Howard R., 301
Townshend, Lord, 67

Ulbricht, Tilo, 70
Usher, Abbott P., 51, 56

Valleau, W. D., 338, 339
Von Humboldt, Wilhelm, 73
Von Liebig, Justus, 68, 72-73, 74, 85, 108,
    250, 332
Von Thunen, Johann Heinrich, 23, 108

Waggoner, Paul A., 256
Wallace, H. A., 300
Warren, G. F., 301
Washington, George, 108
Waugh, Fred, 300
Weston, Sir William, 67
Wholey, Joseph S., 150, 159
Working, Elmer, 300
Working, Holbrook, 300

Young, Arthur, 67, 108

Vernon W. Ruttan is a professor in the departments of economics and of agricultural and applied economics at the University of Minnesota. He has served as an economist with the Tennessee Valley Authority, the Council of Economic Advisors, and the Rockefeller Foundation (at the International Rice Research Institute in the Philippines) and was president of the Agricultural Development Council from 1973 to 1977. Among his many publications are *Agricultural Development: An International Perspective* (with Yujiro Hayami) and *Induced Innovation: Technology, Institutions and Development* (with Hans P. Binswanger).